Principles and Applications of
POSITRON & POSITRONIUM CHEMISTRY

Principles and Applications of
POSITRON & POSITRONIUM CHEMISTRY

Editors

Y C Jean
University of Missouri–Kansas City, USA

P E Mallon
University of Stellenbosch, South Africa

D M Schrader
Marquette University, USA

World Scientific
New Jersey • London • Singapore • Hong Kong

Published by
World Scientific Publishing Co. Pte. Ltd.
5 Toh Tuck Link, Singapore 596224
USA office: Suite 202, 1060 Main Street, River Edge, NJ 07661
UK office: 57 Shelton Street, Covent Garden, London WC2H 9HE

British Library Cataloguing-in-Publication Data
A catalogue record for this book is available from the British Library.

PRINCIPLES AND APPLICATIONS OF POSITRON AND POSITRONIUM CHEMISTRY

Copyright © 2003 by World Scientific Publishing Co. Pte. Ltd.

All rights reserved. This book, or parts thereof, may not be reproduced in any form or by any means, electronic or mechanical, including photocopying, recording or any information storage and retrieval system now known or to be invented, without written permission from the Publisher.

For photocopying of material in this volume, please pay a copying fee through the Copyright Clearance Center, Inc., 222 Rosewood Drive, Danvers, MA 01923, USA. In this case permission to photocopy is not required from the publisher.

ISBN 981-238-144-9

This book is printed on acid-free paper.

Printed in Singapore by Uto-Print

Preface

This book is a follow-up of our previous book, *Positron and Positronium Chemistry,* published in 1988, edited by two of the current editors, D.M. Schrader and Y.C. Jean. The idea of publishing a second comprehensive book on the topic of positronium chemistry began in 1997 when one of the editors, Professor D.M. Schrader, conducted a positron summer school (PSS-97) three days preceding the 11th International Conference on Positron Annihilation (ICPA-11) organized by one of the authors, Professor Y.C. Jean in Kansas City. In PSS-97 about sixty students enrolled in a one-credit hour graduate course, "Special Topics in Physical Chemistry," offered at the University of Missouri-Kansas City (UMKC). Instructors were those scientists most active in positronium chemistry, P. Coleman, J. Duplatre, H. Stoll, D.M. Schrader, R.M. Nieminen, and Y.C. Jean who served as the official instructor at UMKC. During the positron summer school, the idea of publishing a graduate-level textbook in positronium chemistry crystallized. At the same time, we realized that positron and positronium chemistry had advanced significantly during the decade since we published the first book in 1988 and a new comprehensive book was needed for scientists who wish to enter this area of research.

This book is designed mainly for two audiences: (1) for graduate students who are pursing positron and positronium chemistry for their master or doctoral degrees, and (2) for entering scientists interested in pursuing their research career in the area of positron and positronium chemistry. This book can also be readily used as a textbook for a graduate-level course in physical chemistry or chemical physics. Some chapters are provided with exercises and problems along with answers for this purpose. This book could be beneficial for positron scientists who wish to have comprehensive reviews on the current status of positronium chemistry.

Y.C. Jean, Kansas City, Missouri, USA, August, 2002

Acknowledgments

This book could not have been realized without the authors' enthusiastic and persistent contributions of their significant scientific works. In addition to the authors, numerous others have contributed their works to the contents of each chapter. Graduate students are always the most valuable contributors to significant new science. I would like to express my gratitude to all my previous and existing graduate students who have dedicated important years of their lives to research in positron and positronium chemistry. During the past two decades, the students who earned their graduate degrees in positron and positronium chemistry at the University of Missouri–Kansas City (UMKC) are: E. Parsai (1983, M.S. Physics), Dr. Chang Yu (M.S. Physics, 1984), Dr. K. Venkateswaren (Chemistry, 1984), Dr. N. Zhou (M.S. Physics, 1986), S. Budhabhattic (M.S. Physics, 1986), Dr. X. Tang (M.S. Physics, 1987), Dr. R. Ganti (Chemistry, 1987), J. Kyle (M.S. Physics, 1987), L.Y. Hao (M.S. Physics, 1990), S. Sung (M.S. Chemistry, 1991), Dr. T. Mahmood (Chemistry, 1991), Dr. X. Lu (M.S. Physics, 1992), Y. Lyu (M.S. Physics, 1992), G.M. Zhou (M.S. Chemistry, 1992), C.F. Tsai (M.S. Chemistry, 1992), Dr. H. Shi (Physics, 1993), Dr. Q. Deng (Chemistry, 1993), Dr. Q. Zhang (M.S. Chemistry, 1994), Dr. J. Liu (Chemistry, 1994), G.F. Dai (M.S. Physics, 1994), H. Zhang (Physics, 1994), Y.S. Rhee (M.S. Chemistry, 1995), Y. Gu (M.S. Physics, 1996), Dr. X. Hong (Chemistry, 1996), C.Y. Cheng (M.S. Chemistry, 1996), Dr. H.L. Yen (Physics, 1997), Dr. H. Cao (Chemistry, 1998), Dr. J.-P. Yuan (Chemistry, 1999), Dr. C.-M. Huang (Chemistry, 2000), and Dr. Renwu Zhang (Chemistry, 2002). Current students working on their Ph.D. degrees in positron and positronium chemistry and contributing much work to this book are: Hongmin Chen, Ying Li, and Junjie Zhang. I also thank the greater than twenty scientists, who have been associated with me at UMKC for their valuable contributions to positron and positronium chemistry. The family tree started from Prof. D.M. Schrader, who was my Ph.D. advisor in 1970 at Marquette University.

I also especially thank Mrs. Florence Middleton, the senior secretary in the Department of Chemistry, UMKC, for her excellent contributions in proofing and editing the manuscript for this book Mr. John Whitchurch, electronic engineer, and Dr. Robert Middleton for their resourcefulness in developing computer software.

Y.C. Jean

CONTENTS

Preface   v

Acknowledgments   vii

1. Introduction to Positron and Positronium Chemistry   1
Y.C. Jean, P.E. Mallon, and D.M. Schrader

    1.1   A new chemistry: positronium chemistry   1
    1.2   Existing books and articles on positron and
            positronium chemistry and annihilation   6
    Problems   9
    References   10
    Answers to problems   12

2. Compounds of Positrons and Positronium   17
D.M. Schrader

    2.1   Introduction   17
    2.2   Quantum mechanical considerations   18
           2.2.1   Basic physics of mixed electron-positron systems   18
           2.2.2   The calculation of annihilation rates   19
           2.2.3   Quantum mechanical methods   20
    2.3   Current knowledge of bound states   25
           2.3.1   Polyleptons   25
           2.3.2   One-electron atom   27
           2.3.3   Two-electron atoms (excluding He)   29
           2.3.4   The nonmetals   30
           2.3.5   The noble gases   31
           2.3.6   Molecules   33
    Acknowledgment   33
    References   34

3. Experimental Techniques in Positron Spectroscopy 37
P. G. Coleman

3.1 Introduction 37
3.2 Positron sources 39
    3.2.1 Introduction 39
    3.2.2 Radioactive sources for laboratory experiments 39
    3.2.3 Positron sources for facility-based beams 41
    3.2.4 Accelerator-based positron sources for the laboratory 41
3.3 Particle and radiation detectors 42
    3.3.1 Radiation detectors 42
    3.3.2 Particle detectors 45
3.4 Notes on pulse electronics 47
    3.4.1 Transmission of pulses 47
    3.4.2 Elements of circuits used in positron spectroscopy 48
3.5 Lifetime spectrometry 49
3.6 Doppler broadening spectroscopy 52
    3.6.1 Introduction 52
    3.6.2 Experimental set-up 52
    3.6.3 Data analysis 54
    3.6.4 Two-detector technique 55
3.7 Age-momentum correlation (AMOC) 56
3.8 Angular correlation of annihilation radiation (ACAR) 56
    3.8.1 One-dimensional ACAR 56
    3.8.2 Two-dimensional ACAR 58
3.9 Positron beams 59
    3.9.1 Positron moderators 59
    3.9.2 Laboratory-based beams 61
    3.9.3 Facility-based beams 63
    3.9.4 Beam bunching 63
    3.9.5 Polarized positron beams 63
    3.9.6 MeV positron beams 63
    3.9.7 Time-of-flight spectrometry 64
    3.9.8 Positron microscopy 64
    3.9.9 Plasma-generated positron beams 65
Problems 65
References 67
Answers to Problems 70

4. Organic and Inorganic Chemistry of the Positron and Positronium   73

G. Duplâtre and I. Billard

    4.1 Positronium formation in condensed matter   73
          4.1.1 The spur model in polar solvents   74
          4.1.2 The spur model in nonpolar solvents   82
          4.1.3 Quantitative approaches and modeling of Ps formation   84
          4.1.4 Positronium formation in solids   86
    4.2 Positron chemistry   87
    4.3 Positronium states in condensed matter   89
          4.3.1 Ps trapping in liquids: the bubble model   89
          4.3.2 Ps trapping in solids: the free volume model   90
          4.3.3 Ps states in condensed matter: the contact density parameter   91
    4.4 Positronium chemistry in liquids   94
          4.4.1 Positronium reactions   95
          4.4.2 Fundamentals of positronium kinetics   101
    4.5 Some chemical applications of (PAT) in liquids and solids   104
          4.5.1 Stability constants   104
          4.5.2 Polyelectrolyte solutions   106
          4.5.3 PAT applications to primary radiolysis processes   107
          4.5.4 PAT applications to phase transitions   108
          4.5.5 PAT applications to defect formation in solids   109
          4.5.6 Spin cross-over in solid metal complexes   111
    References   112

5. Physical and Radiation Chemistry of the Positron and Positronium   117

Sergey V. Stepanov and Vsevolod M. Byakov

    5.1 Introduction   117
    5.2 Energy deposition and track structure of the fast positron   117
          5.2.1 Ionization slowing down   117
          5.2.2 Thermalization stage   121
          5.2.3 Interaction between the positron and its blob   122

5.3	Positronium formation in condensed media	123
	5.3.1 The Ore model	123
	5.3.2 Quasi-free Ps state	123
	5.3.3 Disappearance of the Ore gap	127
5.4	Recombination mechanism of Ps formation	129
	5.4.1 The spur model	129
	5.4.2 The blob model	131
	5.4.3 "Elementary act" of the Ps formation	132
	5.4.4 Role of localized states of electrons and positron	133
5.5	Quantitative formulation of the blob model	134
	5.5.1 Reactions of hot electrons and positrons	134
	5.5.2 Diffusion-recombination stage	138
5.6	Application of the blob model to aqueous solutions	140
	5.6.1 Claculation of Ps, H_2 and hydrated electron yields	140
	5.6.2 Discussion of the results	143
5.7	Conclusion	143
	Acknowledgments	145
	Problems	145
	References	147
	Answers to the problems	149

6. Positrons and Positronium in the Gas Phase — 151

D. M. Schrader

6.1	Introduction	151
6.2	Overview of positron scattering	151
6.3	Annihilation	153
	6.3.1 The primitive annihilation event	153
	6.3.2 The response of the target	154
	6.3.3 Annihilation on specific molecular sites	159
	6.3.4 Relaxation after annihilation	160
6.4	Bound states	162
	6.4.1 Observation of bound states by resonant capture	162
	6.4.2 Observation of bound states by dissociative attachment	163
	Acknowledgment	164
	References	165

7. Positron Porosimetry — 167

Marc H. Weber and Kelvin G. Lynn

7.1	Introduction	167
7.2	Open versus closed porosity	169
	7.2.1 Detector count rates	170
	7.2.2 3-to-2 photon ratio	171
	7.2.3 Positronium lifetime	171
7.3	Pore connectivity	172
	7.3.1 The concept of the 3-to-2 photon ratio technique	173
	7.3.2 Positronium range—depth profiles	174
	7.3.3 Data and results	176
	7.3.4 Analysis	177
	7.3.5 Skin layers and buried open porosity	179
7.4	Impurity intrusion and chemical effects	180
	7.4.1 All materials are not equal	180
	7.4.2 Chemistry	181
	7.4.3 Uptake of water	182
	7.4.4 Metallic impurities	183
7.5	Pore size distributions	184
	7.5.1 Pore sizes determine lifetimes	185
	7.5.2 The apparatus	186
	7.5.3 Raw data	186
	7.5.4 Data analysis and fitting methods	188
	7.5.5 Average lifetimes — mean pore dimensions	188
	7.5.6 Average lifetimes and 3-to-2 photon ratio	189
	7.5.7 Full analysis — Pore size distributions	191
	7.5.8 From lifetimes to pore sizes	195
	7.5.9 Porosity	196
	7.5.10 Bimodal or not bimodal — critical comments	198
7.6	Beyond count rates, ratios, and lifetimes	202
	7.6.1 High sensitivity Doppler broadening	202
	7.6.2 Angular correlation measurements	203
	7.6.3 Lifetime depth profiles	204
	7.6.4 Combination techniques	204
7.7	Inline diagnostics with positronium	204
References		206

8. Positron Annihilation Studies on Superconducting Materials — 211
C. S. Sundar

8.1 Introduction	211
8.2 Positron annihilation in superconductors	213
8.3 Temperature variation of positron annihilation characteristics in HTSC	214
8.3.1 Positron distribution and the electron-positron overlap	218
8.3.2 Studies on other high-temperature superconductors	220
8.4 Defects in HTSC	222
8.5 Fermi surface studies	225
8.6 Studies on other novel superconductors	227
8.6.1 Fullerenes	227
8.6.2 Borocarbides	228
8.7 Summary	230
Acknowledgments	230
References	230

9. Positronium in Si and SiO_2 Thin Films — 235
R. Suzuki

9.1 Introduction	235
9.2 Amorphous Si	236
9.3 Porous Si	239
9.4 Amorphous SiO_2 thin films	240
9.4.1 Thermally grown SiO_2 on Si	240
9.4.2 Detection of defects in amorphous SiO_2	243
9.5 Porous SiO_2	246
9.6 Summary	249
References	249

10. Application to Polymers 253

P. E. Mallon

10.1 Introduction	253
10.1.1 Free volume in polymers	254
10.2 Theoretical aspects	256
10.2.1 Data analysis of PAL spectra in polymer	258
10.2.2 Free volume fractions	259
10.3 Examples of PAL studies in polymers	260
10.3.1 Phase transition phenomena in polymers	260
10.3.2 Studies in thermoplastics	264
10.3.3 Correlation with mechanical and other properties	268
10.3.4 Chemical sensitivity	271
10.3.5 Polymer blends	271
10.4 Potential problems in polymer studies	272
10.4.1 Exposure to the positron source	272
10.4.2 Inhibition and quenching	275
10.5 Concluding remarks	277
Acknowledgements	277
References	277

11. Applications of Slow Positrons to Polymeric Surfaces and Coatings 281

Y.C. Jean, P.E. Mallon, Renwu Zhang, Hongmin Chen, Y.C. Wu, Ying Li, and Junjie Zhang

11.1 Introduction	281
11.1.1 Depth profiles in polymeric materials	284
11.1.2 Positron and positronium dynamics near the surface	285
11.1.3 Chemical composition and the S parameter	290
11.2 Applications to coatings	291
11.2.1 Depth profile of coatings	291
11.2.2 Exposure to UV and degradation of coatings	293
11.2.3 Durability of commercial coatings and paints	295
11.3 Correlations with other properties	299
11.3.1 ESR	299

11.3.2 Cross-link density	300
11.3.3 Gloss	302
11.3.4 Surface roughness	302
11.3.5 Mechanical loading	303
Conclusion	304
Acknowledgment	304
References	305

12. Positron Annihilation Induced Auger Spectroscopy   309

Shafaq Amdani, Anat Eshed, Nail Fazleev, and Alex Weiss

12.1 Introduction	309
12.2 PAES Mechanism	310
12.3 Theory	313
12.4 Catalog of PAES spectra of slected elements	318
12.5 Applications of PAES	319
Acknowledgements	326
References	326

13. Characterization of Nanoparticle and Nanopore Materials   329

Jun Xu

13.1 Nanoparticle materials	329
13.1.1 Vacancy clusters on the surface of gold nanoparticles embedded in MgO	330
13.1.2 Optical properties affected by surface vacancies	333
13.1.3 Quantum dot — anti dot coupling	335
13.2 Nanopore materials — Ultra-low k	337
13.2.1 Ps out diffusion — interconnectivity of pores	339
13.2.2 Positronium lifetime spectroscopy	342
13.2.3 Dependence on molecular weight of porogen	345
Acknowledgments	346
References	347

14. AMOC in Positron and Positronium Chemistry 349

Hermann Stoll, Petra Castellaz, and Andreas Siegle

14.1 Introduction	349
14.2 Beam-based age-momentum correlation (AMOC)	350
14.2.1 AMOC relief and lineshape function	350
14.2.2 Analysis of AMOC data	352
14.2.3 Analysis of the young age broadening of positronium	354
14.3 Time domain observations in positron and Ps chemistry	355
14.3.1 Chemical reactions of positrons	357
14.3.2 The Model	359
14.3.3 Influence of the halide ion size	360
14.3.4 Positron-molecule lifetimes	361
14.3.5 Reaction rates	361
14.3.6 Existence of the PsF molecule	362
14.4 Observation of positronium slowing down	362
14.5 Evidence for the existence of two kinds of o-Ps in condensed rare gases	365
Acknowledgements	367
References	367
Problem	370
Answer to problem	371

Appendix: Free-volume Data in Polymeric Materials (R.T.)

Ying Li, Renwu Zhang and Y.C. Jean   373

Index   397

Chapter 1

Introduction to Positron and Positronium Chemistry

Y.C. Jean[1], P.E. Mallon[2], and D.M. Schrader[3]

Department of Chemistry, University of Missouri—Kansas City[1], USA

Department of Chemistry, University of Stellenbosch, South Africa[2]

Department of Chemistry, Marquette University, Milwaukee, WI, USA[3]

1.1 A new chemistry: positronium chemistry

Tracing the origins of the human concept on the existence of matter and anti-matter is probably best from the Holy Scriptures, that in the beginning the Universe was created from nothing, *ex nihil*, in Latin or Greek [1]. The disappearance of matter could also be traced in the Holy Scriptures, that the whole Universe will turn into nothing, *an nihil*, which is the root of the word "annihilation" [2]. Chemistry is the study of matter, energy, and their changes. Matter is anything that has mass and takes up space. Energy is the capacity of matter to do work. Albert Einstein formulated the relationship between mass (m) and energy (E) with the speed of light (c) in the famous $E=mc^2$ equation. The direct transformation from mass into energy is manifested when matter encounters its *anti* (which means "against" in Greek) matter. By definition, an anti-particle is the one which has an opposite charge or direction of spin to that of the other particle, while all other physical properties are identical except those related to the sign of the charge and the direction of the spin. The process of transformation from mass of matter and anti-matter pair into energy is called *annihilation* in physics. It is a direct process of converting mc^2 to E according to the Einstein equation. The reverse of the annihilation process, i.e. from E into mc^2, is called the pair production. The concept of annihilation and pair production has been known to scientists since the discovery of the $E=mc^2$ equation at the beginning of twentieth century and several decades before the discovery of anti-matter.

The first anti-particle discovered was the anti-electron, the so-called positron, in 1933 by Anderson [3] in the cloud chamber due to cosmic radiation. The existence of the anti-electron (positron) was described by Dirac's hole theory in 1930 [4]. The result of positron—electron annihilation was detected in the form of electromagnetic radiation [5]. The number and event of radiation photons is governed by the electrodynamics [6, 7]. The most common annihilation is via two- and three-photon annihilation, which do not require a third body to initiate the process. These are two of the commonly detected types of radiation from positron annihilation in condensed matter. The cross section of three-photon annihilation is much smaller than that of two-photon annihilation, by a factor on the order of the fine structure constant, α [8]. The annihilation cross section for two and three photons is greater for the lower energy of the positron—electron pair; it varies with the reciprocal of their relative velocity (v). In condensed matter, the positron—electron pair lives for only the order of a few tenths to a few nanoseconds against the annihilation process.

A positron in an electronic media can pick up an electron and form a neutral atom called Positronium (Ps) [9]. The existence of Ps and its chemical reaction with molecules was detected from annihilation photons in 1951 [10]. Ps is formed in most molecular systems. Due to the different combinations of positron and electron, there are two states of Ps: the para-Ps (p-Ps) from the anti-parallel spin. and the ortho-Ps (o-Ps) from the parallel spin combination. The lifetime and the annihilation events for p-Ps and o-Ps are very different from each other, as given by electromagnetic theory. Figure 1.1 shows basic physical properties of Ps and compares them with the H atom, although it should not be considered an isotope of H (see problems 1.5 and 1.6 and answers at the end of this chapter).

The p-Ps has a shorter lifetime than o-Ps and it annihilates into two photons, while o-Ps annihilates into three photons. The intrinsic lifetime is 0.125 ns and 142 ns for the free p-Ps and o-Ps, respectively. In ordinary molecular media, the electron density is low enough so that Ps can pick off electrons from the media that have anti-parallel spin to that of the positron, and undergo two-photon annihilation. This is called the pick-off annihilation of Ps. The pick-off annihilation of o-Ps not only occurs in the form of two-photon annihilation, it also shortens the o-Ps lifetime from 142 ns (free o-Ps) to a few ns. The pick-off annihilation lifetime of o-Ps in molecular systems is about one order of magnitude greater than in crystalline or metallic media. Experimental determination of o-Ps lifetime is one of the most useful methods for positron and positronium chemistry. This is because o-Ps lifetime contains information about electron density, which governs the basic properties of chemical bonding in molecules. It is also controlled by the physical structure of molecules.

Positron and positronium chemistry is usually considered a branch of traditional nuclear chemistry [11]. However, due to the richness of the positron annihilation process in molecular systems, it has developed into four traditional areas of chemistry discipline: analytical chemistry, inorganic chemistry, organic chemistry and physical chemistry and beyond. Analytical development follows electron spectroscopy but it includes certain specialties and sensitivities which the electron does not have. One of the most beneficial positron chemistry focuses of positron annihilation spectroscopy (PAS) in chemical analysis is that the annihilation photons can be easily detected at a single-event level. The quantity of radioisotopes needed for a practical radiochemical method is on the order of μCi in a typical PAS laboratory. PAS also inherits fruitful knowledge from electron spectroscopy and nuclear technology. The specialty of using the positron and Ps in chemical analysis is mainly due to the nature of their localization into the sites of low electron density, such as a vacancy, the surface, free volumes, holes, pores, and voids. Any of these sites in molecular systems will be the targeted regions for studies using PAS as the dominating annihilation signals. Notable advances of PAS in chemical instrumentations in the last two decades are slow-positron beams [12, 13] and positron-induced Auger electron spectroscopy (PAES) [14], which are potentially powerful for chemical analysis of the outermost layer of surfaces, atomic defects, and oxidation states. This will be discussed in Chapters 3 and 14.

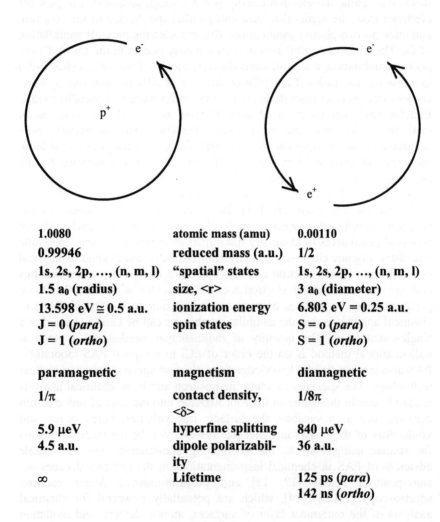

1.0080	atomic mass (amu)	0.00110
0.99946	reduced mass (a.u.)	1/2
1s, 2s, 2p, ..., (n, m, l)	"spatial" states	1s, 2s, 2p, ..., (n, m, l)
1.5 a_0 (radius)	size, $<r>$	3 a_0 (diameter)
13.598 eV ≅ 0.5 a.u.	ionization energy	6.803 eV = 0.25 a.u.
J = 0 (*para*) J = 1 (*ortho*)	spin states	S = 0 (*para*) S = 1 (*ortho*)
paramagnetic	magnetism	diamagnetic
1/π	contact density, $<\delta>$	1/8π
5.9 μeV	hyperfine splitting	840 μeV
4.5 a.u.	dipole polarizability	36 a.u.
∞	Lifetime	125 ps (*para*) 142 ns (*ortho*)

Figure 1.1 Comparison of Positronium (Ps) and H Atoms.

The areas of inorganic and organic positron chemistry deal mainly with material characterization and industrial applications using PAS. Both chemical and electronic industries have found PAS to be a powerful method. In addition to the traditional solution chemistry of the positron and Ps [11], PAS has been developed to determine the free volume Born-Oppenheimer approximation,, such as molecular solids [14] and polymers [15]. The unique localization property of Ps in free volumes and holes has opened new hope in polymer scientific research that determination of atomic-level free volumes at the nanosecond scale of motion is possible. During the last ten years, most positron annihilation research has involved a certain amount of polymer chemistry, polymers and coatings, which will be discussed in Chapters 12 and 13. For inorganic systems, oxides are mostly studied using the positron and Ps. Silicon oxides and zeolites are the most important systems in positron and Ps chemistry. The developments in this area have on the cavity structure and chemical states of inner surfaces. Chapters 8 and 14 will discuss this subject.

Physical chemistry of the positron and Ps is unique in itself, since the positron possesses its own quantum mechanics, thermodynamics and kinetics. The positron can be treated by the quantum theory of the electron with two important modifications: the sign of the Coulomb force and absence of the Pauli exclusion principle with electrons in many electron systems. The positron can form a bound state or scatter when it interacts with electrons or with molecules. The positron wave function can be calculated more accurately than the electron wave function by taking advantage of simplified, no-exchange interaction with electrons. However, positron wave functions in molecular and atomic systems have not been documented in the literature as electrons have. Most researchers perform calculations at certain levels of approximation for specific purposes. Once the positron wave function is calculated, experimental annihilation parameters can be obtained by incorporating the known electron wave functions. This will be discussed in Chapter 2.

Thermodynamics and kinetics of the positron and Ps have been developed to a certain extent, mainly in liquids. It is interesting to observe that the positron and Ps are in fact localized in the bubble or the defect sites of molecular media [17]. Therefore, the motion of the positron and Ps in molecular media is always coupled with the surrounding molecules, i.e. never in the form of a free particle in liquids or solids. Most existing theories of the positron and Ps in condensed media thus still rely on classical thermodynamics and kinetics. This will be discussed in Chapters 5 and 6.

There are other areas of chemistry in addition to the traditional four mentioned above, such as biological chemistry and environmental chemistry, which are basically still untouched by positron and Ps researchers. They are not covered in this book since significant results in these two areas have been presented in our previous book [11].

Ps chemistry stands out by itself, beyond traditional areas of chemistry. For example, when Ps reacts with molecules, it can just become part of the electron and positron orbital in molecules, such as a Ps-molecular bound state formation, or it can travel like a particle, either diffusing through interstitutial regions or localizing in open spaces, such as substitution regions, free volume, holes etc. Because Ps cannot be considered an isotope of the H atom and has zero protons (problem 1.5), it should have its own special place in the Periodic Table of the Elements, as the Period 0 (zero), and Group 1, or 1A (Problem 1.6 and answers at the end of this chapter). Once one adopts Ps as group 0, then one could build the counter part of the Periodic Table built by anti-atoms. The first anti-atom then will be anti-hydrogen, which is made of an anti-proton and a positron. The rest anti-atoms from positrons could be built up similarly as that in the matter of the existing Periodic Table of the Elements. That will be the Periodic Table of the Anti-Elements. However, some people categorize Ps as an exotic atom [18]. In any case, Ps is the lightest atom that exists in this Universe and its study constitutes a new area of chemistry.

1.2 Existing books and articles on positron and positronium chemistry and annihilation

This is the second comprehensive book following our previous *Positron and Positronium Chemistry* in 1988 [11]. Prior to this, there was one excellent book by Green and Lee in 1964 [19]. A series of proceedings have been published tri-annually for Positron and Positronium Chemistry (PPC) Workshops since 1978: PCC1 [20] in Blacksburg, VA; PPC2 [21] in Arlington, TX; PPC3 [22] in Milwaukee, WI; PPC4 [23] in Strasburg, France; PPC5 [24] in Lillafured, Hungary, PPC6 [25] in Tsukuba, Japan. Each proceedings contains the newest results. Another book, one of great value, on Ps by a single author has also been also published [26]. For more general positron, positronium, and positron annihilation research, there is another series of proceedings from the International Conference on Positron Annihilation (ICPA) tri-annually since 1967 [27-35]. Another series of proceedings is the Slow Positron Workshop (SLOPOS), which is held every two to three years since 1990 [36-41]. Many excellent books on positron annihilation in solid states and gases are available [42-47]. Other comprehensive reviews are also available elsewhere and it would not be possible to cite all of them in this chapter.

The first article on the positron was published in 1933 [3]; on Ps, in 1945 [9]. Since then, the number of research articles in this area has grown year-by-year almost linearly [46]. Figure 1 shows the number of articles on positron and Ps per year since 1933, as searched from the Chemical Abstract and Science Abstract database. Certain years show a large number of articles, which reflect important advancements in PAS and Ps chemistry. For example, in the mid-1960s, the positron was discovered to localize in defects of solids; in the 1970's came the discovery of surface positrons; in the 1980's, the observation of Ps localized in free volumes of polymers; in the 1990's, mature developments of the slow-positron beam technique. Currently, there are about one thousand articles per year published in the area of positron, Ps, and positron annihilation in materials. Although there is no clear line to distinguish positron chemistry from other areas of science, about half of the articles published since the late 1990's deal with chemistry or chemistry-related research. For example, in 2000 about 50% of the papers contributed to ICPA-12 were chemistry- and polymer-related research, as shown in Figure 2 [47]. By 2001 the publication rate has grown to about 800 papers per year. From this data, it is obvious that positron and Ps chemistry is a steady and growing area of scientific research.

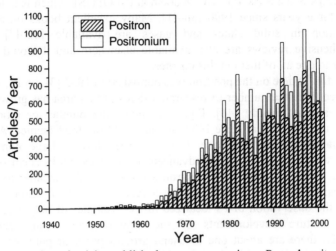

Figure 1.2 Number of articles published per year on positron, Ps, and positron annihilation in materials.

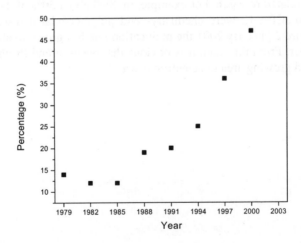

Figure 1.3. Percentage of papers on chemistry and polymers over total papers presented in positron annihilation research (from ICPA data) vs the year [48].

Problems

Problem 1.1 Would you agree that the positron is called the "anti matter" of the electron since it attracts electrons electromagnetically? Do you have any other suggestion for the positron in conjunction with electrons?

Problem 1.2 What is the size of Ps? Is it the same or twice that of H?

Problem 1.3 Justify why the ionization energy of Ps is about half that of H.

Problem 1.4 Would you consider Ps a metal or a non-metal in the Periodic Table of the Elements?

Problem 1.5 Why is it inappropriate to consider Ps a light isotope of H?

Problem 1.6 What is the lightest atom in the Universe? Where would you place it in the the Periodic Table of the Elements?

References

[1] The Bible, Book of Genesis 1:1.
[2] The Bible, Book of Matthew 24:35.
[3] C.D. Anderson *Phys. Rev.* **43**, 491 (1933).
[4] P.A.M. Dirac *Proc. Camb. Phil. Soc.* **26**, 361 (1930).
[5] O. Klemperer *Proc. Camb. Phil. Soc.* **30**, 347 (1934).
[6] For example, see R.F. Feyman, *Quantum Electrodynamics*, Reading, MA: Benjamin/Cummings (1983).
[7] D.C. Lauristsen, J.R. Oppenheimer *Phys. Rev.* **46**, 80 (1934).
[8] P.A.M. Dirac, *Principles of Quantum Mechanics*, Oxford: Oxford University Press (1935).
[9] A.E. Ruark *Phys. Rev.* **68**, 278 (1945).
[10] M. Deutsch *Phys. Rev.* **82**, 455; **83**, 866 (1951).
[11] D.M. Schrader, Y.C. Jean, Eds. *Positron and Positronium Chemistry*, Amsterdam: Elsevier Sci. Pub. (1988).
[12] P.J. Schultz, K.G.Lynn *Rev. Mod. Physics*, **60**, 701 (1989).
[13] P. G. Coleman, Ed. *Positron Beam and Their Applications*, Singapore: World Sci. Pub. (2000).
[14] A. Weiss, *Positron Annihilation Electron Spectroscopy*, in *Handbook of Surface Imaging and Visualization*, A.T. Hubbard, Ed., Chapter 45, pp. 617-632, CRC Press (1995).
[15] M. Eldrup, in *Proceedings of 6th International Conference on Positron Annihilation*, Amsterdam: North-Holland Pub. P.G. Coleman, S.C. Sharma, L.M. Diana, Eds. (1982), p 773.
[16] Y.C. Jean *Microchem. J.* **42**, 72 (1990).
[17] R.A. Ferrell *Phys. Rev.* **108**, 167 (1957).
[18] D. Horváth and R.M. Lambrech, *Exotic Atoms*, Amsterdam: Elsevier Sci. (1984).
[19] J. Green and J. Lee, *Positronium Chemistry*, New York: Academic Press, (1964).
[20] H. Ache, Ed. *Positronium and Muonium Chemistry*, Adv. Chem. Ser. #175, Washington D.C.: ACS Press (1979).
[21] S.C. Sharma, Ed. *International Symposium on Positron Annihilation Studies of Fluids*, Singapore: World Sci. Pub. (1988).
[22] Y.C. Jean, Ed. *Third International Workshop on Positron and Positronium Chemistry*, Singapore, World Sci. Pub. (1990).
[23] I. Billard, Ed. *Fourth International Workshop on Positron and Positronium Chemistry, J. De Physique, Vol. 3, Coll. N 4, Suppl. JP, n 9*, Les Ulis Cedex A, France: Zone Indu. De Couttaoeuf, (1993).
[24] Zs. Kajcsos, B. Levay, K. Suvegh, Eds. *Proceedings of the 5th International Workshop on Positron and Positronium Chemistry, J. Radioanal. Nucl. Chem.*, Akademiai Kiado, Budapest: Elsevier Science, A (1996).

[25] Y. Ito, T. Suzuki, Y. Kobayashi, Eds. *PCC6 Proceedings, Radiation Physics and Chemistry, Vol. 58*, Oxford: Paragon, Elsevier Sci. (2000).
[26] O.E. Morgensen, *Positron Annihilation in Chemistry*, Berlin: Springer-Verlag (1995).
[27] A.T. Stewart, L.O. Roellig, Eds. *Positron Annihilation*, New York: Academic Press (1967).
[28] R.R. Hasiguti, K. Fujiwara, Eds. *Proceedings of 5th International Conference on Positron Annihilation*, Sendai, Japan: Japan Institute of Metals (1979).
[29] P.G. Coleman, S.C. Sharma, L.M. Diana, Eds. *Proceedings of 6th International Conference on Positron Annihilation*, Amsterdam: North-Holland Pub. (1982).
[30] P.C. Gain, R.M. Singru, K.P. Gopinathan, Eds. *Proceedings of 7th International Conference on Positron Annihilation*, Singapore: World Sci. Pub. (1985).
[31] L. Dorikens-Vanpraet, M. Dorikens, D. Segers, Eds., *Proceedings of 8th International Conference on Positron Annihilation*, Singapore: World Sci. Pub. (1988).
[32] Zs. Kajcsos, Cs Szeles, P.G. Coleman, S.C. Sharma, L.M. Diana, Eds. *Proceedings of 9th International Conference on Positron Annihilation, Mater. Sci. Forum*, **105-110**, Aedermannsdorf, Switzerland: Trans Tech Pub. (1992).
[33] Y.-J. He, B.-S. Cao, Y.C. Jean, Eds. *Proceedings of 10th International Conference on Positron Annihilation, Mater. Sci. Forum* **175-178**, Aedermannsdorf, Switzerland: Trans Tech Pub. (1995).
[34] Y.C. Jean, M. Eldrup, D.M. Schrader, R.N. West, Eds. *Proceedings of 11th International Conference on Positron Annihilation, Mater. Sci. Forum*, **255-157**, Aedermannsdorf, Switzerland: Trans Tech Pub. (1997).
[35] W. Trifshäuser, G. Kögel, P. Sperr, Eds. *Proceedings of 12th International Conference on Positron Annihilation, Mater. Sci. Forum*, **363-365**, Aedermannsdorf, Switzerland: Trans Tech Pub. (2000).
[36] P.J. Schultz, G.R. Massoumi, P.J. Simpson, Eds. *Positron Beams for Solids and Surfaces, AIP Conference Proceedings*, **218**, New York: Academic Press (1990).
[37] E. Ottewitte, A.H. Weiss, Eds. *Positron Beams for Solids and Surfaces, AIP Conference Proceedings*, **303**, New York: Academic Press (1994).
[38] M. Doyama, T. Akahane, M. Fujinami, Eds. *Proceedings of the 6th International Workshop on Slow-Positron Beam Techniques for Solids and Surfaces, Appl. Surf. Sci.* **85**, Amsterdam: Elsevier (1995).
[39] W.B. Waeber, M. Shi, A.A. Manuel, Eds. *Proceedings of the 7th International Workshop on Slow-Positron Beam Techniques for Solids and Surfaces, Appl. Surf. Sci.* **116**, Amsterdam: Elsevier (1997).
[40] D.T. Britton, M. Härting, Eds. *Proceedings of the 8th International Workshop on Slow-Positron Beam Techniques for Solids and Surfaces, Appl. Surf. Sci.* **149**, Amsterdam: Elsevier (1999).

[41] G. Bauer, W. Anwand, Eds. *Proceedings of the 9th International Workshop on Slow-Positron Beam Techniques for Solids and Surfaces, Appl. Surf. Sci.* **169**, Amsterdam: Elsevier (2002).
[42] P. Hautojärvi, Ed. *Positrons in Solids*, New York: Springer, (1979).
[43] W. Brandt, A. Dupasquier, Eds. *Positron Solid-State Physics*, Amsterdam: North-Holland (1983).
[44] A. Dupasquier, A.P. Mills, Jr., Eds. *Positron Spectroscopy of Solids*, Amsterdam: ISO (1995).
[45] R. Krause-Rehberg, H.S. Leipner, *Positron Annihilation in Semiconductors*, Berlin: Springer (1999).
[46] C.M. Surko, F.A. Gianturco, Eds. *New Direction in Antimatter Chemistry and Physics*, Dordrecht, The Netherlands: Kluer Acad. Press (2001).
[47] Y.C. Jean, R.M. Lambrecht, D. Horvath, *Positrons and Positronium: A Bibliography 1930-1984*, Amsterdam: Elsevier (1988).
[48] Y.C. Jean, in *Proceedings of 9th International Conference on Positron Annihilation, Mater. Sci. Forum*, **363-365**, Aedermannsdorf, Switzerland: Trans Tech Pub., W. Trifshäuser, G. Kögel, P. Sperr, Eds. (2000), pp. 701-704.
[49] I.N. Levine, *Quantum Chemistry*, 5th ed. Upper Saddle River, NJ: Prentice Hall (2000).

Answers to problems

Problem 1.1 Anti-matter and matter interact via the annihilation process, which is represented by a delta function or zero distance between them (see Chapter 2), while the electromagnetic interaction is represented by the reciprocal of the distance. Electromagetically, the positron attracts an electron; it is suggested to call the positron as the *counter* electron analogy to a counter ion to an ion from chemistry perspective.

Problem 1.2 The size of Ps is about 1.5 Å as the expectation value of the Bohr diameter (Figure 1.1). The size of Ps is the same as that of H (note: in Figure 1.1, the size for Ps is diameter while for H, it is radius) and not twice.

Problem 1.3 According to Bohr's theory of the H atom [49], the energy (E_n) at n levels of H is given as $= -\mu/2n^2$ (in a.u.), where μ is the reduced mass of H. The same Bohr's theory is applicable to Ps, with a reduced mass μ_{Ps} of half mass of the electron or about half that of μ_H. Therefore, the ionization energy (I.E). for the ground state (n=1) of Ps is only about half that of H.

Problem 1.4 Ps could be considered a metalic element instead of a non-metallic element as it is put in the Period Table of the Elements as shown later in the answer of Problem 1.6 (Figure 1.6). Ps has an ionization energy (I.E.) of 6.8 eV, which is in a nice sequence along with the elements of Group 1A: Li (5.4 eV), Na (5.1 eV), K (4.3 eV), Rb (4.2 eV), and Cs (3.9 eV), while H is off-line of the sequence as plotted in Figure 1.4. Similarly, Ps has an electron affinity (E.A.) of 1.1 eV (see Chapter 2), which follows another nice trend with those of Group 1A elements: H (0.76 eV), Li (0.62 eV), Na (0.55 eV), K (0.50 eV), Rb (0.49 eV), and Cs (0.48 eV) as plotted in Figure 1.5. Note that I.E. and E.A. are defined as follows: I.E.= energy of $(X^+ + e^-)$—energy of (X) for the reaction $X \rightarrow X^+ + e^-$ and E.A.= energy of $(X + e^-)$—energy of (X) for the reaction $X + e^- \rightarrow X^-$.

Problem 1.5 By definition, isotopes have the same proton number but a different neutron number. However, Ps and H have a different proton number but the same neutron number (zero). In chemistry, isotopes always have similar chemical reactivities. However, Ps reacts with molecules very differently from H. In physics, although their atomic size is the same, their energy levels are distinctly different from each other. Therefore, it is inappropriate to consider Ps a light isotope of H. It should stay at its own location: Period 0 and Group 1, or 1A as proposed in the revised Periodic Table of the Elements as revised and shown in Figure 1.6. Ps has its own new chemistry.

Problem 1.6 Ps is the lightest atom in the Universe. This atom has no nucleons with a mass of 0.0011 a.u., which is close to zero mass, which should be placed in Period 0. However it has one electron, which belongs to Group 1, or 1A in the existing Periodic Table of the Elements. A revised Periodic Table of the Elements is proposed, as in Figure 1.6, after one includes Ps.

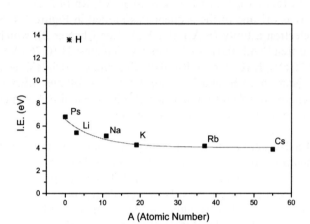

Figure 1.4 Ionization Energy of Elements vs Atomic Number.

Figure 1.5 Electron Affinity of Elements vs Atomic Number.

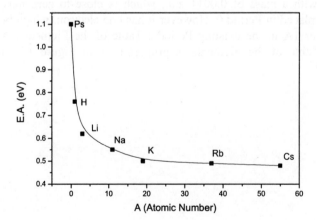

Introduction to Positron and Positronium Chemistry 15

Figure 1.6 Suggested Period Table of the Elements

Chapter 2

Compounds of Positrons and Positronium

D. M. SCHRADER

*Chemistry Department, Marquette University, P.O. Box 1881,
Milwaukee, WI 53201-1881, USA
david.schrader@marquette.edu*

2.1 Introduction

In this chapter we summarize our present knowledge of chemical compounds that contain positrons or positronium. We do not give an exhaustive review of the literature, but rather we try to give information that is correct up to at least January 1, 2002. Since most of our knowledge of bound positronic systems comes from theoretical work, this chapter is primarily concerned with quantum mechanics. The most conspicuous exception is provided by recent observations of vibrational shifts of compounds that have resonantly captured a positron [1], implying the existence of bound states.

We first review some elementary physics that establishes the kind of quantum mechanics that can be profitably applied to mixed electron-positron systems. Next we describe some methods of calculation that have proven to be useful recently. Finally, all the binding energies and annihilation rates that are known for atomic and molecular systems are listed in tables and discussed.

The history of quantum calculations on mixed systems is interesting. Except for positronium, it all began in 1946 with a famous paper on polyleptonic systems by Wheeler [2]. For the next fifty years, papers appeared at the glacial rate of about one every two years. Then two powerful methods appeared at about the same time in the hands of two powerful groups, and the publication rate increased abruptly by an order of magnitude. The number of publications per year shows an interesting pattern:

1946–95	1996	1997	1998	1999	2000	2001
0.4 (ave.)	4	4	12	9	5	6

Perhaps the spike in 1998 indicates that the easy systems have been treated, and from now on more work will be required to continue our exploration of these systems. We show below that for atomic systems, most attention has been paid to the families on the left and right edges of the periodic chart, leaving the interesting middle part less well understood.

2.2 Quantum mechanical considerations

Positrons have the charge of protons and the mass of electrons. Quantum chemists who take an interest in mixed electron-positron systems immediately recognize some interesting consequences of this transparent observation. For example, (a) the familiar Born-Oppenheimer approximation cannot be used for positrons, but rather positrons must be treated as distinguishable electrons; (b) electron-positron correlation is more important, pair by pair, than correlation between leptons of like charge; and (c) there are always core electrons (except for the simplest systems), but positrons congregate in the valence region or beyond.

2.2.1 Basic physics of mixed electron-positron systems

The most interesting quantities of mixed systems that can be both calculated and measured today are binding energies, annihilation rates, and momentum distributions of the annihilation photons. A system has a binding energy in some state if it is *chemically* stable in that state, meaning that it stays in that state until it annihilates. More properly, we say that a system is chemically stable in some state if its annihilation rate is greater than the sum of the rates of all other processes that depopulate that state.

Of course, all mixed systems are unstable in a broader sense because they ultimately annihilate. This implies that each pre-annihilation system is in reality a metastable state or a resonance embedded in a continuum that embraces one less electron, one less positron, and two or three more photons. If we perform calculations on the pre-annihilation system while ignoring the coupling to this continuum and then include the coupling as a second step in the calculation, our energies will be shifted and broadened on account of the coupling. Fortunately, the coupling is quite weak [2], and the shift and broadening are quite small. For the positronium atom the shift for the ground state energy is only 360 μeV [3], and the broadening is no more than $\sim \hbar/125$ ps ≈ 5 μeV. We conclude that the coupling is safely ignored for purposes of calculating wave functions and energy levels.

The positron has the same spin magnetic moment (except for direction) as the electron, so familiar approximations such as the Born-Oppenheimer approximation for nuclei and Russell-Saunders coupling for lepton spin in light atoms are appropriate in applications to electron-positron systems. In other words, we need only make straightforward adjustments in our methods of calculation to accommodate two kinds of electrons, and the full, formidable technology of quantum chemistry is at our disposal. In order to get accurate results, we need to account specifically for electron-positron correlation, which precludes the use of single- particle models such as the SCF (self-consistent field) method. In particular, many-body perturbation theory seems fraught with hazards attending double-counting from using positronium intermediate states [4].

When an accurate wave function is calculated by some effective method, we can calculate an accurate annihilation rate with the use of an effective annihilation operator [5]:

$$\lambda = \langle \Psi | \sum_{e,p} \hat{\lambda}_{e,p} | \Psi \rangle \tag{2.1}$$

$$\hat{\lambda}_{e,p} = 8\pi \sum_{S,M} D_S \, |\sigma_{S,M}(e,p)\rangle \, \delta^3(\boldsymbol{r}_{ep}) \, \langle \sigma_{S,M}(e,p)| \tag{2.2}$$

$$D_0 \approx 7.989 \, \text{ns}^{-1}, \quad D_1 \approx 7.040 \, \mu\text{s}^{-1} \tag{2.3}$$

The sum in Eq. (2.1) is over all electron-positron pairs in the system, and the sum in Eq. (2.2) is over $S = 0, 1$. The projection operators in Eq. (2.2) contain the familiar two-particle spin functions.

2.2.2 The calculation of annihilation rates

A given system in a given state has only one annihilation rate. The terms "two-photon annihilation rate," "three-photon annihilation rate", and "spin-averaged annihilation rate," sometimes seen in the literature, have no operational meaning and are not measurable [5].

Despite this, the spin-averaged annihilation rate is a useful concept. If a positron is immersed in an electron-rich medium in which all electron spins are paired, its wave function at short range resembles that of positronium but with all the electrons participating, one at a time. This is called "virtual positronium." Since all the electrons are equivalent, the positron experiences, in equal measure, electron spins parallel and antiparallel to its own. The parallel alignment can only annihilate by the slow three-photon process, and the antiparallel alignment is equally likely to become quantized as either o-Ps ($m = 0$) or p-Ps at the moment of annihilation. The *para* annihilation process is much faster and dominates. Therefore the annihilation rate for this environment is close to one-fourth that of isolated p-Ps. This is called the spin-averaged annihilation rate, and its ideal value is:

$$\frac{1}{4}D_0 + \frac{3}{4}D_1 \approx 2.003 \, \text{ns}^{-1} \qquad (2.4)$$

While this exact value is not the annihilation rate for any real system, it is remarkable how many diverse closed-shell systems have annihilation rates close to it. Presumably, this is because the buildup of electron density at an embedded positron quickly saturates due to moderating Coulomb repulsion and exclusion effects except in media of extreme electron densities. One estimation of the effective number of electrons gathered around the positron in closed-shell systems is its annihilation rate divided by the rate on the right side above.

2.2.3 Quantum mechanical methods

2.2.3.1 The stochastic variational method (SVM)

A complete description of the stochastic variational method has been given elsewhere [6], and we only briefly sketch the method here.

A trial wave function is expanded in a sum of correlated Gaussian functions:

$$\Psi = \sum_{i=1}^{K} c_i \exp\left(-\sum_{\mu,\nu=1}^{N-1} a_{\mu\nu}^{(i)} \boldsymbol{r}_\mu \cdot \boldsymbol{r}_\nu\right) \qquad (2.5)$$

, where K is the number of terms in the expansion, c_i are linear variational parameters, N is the number of particles considered, $\boldsymbol{r}_\nu$ is the position of ν-th particle relative to one reference particle, and $a_{\mu,nu}^{(i)}$ is an exponential variational parameter. For example, the first term in the sum over i for the helium atom is composed of four terms:

$$c_1 \exp\left(-a_{11}^{(1)} r_1^2 - (a_{12}^{(1)} + a_{21}^{(1)})\mathbf{r}_1 \cdot \mathbf{r}_2 - a_{22}^{(1)} r_2^2\right) \qquad (2.6)$$

Symmetry constraints, including those that serve to define the spin state of the system, fix the number of independent exponential parameters at two in this example. For positronium hydride, PsH, each term in the sum over i is composed of nine terms in the sum over μ and ν, of which four are independent.

One can see in Eq. (2.6) that interparticle correlation is treated directly. One also notices that both the short- and long-range behaviors of the basis functions are incorrect. To compensate for these inadequacies, basis sets must be large. The number of basis functions required for good accuracy is so large that the exponential parameters are in practice indeterminable by direct minimization of the energy, and instead are found by a stochastic process described below. The disadvantage of dealing with large basis sets is compensated for by the ease with which the matrix elements that appear in the secular equation are calculated.

The wave function is built up in steps. The process is initiated by choosing a small basis set of $K' \ll K$ members with exponential parameters determined by experience, guess, or a direct minimization. The linear parameters are calculated by solving a secular equation.

At each step, one basis set is added to the set already in use, and a sequence of calculations is performed, each with the added exponential parameters determined by random selection. The set of added parameters that give the lowest energy defines the new basis function, which then becomes a permanent member of the basis set provided the energy improvement is greater than a predetermined increment. This increment is adjusted during the calculation to control convergence.

The process is terminated when the energy converges satisfactorily. Computers presently in use permit a maximum of about 1000 linear parameters and about 10,000 exponential parameters. Up to six particles have been treated explicitly by this method, which can be extended to larger systems by use of effective core potentials. The accuracy of calculated binding energies can be substantially less than 1 meV.

The SVM method gives the energy directly, and the wave function provided by SVM is well-suited to the calculation of annihilation rates.

2.2.3.2 The quantum Monte Carlo (QMC) method

There are several distinct methods that that are referred to as "QMC" methods. Here we consider the "fixed-node diffusion" QMC method, or the "DMC" method; other members of the QMC family are described in several recent reviews [7, 8, 9]. DMC has been widely applied to ground state systems in the Born-Oppenheimer approximation [10], but is not limited to such cases.

In QMC, the time-dependent Schrödinger equation (we use atomic units throughout this chapter),

$$-\frac{1}{i}\frac{\partial}{\partial t}\Psi = -\frac{1}{2}\nabla^2\Psi + V\Psi \tag{2.7}$$

is converted to a classical diffusion equation by the substitution $\tau = it$, giving

$$\frac{\partial}{\partial \tau}\Psi = \frac{1}{2}\nabla^2\Psi - V\Psi. \tag{2.8}$$

These two equations are similar but have very different interpretations. Equation (2.7) describes an n-particle system in 3-dimensional space. Equation (2.8) can be considered to describe the diffusion of a collection of particles (called "walkers" here to avoid confusion) in $3n$-dimensional space with a source or sink. The function Ψ gives the distribution of the walkers in this space. Each walker is a point in $3n$-dimensional space. Equation (2.8) can be solved by permitting each of the walkers to make a $3n$-dimensional random walk, and for the population of particles to grow or shrink according to the potential term.

The solutions of classical diffusion equations are necessarily positive, and the nodes of quantum mechanical wave functions required by the exclusion principle render a direct solution of Eq. (2.8) impossible. Instead, a new function F is defined as the product of the unknown function Ψ and an approximation to it:

$$F = \Psi\tilde{\Psi}, \tag{2.9}$$

The function $\tilde{\Psi}$ is not provided by the QMC method. It need not be highly accurate in all regions in space, but it must have accurate nodal surfaces in $3n$-dimensional space, and it must be accurate near singularities of V. In practice, simple SCF wave functions provide sufficiently accurate nodal surfaces, and symmetrical Jastrow factors [11] that serve to satisfy cusp conditions that can be appended to an SCF wave function to give a serviceable $\tilde{\Psi}$. The positronic part of $\tilde{\Psi}$ can usually be inferred by simple structural considerations.

The new unknown, F, must satisfy an equation slightly more complicated than Eq. (2.8). Ψ is assumed to have the same nodes as $\tilde{\Psi}$, and F is therefore non-negative. Typical calculations embrace 1000 walkers, each of which take 10^4 to 10^5 steps, each of length 10^{-2} to 10^{-4} atomic units of time. The guided random walks are run until a steady distribution of walkers is reached. A solution consists of a table of set $3n$ coordinates for each walker, and the corresponding distribution of walkers replicates F.

Sources of imprecision are an insufficient number of particles and steps, too large a step length, and inaccuracies in $\tilde{\Psi}$ other than the position of its nodes. Errors in the nodal surfaces of $\tilde{\Psi}$ translate into positive errors in the calculated energy, so an upper bound is obtained. Ideally, the exact solution Ψ of the Schrödinger equation is provided within the loges bounded by the (approximate) nodal surfaces.

The QMC method is ideally suited for mixed systems because electron-positron correlation, which is difficult to treat with CI methods, is automatically treated correctly. Systems of up to a bit more than ten leptons are routinely treated. Effective core potential methods can be used to extend the method to larger systems. Expectation values of local operators for the distribution $|\Psi|^2$ are calculated by straightforward sampling procedures, but nonlocal operators, such as those for the annihilation rate, are problematic and are under active investigation [12].

QMC was first applied to electron-positron systems in 1992 [13].

2.2.3.3 Other methods

In the configuration interaction (CI) method, basis functions are centered on nuclei, which requires that a great many functions are needed to satisfactorily simulate the important electron-positron cusps in the exact wave function. Both binding energies and annihilation rates converge slowly. Nonetheless, owing to steady improvements in computing resources, CI methods have lately become competitive, in particular for systems that do not have free positronium in their lowest dissociation thresholds. Indeed, the first reliable predictions of positron binding to Ca, Sr, and Cd were made from CI calculations (see Table 2.4).

The simple Hartree-Fock (HF) method, which is essentially a CI method

Table 2.1 Explanation of the tables.

Symbol	Meaning
AR	Annihilation rate, in ns^{-1}, in italics in Table 2.8.
BE	Binding energy, in eV.
CI	Configuration interaction method.
CIcp	Configuration interaction method with a core potential.
cp	Core potential.
DFT	Density functional theory.
HF	Hartree-Fock method.
Hyll	Hylleraasian basis expansion method.
MBPT	Many-body perturbation theory.
nbr	The number of leptons explicitly treated.
o	*ortho*, the high-spin member of a set of states.
p	*para*, the low-spin member of a set of states.
QMC	Quantum Monte Carlo method.
QMCcp	Quantum Monte Carlo method with a core potential.
SVM	Stochastic variational method.
SVMcp	Stochastic variational method with a core potential.
(15)	Uncertainty in the decimal immediately preceding.

with only one configuration, would seem to be an unpromising method. However, there is a class of compounds for which even this ancient approach is able to give reasonable binding energies. These are simple polar molecules such as diatomic molecules for which the bonding is predominantly ionic. Since a simple dipole moment of as little as 1.625 debye can bind an electron or a positron, real molecules with dipole moments rendered larger than that by a HF calculation are guaranteed to bind a positron in that approximation, even without electron-positron correlation. A proper treatment of electron-positron correlation is required for accuracy in calculated annihilation rates, so this approach will not give satisfactory results for that quantity.

Density functional theory (DFT) has found useful application to positrons that are localized in extended structures [14], and to the construction of

Table 2.2 The purely leptonic systems.

System	Binding Energy	Threshold	Annihilation Rate
p-Ps	$6.8032081(3)^{(a)}$	$e^+ + e^-$	$7.989476(13)^{(b)}$
o-Ps	$6.8023628(3)^{(a)}$	$e^+ + e^-$	$7.039970(10)\times 10^{-3,\,(c)}$
$e^\pm$Ps	$0.32667457(1)^{(d)}$	$e^\pm + $ Ps	$2.0861223^{(e)}$
Ps$_2$	$0.435485^{(f)}$	2Ps	$4.446^{(g)}$
$e^\pm$,Ps$_2$	unbound$^{(h)}$	$e^\pm + $ Ps$_2$	

(a)[17, 18, 19, 20]. (b)[21, 22]. (c)[21, 23, 24, 25]. (d)[2, 26]. (e)[27, 28]. (f)[2, 29]. (g)[29]. (h)[30].

correlation potentials for scattering calculations [15]. Recently, DFT has been applied to the binding of positrons and positronium to simple atoms with encouraging results [16], although this method has not yet proven to be superior to more fully *ab initio* approaches for the simple systems to which it has been applied.

2.3 Current knowledge of bound states

Tables of data are given here, meant to be complete through January 1, 2002. For each species, we list only the "best" results, which we have taken from the most nearly pure *ab initio* calculations. Some authors present extrapolations beyond calculated results; these are not given here. The symbols used in the data tables are explained in Table 2.1.

2.3.1 *Polyleptons*

Binding energies and annihilation rates for polyleptons are given in Table 2.2. The current values for positronium are listed for completeness. Since Wheeler's seminal 1946 paper, the 3- and 4-particle polyleptons have been the subject of many studies, and their properties are well understood today. The annihilation rate of diatomic positronium, Ps$_2$, is about twice the spin-averaged rate in Eq. (2.4) because there are two positrons and each of them sees spin-paired electrons. Recently a calculation of the 5-particle

Table 2.3 Hydrogen, and the alkali and coinage metals.

System	BE/eV	Method,nbr	Threshold	AR/ns^{-1}	Refs.
p,o-(e$^+$,H)	unbound		e$^+$ + H		[32]
e$^+$PsH	0.5728	SVM,4	e$^+$ + PsH	2.7	[30]
PsH$^{(a)}$	1.0666	Hyll,3	Ps + H	2.443	[33, 34]
p-e$^+$Li	0.0672	SVM,4	Ps + Li$^+$	6.967	[35]
o-e$^+$Li	0.0672	SVM,4	Ps + Li$^+$	7.83×10^{-3}	[35]
Ps$_2$Li$^+$	0.3647	SVMcp,4	Ps$_2$ + Li$^+$	3.881	[30]
PsLi	0.3306	SVMcp,3	Ps + Li	2.151	[36]
p-e$^+$Na	0.0128	SVMcp,2	Ps + Na$^+$	7.586	[37]
o-e$^+$Na	0.0128	SVMcp,2	Ps + Na$^+$	8.49×10^{-3}	[37]
Ps$_2$Na$^+$	0.1714	SVMcp,4	Ps$_2$ + Na$^+$	4.044	[30]
PsNa	0.2291	SVMcp,3	Ps + Na	2.085	[36]
p,o-(e$^+$,K)	unbound		Ps + K$^+$		[38]
PsK	0.0875	SVMcp,3	Ps + K	1.994	[39]
p-e$^+$Cu	0.1502	SVMcp,2	e$^+$ + Cu	2.211	[40]
o-e$^+$Cu	0.1502	SVMcp,2	e$^+$ + Cu	3.59×10^{-2}	[40]
PsCu	0.2920	CIcp,3	Ps + Cu	1.248$^{(c)}$	[41]
p,o-(e$^+$,Rb)	unbound		Ps + Rb$^+$		[38]
p-e$^+$Ag	0.158	SVMcp,2	e$^+$ + Ag	2.331	[35]
o-e$^+$Ag	0.158	SVMcp,2	e$^+$ + Ag	4.46×10^{-2}	[35]
p,o-(e$^+$,Cs)	unbound		Ps + Cs$^+$		[38]
p,o-(e$^+$,Au)	unbound	CIcp,2$^{(b)}$	e$^+$ + Au		[42]

$^{(a)}$Polarizabilities have been calculated [43].
$^{(b)}$The 78-electron core of Au is treated by MPBT.
$^{(c)}$Unextrapolated value. The authors reported extrapolations that are not given here.

polylepton was attempted by SVM, but a bound state was not found.

Table 2.4 Alkaline earths and the zinc family.

System	BE/eV	Method,nbr	Threshold	AR/ns^{-1}	Refs.
e^+Be	0.04591	SVM,5	e^+ + Be	0.334	[36]
e^+Mg	0.4248	SVMcp,3	e^+ + Mg	0.955	[36]
e^+Ca	0.3363	CIcp,3	Ps + Ca$^+$	0.747$^{(a)}$	[45]
e^+Zn	0.0827	CIcp,3	e^+ + Zn	0.263$^{(a)}$	[45]
e^+Sr	0.1325	CIcp,3	Ps + Sr$^+$	0.764$^{(a)}$	[45]
e^+Cd	0.1367	CIcp,3	e^+ + Cd	0.327$^{(a)}$	[45]
e^+Hg	0.045	MBPT	e^+ + Hg		[4]

$^{(a)}$Unextrapolated value. The authors reported extrapolations that are not given here.

2.3.2 One-electron atoms

Our knowledge of positron and positronium interactions with atomic hydrogen and the *quasi*-one-electron atoms is summarized in Table 2.3. For hydrogen, the first calculation that established the chemical stability of PsH was done in 1951 [31]. Since then this system has been much studied, and its properties are now well understood. Quite recently, this molecule was found to bind a second positron as well [30].

Lithium was the first neutral atom shown to bind a positron, and this as recently as 1997 [44]. Now eleven examples are known. Since the static potential provided by a neutral atom to an approaching positron is repulsive everywhere, this type of binding is due entirely to electron-positron correlation. For atoms that have ionization potentials much less than 6.8 eV, the binding energy of positronium, the positron is more attractive to the valence electron than to the atomic core. The result is a well-defined positronium atom distorted by and loosely bound to the cationic core. Such species have *para*- and *ortho*-states, as positronium, and their lifetimes are similar to those of the corresponding states of positronium. The coinage metals, with ionization potentials greater than 6.8 eV, have lifetimes of the *para*-states closer to the spin-averaged annihilation rate. This suggests that the positron is bound to the whole atom, and is correlating strongly not only the highest *s*-electron but with the higher *d*-electrons also.

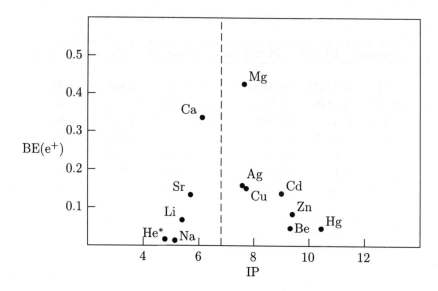

Fig. 2.1 Correlation between ionization potentials (IP) and positron binding energies (BE(e^+)) for atoms, in eV. The dashed vertical line is at 6.8 eV.

The annihilation rates of all the two-electron systems except PsCu are close to the spin-averaged rate, 2 ns^{-1}, and show only a weak monotonic dependence on the binding energy. The only exception, PsCu, is also the only bound system in Table 2.3 that was studied by CI. We have already noted the slow convergence of annihilation rates for this method.

The calculated annihilation rate of Ps$_2$Li$^+$ is close to twice the spin-averaged rate, suggesting a structure in which a relatively well-defined diatomic positronium molecule is bound to the Li$^+$ core.

Most of our knowledge of the atoms in Table 2.3 comes from the application of the SVM method, and most of that from one group, that of Dr. Mitroy. SVM is superior to other methods for these systems.

Table 2.5 The B, C, N, and O families, and the halogens.

System	BE/eV	Method,nbr	Threshold	AR/ns^{-1}	Refs.
p,o-(Ps,B)	unbound	DMC,7	Ps + B		[46]
p,o-PsC	0.48(2)	DMC,8	Ps + C		[46]
Ps$_2$O	1.27(8)	DMC,12	Ps + PsO	4.6	[47]
p,o-PsO	0.43(2)$^{(a)}$	DMC,10	Ps + O		[46]
p,o-(e$^+$,F)	unbound	cp,1	e$^+$ + F		[48]
PsF	2.85(2)	DMC,11			[46]
p,o-(e$^+$,Cl)	unbound	cp,1	e$^+$ + Cl		[48]
PsCl	1.9(2)	DMCcp,9	Ps + Cl		[13]
p,o-(e$^+$,Br)	unbound	cp,1	e$^+$ + Br		[48]
PsBr	1.14(11)	DMCcp,9	Ps + Br		[49]

$^{(a)}$A seemingly equivalent calculation [47] gives a binding energy of 0.8 ± 0.2 eV.

2.3.3 Two-electron atoms (excluding He)

Known data is shown in Table 2.4. The relationship between the ionization potentials and positron affinities of neutral atoms shown in Fig. 2.1 confirms the conjectures of several that atoms with ionization potentials near 6.803 eV should have a large positron affinity. For an atom with this exact ionization potential, there would be an accidental degeneracy between the thresholds (e$^+$ + atom) and (Ps + cation), giving the largest quantum mechanical resonance effect. The atoms whose ionization are closest to 6.803 eV are hafnium (IP = 6.825), and titanium (IP = 6.828 eV). e$^+$Ti and e$^+$Hf can be treated as 5-particle systems, but they have not yet been studied for their positron affinities, to our knowledge.

Atoms in our tables that do not bind a positron only corroborate this notion: Their ionization potentials are either less than that of the left-most atom (2^{3}S He – 4.77 eV), or greater than the right-most (Hg – 10.44 eV). The lone exception is gold, Au, a bound state for which was not found by the CI method.

Clearly, positron and positronium binding follow quite different trends.

Table 2.6 The noble gases.

System	BE/eV	Method,nbr	Threshold	AR/ns^{-1}	Refs.
e^+,He	unbound	(a)	e^+ + He		[50, 51]
p-e^+He*[b]	0.0161	SVM	Ps + He$^+$	5.67	[52]
o-e^+He*[b]	0.0161	SVM	Ps + He$^+$	7×10^{-3}	[52]
e^+,Ne	unbound	(a)	e^+ + Ne		[50, 51]
e^+,Ar	unbound	(a)	e^+ + Ar		[50, 51]
e^+,Kr	unbound	(a)	e^+ + Kr		[50, 51]
e^+,Xe	unbound	(a)	e^+ + Xe		[50, 51]

[a] Similar conclusions are reached with a 1-particle model potential method [50] and a model relating Z_{eff} and the scattering length [51].
[b] The electrons in this system have parallel spins ($S_e = 1$).

Positron binding is common among one- and two-electron atoms, except for the heaviest few. (Helium is included in the noble gas group, not two-electron atoms, in this chapter.) Positronium binding seems to be common for the alkali metals and halogens, plus some of the lighter nonmetals (C and O) and coinage metals (Cu). Trends are difficult to interpret because there are large gaps in the data.

2.3.4 The nonmetals

The nonmetals (Table 2.5) that have been studied are remarkably uniform in their inability to bind positrons, and their ability to bind positronium (boron is an exception–it apparently does not bind positronium). There is a strong correlation between the Ps–atom and H–atom bond strengths for these atoms, and this correlation extends to almost all examples of positronium binding shown in this chapter.

QMC methods have proven to be the most powerful for these systems, which have too many particles to be amenable to attack by SVM.

Table 2.7 Knowledge of molecules from theory.

System	BE/eV	Method,nbr	Threshold	AR/ns^{-1}	Refs.
e^+LiH	0.615	SVM,5[a]	PsH + Li$^+$	1.69	[53]
PsCH	0.44(2)	QMC,9	Ps + CH		[54]
PsOH	0.28(3)	QMC,11	Ps + OH		[54]
e^+HF	0.04(4)	DMC,11	e^+ + HF		[55]
e^+BeO	0.76(5)	DMC,13	e^+ + BeO		[55]
e^+LiF	0.47(4)	DMC,13	e^+ + LiF		[55]
e^+NaH	0.285	HF,13	PsH + Na$^+$		[56]
PsCN	0.83	HF,15	Ps + CN		[57, 58]
e^+KH	0.592	HF,21	PsH + K$^+$		[56]
e^+NaF	0.49	HF,21	e^+ + NaF		[59]
e^+RbH	0.824	HF39	PsH + Rb$^+$		[56]
e^+LiRb	0.024	HF,41	e^+ + LiRb		[56]
e^+NaRb	0.009	HF,49	e^+ + NaRb		[56]
PsN$_2$	0.15(2)	QMC,11	Ps + NH$_2$		[54]
e^+H$_2$O	0.043(35)	DMC,11	e^+ + H$_2$O		[55]
PsN$_3$	0.14	HF,23	Ps + N$_3$		[58]
e^+,C$_2$H$_2$	unbound	(b),15	e^+ + C$_2$H$_2$		[60]
e^+HCNO	0.017	HF,23	e^+ + HCNO		[56]
PsNO$_3$	~0.3	HF,33	Ps + NO$_3$		[61]

[a] This is a non-adiabatic calculation involving 5 light particles and the nuclei.
[b] Schwinger multichannel method, 23,112 configurations.

2.3.5 The noble gases

The noble gases (Table 2.6) are uniform in their inability to bind a positron. The stability of the system of a positron and an excited helium atom in its metastable 2^3S state is interesting. This state has an optical lifetime of about 20 ns. The excited electron in a 2s-orbital is far enough from the nucleus to attract and bind a positron. Similar states of other atoms must be common.

Table 2.8 Knowledge of atoms and molecules from experiment.

System	Year	BE or AR	Method
PsO_2	1959	> 2.3	Positron lifetime spectra in liquid oxygen [62]; gas-phase quenching data with a Born cycle interpretation [63].
PsCl	1969	2.0(5)	Enhancement of annihilation rates in Cl_2–Ar mixtures [64].
OPs	1969	2.2(5)	Positron lifetime data and thermodynamic argument based on a proposed mechanism involving the oxidation of Ps by H^+ [64].
PsF	1969	2.9(5)	Interpretation of positron lifetime spectra in liquid C_6H_6 vs C_6H_5F [64].
PsOH	1969	< 1.5	Same as for OPs [64].
$PsNH_2$	1969	~ 0	Analogy with the argument for OPs [64].
$PsCH_3$	1969	~ 0	Analogy with the argument for OPs [64].
$PsNO_2$	1972	> 0	Positron lifetimes in gaseous NO and NO_2 [65].
e^+Cl_2	1972	> 0	Positron lifetime spectra of Cl_2–Ar mixtures inside silica gel pores [66].
$e^+C_6H_5NO_2$	1974	> 0	A feature in the positron lifetime spectrum associated with the shrinking of positronium traps attending compound formation [67].
$e^{\pm}Ps$	1983	*2.09(9)*	Time of flight [68].
PsH	1968	*2.35(11)*[a]	Extrapolation to zero density for a sequence of alkali hydride crystals [69].
PsH	1992	1.0(2)	Dissociative attachment to CH_4 in a positron beam [70].
$e^+CH_nF_{4-n}$	2000	(b)	A model relating measured Z_{eff} to the scattering length [51].
$e^+C_2H_6$	2002	0.000(15)	Vibrational shifts measured by annihilation rates [1].
$e^+C_3H_8$	2002	0.020(15)	Vibrational shifts measured by annihilation rates [1].
$e^+C_4H_{10}$	2002	0.030(15)	Vibrational shifts measured by annihilation rates [1].

[a]Lifetime measurements hydrogen-laden thermochemically reduced MgO single crystals suggest an annihilation rate of 1.75(15) ns^{-1} [71].
[b]Reported binding energies are: 0.03 eV for CH_4 and 0.0003 eV for CH_3F. The species for $n = 2, 3$, and 4 are predicted not to be bound.

2.3.6 Molecules

Our present theoretical knowledge is summarized in Table 2.7. There is a strong correlation between positron affinities and dipole moments, and between Ps affinities and strengths of bonds with atomic hydrogen.

The simple Hartree-Fock method shows binding for a number of polar diatomics as expected (see Sec. 2.2.3.3), but without electron-positron correlation these results cannot be regarded as highly accurate.

The data from experiments (Table 2.8) can be divided into two groups: earlier data that are based on indirect inferences (sometimes while ignoring the influence of surrounding condensed phases), and results from the 1980's and later that rely on more direct experimental evidence. Nonetheless, the agreement of some of the older data with modern calculations is uncanny, e.g., the binding energies of PsCl and PsF and the annihilation rate of PsH. The binding energies for PsOH and $PsNH_2$ in Table 2.8 are at least consistent with modern calculations. Only the OPs prediction is inaccurate. On this basis, a calculation of positronium binding to molecular oxygen will be of great interest.

Acknowledgments

The author is pleased to acknowledge instructive discussions with J. Mitroy. We are grateful to Dr. Mitroy and to Dr. M. Mella for providing us with pre-publication results, and for offering critical comments on an earlier version of this chapter. All errors that remain are the author's.

References

[1] S. J. Gilbert, L. D. Barnes, J. P. Sullivan, and C. M. Surko, Phys. Rev. Lett. **88**, 043201, 079901(E) (2002).
[2] J. A. Wheeler, Ann. N. Y. Acad. Sci. **48**, 219 (1946).
[3] J. Pirenne, Arch. sci. phys. et nat. **29**, 207 (1947). This shift, called the "radiative shift," applies mostly to the *ortho*-state. A larger source of the splitting comes from the spin-spin interaction, hence the term "hyperfine splitting." The total splitting is 845 μeV.
[4] V. A. Dzuba, V. V. Flambaum, G. F. Gribakin, and W. A. King, Phys. Rev. A **52**, 4541 (1995).
[5] D. M. Schrader and J. Moxom, *Radiat. Phys. Chem.* **58**, 649 (2000).
[6] Y. Suzuki and K. Varga, *Stochastic Variational Approach to Quantum-Mehanical Few-Body Problems* (Springer, New York, 1998).
[7] A. Luchow and J. Anderson, Annu. Rev. Phys. Chem. **51**, 501, 527 (2000).
[8] D. Bressanini and P. J. Reynolds, Adv. Chem. Phys. **105**, 37 (1998).
[9] P. H. Acioli, THEOCHEM **394**, 75 (1997).
[10] *Recent Advances in Quantum Monte Carlo Methods*, by W. A. Lester, Jr., Ed. (World Scientific Publishing, Singapore, 1997).
[11] R. Jastrow, Phys. Rev. **98**, 1479 (1955).
[12] M. Mella, S. Chiesa, and G. Morosi, J. Chem. Phys. **116**, 2852 (2002).
[13] D. M. Schrader, T. Yoshida, and K. Iguchi, Phys. Rev. Lett. **68**, 3281 (1992).
[14] B. Barbiellini, in *New Directions in Antimatter Chemistry and Physics*, edited by C. M. Surko and F. A. Gianturco (Kluwer Academic Publishers, The Netherlands, 2001), p. 127.
[15] F. A. Gianturco, P. Paioletti, and J. A. Rodriguez-Ruiz, Z. Phys. D **36**, 51 (1996); F. A. Gianturco and P. Paioletti, Phys. Rev. A **55**, 3491 (1997).
[16] D. G. Kanhere, A. Kshirsagar, and V. Bhamre, Chem. Phys. Lett. **160**, 526 (1989); K. Kim and J. G. Harrison, J. Phys. B **29**, 595 (1996).
[17] S. Mohorovičić, Astron. Nacht. **253**, 94 (1934).
[18] A. E. Ruark, Phys. Rev. **68**, 278 (1945).
[19] E. U. Condon and G. H. Shortley, *The Theory of Atomic Spectra* (Cambridge University Press, 1959).
[20] M. W. Ritter, P. O. Egan, V. W. Hughes and K. A. Woodle, Phys. Rev. A **30**, 1331 (1984).
[21] B. A. Kniehl and A. A. Penin, Los Alamos Natl. Lab., Prepr. Arch., High Energy Phys.–Phenomenol. (2000) 1-8, arXiv:hep-ph/0010159, 16 Oct 2000. URL: http://xxx.lanl.gov/pdf/hep-ph/0010159.
[22] A. H. Al-Ramadhan and D. W. Gidley, Phys. Rev. Lett. **72**, 1632 (1994).
[23] C. I. Westbrook, D. W. Gidley, R. S. Conti, and A. Rich, Phys. Rev. A **40**, 5489 (1989).
[24] J. S. Nico, D. W. Gidley, A. Rich, and P. W. Zitzewitz, Phys. Rev. Lett. **65**, 1344 (1990).

[25] S. Asai, S. Orito, and N. Shinohara, Phys. Lett. B **357**, 475 (1995).
[26] A. M Frolov, Phys. Rev. A **60**, 2834 (1999).
[27] H. Cox, P. E. Sinclair, S. J. Smith, and B. Y. Sutcliffe, Mol. Phys. **87**, 399 (1996).
[28] Y. K. Ho, Phys. Rev. A **48**, 4780 (1993).
[29] J. Usukura, K. Varga, and Y. Suzuki, Phys. Rev. A **58**, 1918 (1998).
[30] J. Zs. Mezei, J. Mitroy, R. G. Lovas, and K. Varga, Phys. Rev. A **64**, 032501 (2001).
[31] A. Ore, Phys. Rev. **83**, 665 (1951).
[32] I. Aronson, C. J. Kleinman, and L. Spruch, Phys. Rev. A **4**, 841 (1971).
[33] Z.-C. Yan and Y. K. Ho, Phys. Rev. A **59**, 2697 (1999).
[34] K. Strasburger and H. Chojnacki, J. Chem. Phys. **108**, 3218 (1998).
[35] J. Mitroy, M. W. J. Bromley, and G. G. Ryzhikh, in *New Directions in Antimatter Chemistry and Physics*, C. M. Surko and F. A. Gianturco, Eds. (Kluwer Academic Publishers, The Netherlands, 2001). p. 199.
[36] J. Mitroy and G. G. Rytzhikh, J. Phys. B **34**, 2001 (2001).
[37] G. Ryzhikh, J. Mitroy, and K. Varga, J. Phys. B **31**, L265 (1998).
[38] Dr. Mitoy's website at http://www.cs.ntu.edu.au/homepages/jmitroy/j_mitroy.html, click on Atomic Physics at the NTU, Positronic Atoms, scroll to Current list of atoms binding positrons.
[39] J. Mitroy and G. Ryzhikh, J. Phys. B **32**, 3839 (1999).
[40] G. G. Ryzhikh and J. Mitroy, J. Phys. B **31**, 4459 (1998).
[41] M. W. J. Bromley and J. Mitroy, preprint (2002).
[42] V. A. Dzuba, V. V. Flambaum, and C. Harabati, Phys. Rev. A **62**, 042504 (2000).
[43] M. Mella, D. Bressanini, and G. Morosi, Phys. Rev. A **63**, 024503 (2001).
[44] G. G. Ryzhikh and J. Mitroy, Phys. Rev. Lett. **79**, 4124 (1997).
[45] M. W. J. Bromley and J. Mitroy, Phys. Rev. A, under review (2002).
[46] D. Bressanini, M. Mella, G. Morosi, J. Chem. Phys. **108**, 4756 (1998).
[47] N. Jiang and D. M. Schrader, Phys. Rev. Lett. **81**, 5113 (1998). [Erratum, *ibid.* **82**, 4735 (1999)] .
[48] D. M. Schrader, Phys. Rev. A **20**, 933 (1979).
[49] D. M. Schrader, T. Yoshida, and K. Iguchi, J. Chem. Phys. **98**, 7185 (1993).
[50] D. M. Schrader, Phys. Rev. A **20**, 918 (1979).
[51] G. F. Gribakin, Phys. Rev. A **61**, 022720 (2000).
[52] G. Ryzhikh and J. Mitroy, J. Phys. B **31**, 3465 (1998).
[53] J. Mitroy and G. G. Ryzhikh, J. Phys. B **33**, 3495 (2000).
[54] D. Bressanini, M. Mella, and G. Morosi, J. Chem. Phys. **109**, 5931 (1998).
[55] D. Bressanini, M. Mella, and G. Morosi, J. Chem. Phys. **109**, 1716 (1998).
[56] M. Tachikawa, I. Shimamura, R. J. Buenker, and M. Kimura, in *New Directions in Antimatter Chemistry and Physics*, C. M. Surko and F. A. Gianturco, Eds. (Kluwer Academic Publishers, The Netherlands, 2001), p. 437.
[57] H. A. Kurtz and K. D. Jordan, Int. J. Quantum Chem. **14**, 747 (1978).

[58] C.-M. Kao and P. E. Cade, J. Chem. Phys. **80**, 3234 (1984).
[59] H. A. Kurtz and K. D. Jordan, J. Chem. Phys. **75**, 1876 (1981).
[60] E. P. da Silva , J. S. E. Germano, M. A. P. Lima, Phys. Rev. Lett. **77**, 1028 (1996).
[61] A. Farazdel and P. E. Cade, Chem. Phys. Lett. **72**, 131 (1980).
[62] D. A. L. Paul, Can. J. Phys. **37**, 1059 (1959).
[63] V. I. Goldanskii, At. En. Rev. **6**, 3 (1968).
[64] S. J. Tao and J. H. Green, J. Phys. Chem. **73**, 882 (1969).
[65] S. J. Tao, S. Y. Chuang, and J. Wilkenfeld, Phys. Rev. A **6**, 1967 (1972).
[66] V. I. Goldanskii, A. D. Mokrushin, A. O. Tatur, and V. P. Shantarovich, Kinet. Katal. **13**, 961 (1972) [Kinet. Catal. **13**, 861 (1972)] .
[67] V. I. Goldanskii and V. P. Shantarovich, Appl. Phys. **3**, 335 (1974).
[68] A. P. Mills, Jr., Phys. Rev. Lett. **50**, 671 (1983).
[69] A. Gainotti, C. Ghezzi, M. Manfredi, and L. Zecchina, Nuovo Cimento B **56**, 47 (1968).
[70] D. M. Schrader, F. M. Jacobsen, N.-P. Frandsen, and U. Mikkelsen, Phys. Rev. Lett. **69**, 57 (1992).
[71] R. Pareja, R. M. de la Cruz, M. A. Pedrosa, R. González, and Y. Chen, Phys. Rev. B **41**, 6220 (1990).

Chapter 3

Experimental Techniques in Positron Spectroscopy

P.G. Coleman

Department of Physics, University of Bath, Bath BA2 7AY, UK

A broad overview of traditional methods and recent developments in experimental positron spectroscopy is presented. A discussion of the generation and detection of positrons and their annihilation radiation is followed by a survey of techniques used for positron lifetime measurement, Doppler broadening spectroscopy and angular correlation of annihilation radiation, and the opportunities presented by combining these methods (e.g. in age-momentum correlation) and/or extending their capabilities by the use of monoenergetic positron beams. Novel spectroscopic and microscopic techniques using positron beams are also described.

3.1 Introduction

This chapter is intended to introduce the reader to the basic experimental techniques used in low-energy positron studies of solids, liquids and gases, as well as in fundamental studies involving the positron itself. They are not an exhaustive review of research work in the field—in addition to this work there are a number of books to which reference can be made (e.g., [1-5]) and so specific pieces of research will only be referred to when they provide an illustration of a particular technique or device.

The history of the use of slow positrons to study matter spans the second half of the twentieth century; it is remarkable that the pioneering experiments of DeBenedetti et al [6] and Deutsch [7] followed the prediction and discovery of the positron after only sixteen years (six of which were blighted by world war). It is also remarkable that forty years have passed since Cherry [8] first observed the emission of slow positrons from a solid surface.

As so often in scientific research, steps forward in experimental positron science have resulted from a combination of technological advances and the foresight and ingenuity of individuals. Between 1955 and 1965 the groundwork was laid for two of the primary positron spectroscopies— angular correlation of annihilation radiation (ACAR) and lifetime measurement; a third—Doppler broadening spectroscopy—came along a decade later, with the development of high-resolution semiconductor detectors. It is always salutary to remember that the study of defects by positron annihilation—a field which burgeoned in the seventies and is still strong today—grew from apparent 'inconsistencies' in early positron measurements of solids which for a while made it seem as though the technique had no future.

All three spectroscopies mentioned in the previous paragraph were originally applied to the study of bulk properties of solids, each involving the detection of gamma radiation from the annihilation of beta positrons implanted into the samples under study. All three can now be carried out at or near the surface of solids by controlling the energy of the implanted positrons. To the traditional three we can now add a range of spectroscopies associated with the re-emission of slow positrons and positronium from solid surfaces, usually (but not always) requiring the detection of the emitted particle.

Along the way there have been several landmark experiments of fundamental interest involving positrons and positronium.

This chapter will begin by looking at some of the hardware requirements for positron-based experiments and then move on to their application in the measurement of angular correlation, positron lifetimes and Doppler broadening parameters. We shall then look at the generation and application of beams of mono-energetic positrons.

An interesting historical perspective is provided by *Thoughts at a Dinner: reflections on the field of positron science* by Prof. Alec Stewart, in *Positron Annihilation*: Proceedings of the 10th International Conference on

Positron Annihilation, edited by Y-J He, B-S Cao and Y C Jean (Trans Tech, Switzerland, 1995).

3.2 Positron sources

3.2.1 Introduction

The old units of curies (Ci) for radionuclide activities have been replaced by Becquerels (Bq) throughout much of the world in recent years. 1 Bq = 1 nuclear disintegration per second, and 1 Ci = 37 GBq. Safety considerations are always paramount in a positron laboratory; in a standard university laboratory environment it is usually sufficient to ensure that adequate lead shielding is provided around the source region, and that the need for close proximity to this region is minimised (the inverse square law is a great help). In reactor or LINAC facilities much greater shielding precautions are usually necessary, in light of the much higher ambient radiation levels. Needless to say, dosimetry regulations should always be adhered to.

3.2.2 Radioactive sources for laboratory experiments

In the laboratory the choice of radioactive sources for positron experiments has always been a compromise between application and cost. The most common source by far has been ^{22}Na, bought in the form of a ^{22}Na salt either in solution (for in-house source preparation by evaporative deposition - usually for activities ≤ 37MBq) or as a ready-made sealed source (for activities up to 3.7GBq). In the latter case the source, usually in capsule form, is required to emit as large a fraction of the beta positrons as possible. This is achieved by having a high-Z backing material and a thin low-Z foil cover (usually 5-10µm Ti). [Note: we shall see later that the high-Z backing is not desirable for polarised positron experiments.]

^{22}Na is most widely used because of the benefits of long half life (2.6y) coupled with reasonable price. The positron flux over a typical experimental run can be regarded as essentially constant. Furthermore, the ^{22}Na nucleus decay to an excited state of ^{22}Ne, which then decays to its ground atomic state with the emission of a 1.274 MeV gamma ray. Detection of this gamma ray provides the means for measuring time zero- i.e., the time of birth of the positron, with high time resolution. This is invaluable in conventional positron lifetime spectroscopy. In other timing applications—such as beta-gamma delayed coincidence timing, where time zero is, for example, measured by passing beta positrons through a thin scintillator and the time of the positron's death is recorded by detecting an annihilation gamma ray—or in standard Doppler broadening spectroscopy—the time-zero gamma ray is both unnecessary and even disadvantageous, increasing the background in the stop (gamma) detector (see, e.g., [9]). In this case another source, such as ^{68}Ge, is preferable, which has little gamma emission associated with positron emission. However, there are problems in depositing ^{68}Ge and its half-life is only 275 days. ^{58}Co has been a popular choice for high-flux beam measurements in the laboratory (e.g. in the positron diffraction and microscopy measurements at Brandeis); it is affordable because it has a short half-life of 71 days; however, the experiment which needs a high positron flux needs to be in good working order (checked with a lower-activity source if possible) before such a source is installed, to avoid time wastage. ^{58}Co is not suitable for standard lifetime or Doppler-broadening spectroscopies as it emits many gamma-rays per positron, and most of these are not time-related to the positron [10].

One problem in traditional lifetime and Doppler-broadening spectroscopies has been the detection of components associated with positron annihilation in the source (the source being both the radionuclide material itself and its thin foil envelope). The intensity of the resultant source component in the measured spectra can be reduced by choosing an isotope with higher-energy beta emission, thereby reducing the fraction of positrons stopping in the source assembly. An excellent choice for lifetime experiments would be ^{44}Ti (half-life 47y, maximum energy 1.47 MeV, 1.16 MeV time-zero gamma ray), but the higher cost of this isotope explains why it loses out to ^{22}Na for these experiments.

Past difficulties in the availability of high-quality, high-emissivity ^{22}Na sources from commercial suppliers has led researchers to investigate the in-house production of suitable positron sources. For example, Weng et al. [11] have constructed an in-house apparatus for drop-depositing up to 740MBq ^{22}Na from ^{22}NaCl solution and sealing with 3μm Ti; Britton et al. [12] have produced ^{22}NaCO$_3$ at the N.A.C., South Africa.

3.2.3 Positron sources for facility-based beams

The first reactor-based slow positron beam was developed by Lynn at Brookhaven [13]. In this system, housed in the reactor building, a copper ball was irradiated in the reactor core (^{63}Cu(n,γ)^{64}Cu) and transferred automatically into the source chamber. It was dropped into a crucible and evaporated on to a tungsten backing. The strong ^{64}Cu source therefore acted as a self-moderator for the production of slow positrons. At the Munich reactor positrons are produced by pair production by gamma rays emitted in the reaction ^{113}Cd(n,γ)^{114}Cd [14].

It is of great interest to note that after Cherry's observation of positron emission from a solid surface the first 'beam' system—developed in the late 60's—was based at a LINAC facility [15]. Bremsstrahlung gamma radiation from the energetic (50MeV) LINAC electrons create electron-positron pairs in a Ta target; the fast positrons thus created are then moderated (see section 8) to form the slow positron beam. The efficiency of this process clearly depends on the LINAC electron energy and the thickness of the converter.

3.2.4 Accelerator-based positron sources for the laboratory

The principal option to date for strong primary sources in the laboratory has been the purchase of very active but short-lived ^{58}Co. With the advent of table-top cyclotrons for medical uses this may change. Canter [16] and Hirose et al. [17] have proposed schemes for the use of such systems for the quasi-continuous production of fast positrons. The latter system produces about 2×10^6 slow positrons using 18MeV protons at a current of 30A. The advantages of such systems is that the strong positron source can be turned off at will; disadvantages are the need for continual maintenance of the cyclotron, and (currently) the high initial cost. (Note that after the initial capital outlay the cost per positron competes well with radioactive sources).

3.3 Particle and radiation detectors

3.3.1 Radiation detectors

3.3.1.1 Scintillation detectors
Organic phosphors (commonly loaded plastics) exhibit luminescence, of which the fastest process is fluorescence; a gamma ray releases an energetic electron which then excites molecules in the phosphor; these lose energy in decaying to an intermediate state and then de-excite with photoemission.

Inorganic phosphors can have their luminescent efficiency improved by the addition of an activator, e.g., 0.1% Tl in NaI (then referred to as NaI(Tl); excitation energy of NaI molecules is transferred to impurity centres in the crystal lattice, which de-excite with photoemission characteristic of the impurity. The higher atomic number of these phosphors improves the probability of photoelectric effect events. NaI is deliquescent and has to be sealed in an Al pot with glass window. The useable light output is improved by painting white to improve reflection from the surface.

The light output of NaI(Tl) is about 10^4 photons, of wavelength 410 nm, per MeV of incident gamma energy. Loaded plastic scintillators yield about 4000 photons per MeV in a wavelength range 350—480 nm.

Phosphors are coupled to photomultiplier (PM) tubes with optical coupling grease (compatible refractive index, good optical coupling). Photons from the phosphor fall on the PM photocathode, releasing electrons (with 10-20% efficiency) which travel to an anode via several electrodes (dynodes); for each electron leaving the photocathode 10^6—10^8 may eventually reach the anode, creating a current pulse of useable size. The photocathode efficiency of a typical PM peaks at around $\lambda \approx 400$ nm (this is why photons of this wavelength are desired from a phosphor). At least 100 photons are required to produce a useable pulse from the PM anode. If all the energy of an annihilation photon (i.e., 511keV) is deposited in NaI(Tl), about 5000 photons release ~1000 photoelectrons from the cathode. This last number is subject to Poisson statistics and hence an fractional uncertainty of $1/\sqrt{N} \approx 3\%$ (5% for plastic), leading to a spread in resultant pulse height which is increased by a further 20-25% at the first dynode. The statistical spread is reduced considerably by using a photocathode and dynodes of a material with much higher secondary electron coefficients η, such as GaP

($\eta \approx 50$). A second advantage is that fewer dynodes (e.g. 5 instead of 12: $50^5 \approx 10^8$) are required, decreasing the electron transit time in the PM. The relatively good resolution of NaI(Tl) is not maintained if the crystal is large, because of reflection and absorption effects.

The rate of emission of photons from a scintillator is $\lambda N_0 e^{-\lambda t}$, where N_0 is the total number of photons emitted and λ is the decay constant of the phosphor. For NaI(Tl) the decay time τ ($=\lambda^{-1}$) $\approx$ 220ns, whereas for 'fast' plastic $\tau \approx$ 2ns (i.e., in the first ns 40% of the light from a plastic scintillator is typically emitted, whereas only 0.5% is emitted from NaI(Tl)).

The distribution of heights of pulses generated by a scintillation detector depends on the range of gamma ray energies dumped in the phosphor. In plastic scintillators the probability of photoelectric emission is negligible and all electrons ejected are from Compton scattering events. Thus the pulse height spectrum from plastic scintillators is roughly uniform up to what is known at the *Compton edge*, corresponding to the maximum energy which can be transferred to an electron in a Compton interaction (when the gamma ray is scattered through 180°).

In the case of NaI(Tl), however, the total gamma ray energy can be deposited in the crystal via the photoelectric effect; this leads to the observation of *photopeaks* on the pulse-height spectrum. For 511keV annihilation gamma rays entering a 25x25mm crystal, the ratio of photoelectric to Compton events is approximately 1:4; this ratio increases as the crystal size increases because of the increased probability that all the gamma energy will be dumped in the crystal (and because the probability of photoelectric emission increases as the electron energy decreases). The total absorption coefficient of NaI(Tl) for 511keV gamma radiation is 0.3 cm^{-1}, and for plastics 0.12 cm^{-1}. A comparison of pulse height spectra for ^{22}Na obtained with NaI(Tl), plastic scintillator and semiconductor (Ge) detectors (see next section) is represented by Figure 3.1.

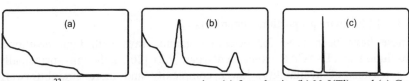

Figure 3.1 ^{22}Na gamma energy spectra using (a) fast plastic, (b) NaI(Tl), and (c) Ge.

BaF$_2$ scintillators have been developed in recent years for use in positron lifetime systems. BaF$_2$ combines the high stopping power/detection efficiency of NaI(Tl) (density 4.9 gcm^{-3}) with the fast pulse characteristics of plastics (decay time ~ 700ps, less than half that of fast plastics). It also exhibits strong photopeaks and so is well suited to energy discrimination and, unlike NaI, is not hygroscopic. BaF$_2$ scintillates in the ultra-violet (220nm) and so requires matched PM tubes with quartz windows. With such scintillators high-resolution measurements (e.g. full-width at half maximum ~ 220 ps) can be made with counting rates about five times bigger than those possible with organic detectors [18].

Scintillators like NaI(Tl) can be made position-sensitive for applications like 2D-ACAR (section 7) and medical imaging [19]. Light from a large-diameter crystal is detected by a close-packed array of PM tubes.

3.3.1.2 Semiconductor detectors

The semiconductor detector most commonly used for annihilation gamma spectroscopy has at its heart a hyper-pure Ge crystal, cooled to liquid nitrogen temperatures while in use to reduce significantly thermal noise. Incident photons release electrons which then create electron-hole pairs; the electrons are swept to an anode by potential of ~ 2—4kV. As the energy required per pair ~2eV approximately 2.5x10^5 electrons are released following a photoelectric event, with consequently high resolution ($1/\sqrt{N} \approx$ 0.2%, cf 5% for scintillators). The probability for photoelectric events is high, and the pulse spectrum from a Ge detector is thus dominated by very narrow peaks corresponding to the energies of the incident photons. Ge detectors are thus very much suited to annihilation Doppler-broadening spectroscopy. Typical detection efficiencies range from 25 to > 100% (compared to NaI(Tl) in standard geometry). The energy resolution of a small detector may be $\leq$ 1.2 keV at 511 keV, comparable to the Doppler broadening due to positrons annihilating electrons in solids.

3.3.1.3 Multiwire proportional chambers

These have been used for detecting gamma rays with high positional resolution in 2D-ACAR experiments (see 7.2) [20], although—principally for reasons of availability, lower complexity and more straightforward maintenance—scintillation detectors are more common. Very briefly, a 'spark' chamber is modified to detect gamma radiation with positional sensitivity first by converting the gamma rays to electrons (via secondary electron emission in an array of holes similar in principle to the CEMA plate (see below); the electrons then enter the multiwire chamber and an avalanche occurs between two orthogonal sets of closely-spaced parallel wires; the spatially confined electron pulse travels to the wires and the 'centre of gravity' of the induced pulse distributions—i.e., the point of impact of the gamma ray—is determined by computer.

3.3.2 Particle detectors

3.3.2.1 Scintillators

Direct detection of fast positrons is possible with scintillator detectors; as the processes involved are the same as for gamma rays without the initial Compton or photoelectric event. In positron experiments the detection of fast positrons by thin plastic scintillators has been put to best use in time-of-flight and lifetime studies [21, 22]. Here almost 100% of the positrons from a radioactive source or MeV beam which pass through a thin scintillator disc deposit enough energy ($\geq$ 20keV) to create photons which produce a useable time-zero tagging pulse in a PM. (See secs. 3.5).

3.3.2.2 Surface-barrier detectors

These detectors are thin ($\geq$500μm) discs of passivated, implanted planar Si (known also as PIPS detectors); for beta and high-energy positron detection they have extremely thin entrance 'window' regions, not deposited metallic films but an ion-implanted region. A detector of active area 50mm^2 can detect positrons above 18 keV, with an energy resolution of ~ 6 keV [23].

3.3.2.3 Channel electron multipliers

Pendyala *et al.* [24] found that the efficiency of single-channel electron multipliers (CEMs) (also called channeltrons or spiraltrons) was about the same for positrons and electrons of the same incident energy.

The CEM is manufactured from high-resistivity glass with a high secondary electron coefficient (Figure 3.2). Incident projectiles release secondary electrons which are then accelerated towards the anode at the other end of the glass tube, interacting with the walls on many occasions on the way and releasing more electrons each time—causing an avalanche and a current pulse collected at the anode. The tube is usually coiled to prevent positive ions from travelling in the opposite direction and creating unwanted pulses. CEMs often have a conical front end, of maximum diameter ~1cm, to increase the collection efficiency. For fast timing applications some CEMs have an isolated anode, with a potential difference of ~10^2eV between it and the tube end, to pull the electron shower together to the anode. Whilst being sensitive to soft x-radiation, CEMs are quite insensitive to gamma radiation, which makes the shielding of source from detector unnecessary.

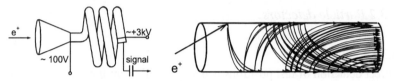

Figure 3.2 (left) typical CEM: (right) CEM operation (secondary electron emission).

The positron/electron detection efficiency η of CEMs rises rapidly with incident projectile energy to ≈ 90% at 30eV and 100% at 100eV, and starts to decline slowly at keV energies. (Ions require keV energies to be detected.) It is usual to hold the front end of the CEM at a small negative potential for positrons and a small positive potential for electrons.

The CEM array (CEMA) is constructed of thousands of microtubes, each some 25 µm diameter and 500µm long, formed into a disc of diameter 25mm upwards. The maximum detection efficiency is ~70%, limited by the open area on the front face of the detector plate. Because with ~1kV across a plate the gain is ~10^4, it is usual to use two closely-separated plates in tandem. The channels in each plate are angled at ~10° to the plate surfaces and the two plates are arranged to achieve a chevron channel geometry—again to reduce problems associated with ion feedback. The large collection area of CEMAs has found application in many positron experiments—particularly those requiring position-sensitive detection (e.g. see [25, 26]).

3.4 Notes on pulse electronics

3.4.1 Transmission of pulses

Signals in most if not all positron experiments are transmitted along coaxial cables in the form of current pulses—i.e., a charge dq in time dt, where dt is in the range 10^1-10^3 ns. The pulses are used for counting, to locate events, to measure energy, as time markers, and so on.

A coaxial transmission line consists of a central (core) wire surrounded by polythene (relative dielectric constant K ~ 2.5) and an outer, cyclindrical conducting sheath. The line parameters are series capacitance C_0, inductance L_0 and resistance R_0, and shunt dielectric loss G_0 (conductance). The quality of the dielectric may be such that R_0 and G_0 are negligible (the line is said to be 'less lossy'; transmission is faster and pulse distortion is minimised). Typically $C_0 \approx 100\text{pFm}^{-1}$ and $L_0 \approx 0.25\mu\text{Hm}^{-1}$. For this cable we specify a *characteristic impedance* Z_0 which, for zero R_0 and G_0, is $(L_0/C_0)^{1/2} = 50\Omega$. The line is commonly referred to as '50Ω cable'. If a finite length of such cable is terminated in 50Ω it appears infinite—pulses are not reflected. If the load impedance (termination) is Z_L then R_v, the ratio of the voltage amplitudes of the reflected pulse to that of the incident pulse, is (Z_L-Z_0)/(Z_L+Z_0), so that for an open-ended cable ($Z_L = \infty$) $R_v = 1$ and the pulse is reflected. If the cable is short-circuited at the end then $R_v = -1$ and the pulse is reflected and inverted. If a pulse generator has a high-impedance output it should be terminated with Z_0 to avoid source reflections. One must also terminate a coaxial cable correctly at the input of an oscilloscope when viewing fast (say 20ns) pulses-otherwise one sees an apparently long, slowly-decaying pulse made up of the original pulse and a large number of reflections, each a little delayed and attenuated by travelling along the cable to the source and back. Matching impedances is important also if a pulse is to be split or attenuated by a simple resistor circuit placed in-line.

Coaxial cables are commonly used as low-distortion, low-jitter delay lines in timing circuits. The velocity of a current pulse along an ideal cable (R_0, $G_0 = 0$) is given by $c = (L_0C_0)^{-1/2} \approx 2\times10^8$ ms^{-1}. Hence a pulse takes almost exactly 5ns to travel along 1m of cable-a handy figure to remember.

3.4.2 Elements of circuits used in positron spectroscopy

The electronic units traditionally used in positron spectroscopy experiments were developed commercially for the huge endeavours associated with high-energy physics. Nuclear Instrumentation Module (NIM) systems have their own standard specifications: the modules fit into standard power supply bins (±24, 12 and 6V lines to standard sockets), with standard fast and slow logic pulses (fast pulses are negative 700mV into 50Ω, 20ns wide: slow logic pulses are positive 5V and 500 ns long). Fast pulses are used for timing and fast counting applications; slow logic pulses for energy analysis, gating and other logic processes. In recent times it has been possible to replace the analogue electronics units described below by their digital counterparts; however, the principles of application remain unchanged.

3.4.2.1 Discriminators

There is commonly a need to discriminate against small (e.g.,noise) pulses. The simplest method to reject unwanted pulses is by *leading edge discrimination*. An output pulse is only generated when an input pulse crosses a chosen threshold voltage. However, because the rise time of pulses is not negligible, larger pulses cross the threshold and trigger an output pulse earlier than small pulses, and timing jitter results

This problem can be improved by the use of *zero-crossing discriminators*, which reshape the input pulse into a bipolar pulse (by delaying, inverting and adding) and generate output pulses (a) if the input pulse crosses the preset threshold and (b) at the time when the bipolar pulse crosses the zero level. The best timing has been found to follow from ensuring that the zero crossing occurs at a time after the start of the pulse when a predetermined fraction of the original pulse height has been reached. Timing jitter is reduced significantly because the zero-crossing point is close to the same time after the start of each pulse, whatever the pulse height [27].

3.4.2.2 Amplifiers

For energy spectroscopy, where measurement of 'true' pulse heights is important, pulses are pre-amplified as close as possible to their source and can then be transported over several meters to the amplifier. Modern amplifiers are designed to operate at high throughput (several x 10^3 s^{-1}) without degradation of resolution, with a wide range of amplification settings and pulse shapes, depending on the application. *Pulse pile-up rejection* reduces distortion of pulses by pile-up effects by discarding any output pulses when two or more input pulses arrive within a preset time interval. *Biased amplifiers* set an arbitrary zero level and then amplify further any pulses of heights exceeding that level.

3.4.2.3 Single and multi-channel analyzers

Discriminators with upper and lower levels, giving an output pulse only when the input pulse height lies between the two) are termed *single channel analysers* (SCA's). They are most often used to select pulses generated by gamma radiation within a certain energy range. An oft-used fast timing version of the SCA is the *constant fraction differential discriminator* (CFDD) which combines fast low-jitter timing with pulse height selection. The *multichannel analyser* (MCA), a PC plug-in card, sorts pulses according to their height into several thousand channels or bins. MCAs can be multiparameter, collecting for example spectra simultaneously in two dimensions and storing intensity contour plots.

3.5 Lifetime spectrometry

The measurement of the mean lifetimes of positrons in matter has been one of the cornerstones of positron science over the past half-century. The lifetime of a positron in matter—gas, liquid or solid—will depend on the electronic environment in which it finds itself, and this in turn tells us much about the submicroscopic nature of the material. In condensed matter a positron will approach thermal energies within about 1ps, so that measured lifetimes are essentially those of a thermal positron in the material under study. In some gaseous environments—particularly in the noble gases—the time taken for a positron to come to thermal equilibrium with its surroundings is much longer—10^0-10^2 ns—and this 'thermalisation time' has to be taken into account in the analysis of time spectra.

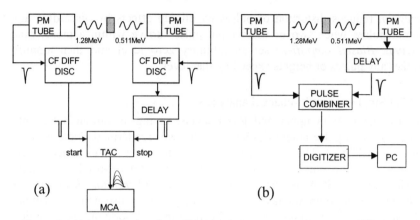

Figure 3.3: (a) standard positron lifetime set-up: PM TUBE: photomultiplier tube/ scintillator assembly: CF DIFF DISC: constant fraction differential discriminator: DELAY: delay box or fixed length of 50Ω cable: TAC: time-to-amplitude converter: MCA: multichannel analyser. (b) digital version of (a) [Rytsölä et al, preprint, 2001].

The principle of operation of most lifetime spectrometers is to measure the spectrum of time intervals between start signals, generated by detecting prompt gamma rays following the emission of positrons, and stop signals from one of the annihilation gamma photons. The simplest positron lifetime set-up using a ^{22}Na source and traditional electronics units would look as shown in Figure 3.3(a), and modified to use digital collection and manipulation of pulses as in Figure 3.3(b).

A drop-deposited ^{22}Na source sandwiched between two thin foils (e.g., 0.5 μm Ni) which is in turn sandwiched between two pieces of sample under study. This assembly is usually mounted in an evacuated chamber to reduce any risks of contamination of the spectrum by long lifetime components associated with air, and to allow heating and/or cooling of the sample without surface contamination occurring. It is common to mount the detectors at 90° to each other, with lead shielding between, to reduce the number of unwanted prompt-related pulses (by detection of both 0.511 MeV gamma rays and/or by backscattering from one detector to the other).

The time scale of the MCA spectrum has to be calibrated. In analogue systems this can be done by (a) using a commercially available time calibrator: (b) stopping the TAC-MCA clock with a delayed start pulse, then introducing an extra known delay and measuring the change in channel number of the centroid of the resultant peaks: or (c) starting and stopping the TAC with trains of unrelated pulses. The (random) start and (regular) stop pulses should also be fed into scalers and the total numbers of pulses—N_1 and N_2, respectively, recorded in the run time T. The number of 'random' events N_i in channel i (time width Δt_i) is $N_1 N_2 \Delta t_i / T$, and Δt_i is determined for all channels i. Corrections to N_1 must be made for any losses due to the system dead time. T/N_2 can have any value up to the time range used.

Bauer et al. [22] measured fast lifetimes using the $\beta^+\gamma$ coincidence technique: fast positrons pass through ~ 1mm fast plastic scintillator to produce the start pulse. This eliminates background because every positron entering the specimen is detected; high signal counting rates are obtained. The same idea was exploited by Coleman et al. [28] to measure the much longer positron lifetimes in gases; a ^{22}Na source was covered in a thin plastic scintillator disc (see sec. 3) to provide start pulses (which, because of the extremely high count rate, were delayed until after the true 'stop' pulses (thus producing a time-reversed spectrum) to avoid high dead-time losses.

Figure 3.4 illustrates two lifetime spectra collected by methods similar to those outlined above. (a) exhibits the non-exponential 'shoulder' region associated with the annihilation of non-thermalised positrons. After thermalisation (essentially at time zero for condensed matter) the spectra are sums of exponential components associated with each decay mode, and a background component B_i: $N_i = \sum A_i \exp(-\lambda_i t_i) + B_i$. For long lifetime components ($\geq$ 1ns) each λ can be extracted by non-linear least squares fitting. For short λ values characteristic of condensed matter, however, a

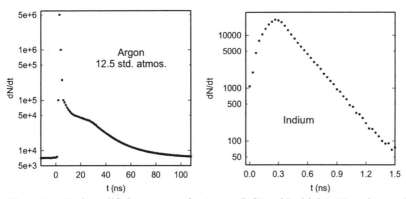

Figure 3.4 Positron lifetime spectra for Ar gas (left) and In (right). Note time scales.

more sophisticated analysis procedure must be followed, and programs such as POSITRONFIT, CONTIN or MELT are employed [29]. Correction for annihilations in the source must be made.

3.6 Doppler broadening spectroscopy

3.6.1 Introduction

The component of the momentum of a positron-electron pair p in the direction of emission of the annihilation photon gives rise to a Doppler shift in the photon energy; we can employ the small-angle approximation to find that the energies of the two photons are $E = m_0c^2 \pm cp/2$, where m_0 is the positron/electron rest mass and c is the speed of light [1]. If p corresponds to an energy E of 5eV, $cp/2 = (c/2)(2mE)^{1/2} = (mc^2E/2)^{1/2} = 1130$ eV. This is of the same order as the energy resolution of a Ge gamma detector. The method is based on measurement of the width of the annihilation gamma photon line, centred at 511keV, by such a detector. Changes of up to a few percent on the annihilation linewidth may be measured reliably; although the technique provides a relatively low-resolution reflection of the average electron momentum, it has found wide application in the study of defects in materials. The high signal rate is particularly suited to following defect dynamics—for example as a function of temperature- and the experimental requirements are straightforward.

3.6.2 Experimental set-up

A typical system for Doppler broadening studies is sketched in Figure 3.5.

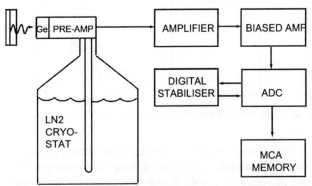

Figure 3.5 Simple Doppler-broadening spectrometer.

The crystal must be cooled during data collection (i.e., if the high voltage is applied), by liquid nitrogen or by an electrical cooler. The biased amplifier (see section 4) allows expansion of the photopeak into a larger number of MCA channels. It is essential that the spectrum be stabilised; this is done by controlling the ambient temperature and by using a digital stabiliser. This module measures the centroid of a chosen peak on the spectrum and adjusts the height of pulses to be sorted by the MCA so that the centroid remains in the same channel. The preferred option is to use a reference line for this purpose, e.g. the ^{7}Be 478 keV line [30]. Digital self-contained systems have been developed which house all the elements shown in Figure 3.5.

Typical photopeak spectra are shown in Figure 3.6 for (a) positron annihilation in a metal, and (b) the (shifted) 478 keV peak for ^{7}Be. The latter is not broadened and reflects the system resolution.

Samples for Doppler broadening spectroscopy should be mounted in an evacuated chamber to avoid sample contamination on

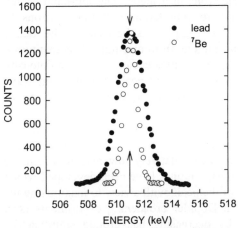

Figure 3.6 (Low-statistics) photopeak spectra: (a) positron annihilation in Pb(●) and (b) 478keV gamma radiation from ^{7}Be, shifted by 33keV (O).

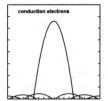 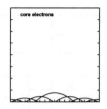 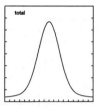

Figure 3.7 conduction and core electron contributions to 0.511keV photopeak.

heating or cooling, and data contamination by annihilation signals in air. A typical sample set-up would be similar to that for lifetime spectrometry. The sample may be mounted *in vacuo* on a cold finger (e.g., a solid copper rod extending into liquid nitrogen, or of a commercial closed-cycle refrigerator unit) and its temperature controlled. The evacuated environment then also acts as a good thermal insulator. Measurements with positron beam systems are uncontaminated by source components.

3.6.3 Data analysis

Figure 3.7 sketches contributions to the electron momentum distributions from conduction band and core electrons in a metal. The left-hand sketch shows a dominant parabolic peak due to annihilations with free electrons. In addition there are higher-momentum components because of the lattice periodicity: there are not only contributions from electrons at momentum k but also at $k+G$, where G is the reciprocal lattice vector. These low-intensity components add to give a long, low tail to the peak. The positron wave function can also overlap with core electrons and this introduces a number of slowly-varying higher-momentum components with combine to give a broadly gaussian distribution, shown diagrammatically in the central sketch.

These sketches serve to illustrate that there is structure in the photopeak which can yield information, albeit with relatively low resolution, on the details of electronic structure. Analysis routines have been developed which take account of the detailed shape of the peak, with a view to maximising the amount of information obtained in Doppler broadening experiments (see, e.g., [31]). We shall later that by decreasing background significantly, detailed analysis of the peak shape can yield considerable fruit.

For the moment we shall consider the standard method for describing the Doppler-broadened linewidth—i.e. by using a simple lineshape parameter. By far the most common parameters used—called S and W—are defined

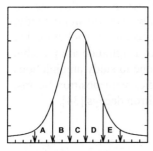

Figure 3.8 Regions chosen for S and W.

using Figure 3.8. Six MCA channels (i.e., gamma energies) are chosen, symmetrically about the peak centroid, to define the *regions of interest* A, B, C, D and E. The limits of regions A, C and E are chosen such that (a) the areas of the wing regions A and E are approximately equal and (A+E)/T ≈ 0.25, and (b) C/T ≈ 0.5, where T is the total area (A + B + C + D + E).

As most experimenters want to know not the absolute value of the linewidth but how it changes as a function of physical parameters, these ratios have been taken up as the simplest way of describing the linewidth. C/T is called the S (sharpness) parameter, and (A+E)/T is W, the wing parameter. As we can induce from Figure 3.7, S and W should be sensitive to changes in the momentum density of lower- and higher-momentum electrons, respectively. Positron annihilation in open volume defects thus typically leads to an increase in S and a decrease in W.

Neither S nor W have absolute values (they are arbitrarily chosen to be ~ 0.5 and 0.25, although these figures are not necessarily optimum). However, it is becoming more common to endeavour to express the parameters in reduced form, e.g. S/S_{bulk}, where is the parameter associated with the defect-free bulk material being studied, measured with the same apparatus—especially in the case of Si.

The choice of which parameter to use, and the limits of regions A to E, should be investigated first for each system studied. The figure of merit here is the difference between S (W) and S_{bulk} (W_{bulk}), expressed as a number of statistical standard deviations.

If S_f and W_f are the S and W parameter characteristic of free positrons, then we can define a new parameter $R, = |(S-S_f)/(W-W_f)|$, which depends only on the nature of the defect, and not on its concentration. If R is found to be constant then this points to the existence of only one kind of defect [32].

3.6.4 Two-detector technique

This technique was proposed many years ago [33] but has recently experienced a renaissance. If the sample in Figure 3.5 is also viewed by a second gamma photon detector, and the pulses from this second detector are used in coincidence mode to gate the pulses from detector 1 on their way to the MCA, the background under the photopeak can be reduced significantly. If the second detector is, say, NaI(Tl), a tenfold reduction in background is achieved; an improvement of two or more orders of magnitude are possible if a second Ge detector is employed. This enables one to study annihilations with core electrons and, in the case of recent studies with positron beams, identify the chemical environment in which the positron decays [34].

3.7 Age-momentum correlation (AMOC)

Simultaneous measurement of positron lifetime and the momentum of the annihilating pair can give information on thermalisation and transitions between positron states (and hence on chemical reactions of positrons or Ps). The most recent version uses MeV positron beams [35]. A full description of AMOC can be found elsewhere in this volume.

3.8 Angular correlation of annihilation radiation (ACAR)

3.8.1 One-dimensional ACAR

Two-gamma annihilation in the centre of mass frame sees the two photons leaving the site of the annihilation with identical energies (mc^2, 0.511 MeV) at in exactly opposite directions (to conserve momentum). This is sketched in Figure 3.9(a). In the laboratory frame [Figure 3.9(b)] the momentum of the annihilating pair (velocity v, $\ll c$, to the right) means that $\theta_1 \neq \theta_2 \neq \theta_0$.

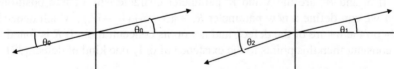

Figure 3.9 (a) 2γ annihilation in centre-of-mass frame; (b) in the laboratory frame.

Relativistic transformation from one frame to the other yields the results for $\tan\theta_1$ and $\tan\theta_2$. Then $\delta\theta \approx \tan(\theta_1-\theta_2)$ which, ignoring terms in $(v/c)^2$, reduces to $(2v/c)\sin\theta_0$. Now if we set $\theta_0 \approx \theta_1 \approx \theta_2 = \theta$, then $\delta\theta \approx 2mv\sin\theta/mc = p_t/mc$, where p_t is the component of momentum of the annihilating pair in a direction transverse to the gamma emission. Measurement of $\delta\theta$ thus directly yields information on p_t.

A schematic diagram of a one-dimensional angular correlation apparatus is shown in Figure 3.10. In order to achieve good angular resolution and reasonable counting rates the 'long slit' geometry is adopted; the lead slits in front of the two gamma detectors extend symmetrically in and out of the paper by a distance exceeding a typical angular distribution. Only one component of the transverse momentum of the annihilating pair is therefore defined. A strong (~ GBq) radioactive source is placed a few cm to one side of the sample and lead blocks shield the source from the two detectors (e.g., long NaI(Tl) crystals). Positrons from the source are constrained to hit the target (mounted on a cold finger in an evacuated chamber) by a magnetic field (~1T). A sample tilt of ~50mr minimises an asymmetric detector response caused by positron penetration into the sample and asymmetric attenuation of the annihilation radiation, and (in the case of single crystal samples) the effects of gamma ray diffraction.

One detector is held stationary and the other is rotated about the sample axis; the coincidence count rate is recorded by the scaler at each angle chosen. A typical time required to obtain an angular correlation curve with good statistics and with an angular resolution of $\sim 10^{-1}$ mr is 100h.

One problem with the long-slit 1D-ACAR geometry is the projection of high-momentum events on to the small-angle parts of $I(\theta)$—where Fermi surface structure is most apparent. This problem is countered somewhat by the use of 'short' slit geometry, which restricts the detection of higher momentum components and enhances any structures associated with the Fermi surface. It is clear that point geometry would be ideal (determining

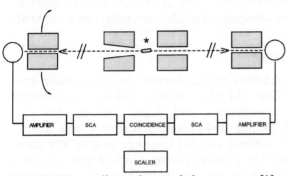

Figure 3.10 Long-slit angular correlation apparatus [1].

both p_x and p_y); the problems with higher momentum contributions from conduction electrons and from core electron annihilation are still present, but can be dealt with. Point gamma detection with 1D-ACAR system would take an unfeasibly long time; two-dimensional parallel data collection is required.

3.8.2 Two-dimensional ACAR

Two-dimensional position-sensitive gamma detectors have been made from an array of discrete scintillator-PM tube detectors, multiwire proportional counters, position-sensitive phototubes, and Anger cameras. The last, manufactured for medical imaging applications, consist of a large scintillation crystal (e.g., NaI(Tl)) optically coupled to a large array of PM tubes. Light ($\sim 10^3$ photons) produced by a gamma photon is picked up by a number of PM tubes and the centroid of the light pulse intensity is evaluated electronically. The linearity of a camera's response is checked and adjusted by irradiating the camera face with a radioactive source placed centrally 1m away through a precisely-constructed lead slit template with grid geometry.

The experimental system is sketched in Figure 3.11. The cameras are placed about 14m on either side of the sample on firm mountings; the distance is chosen to give the best resolution possible with an acceptable gamma photon count rate ($\sim 100s^{-1}$ for a 780MBq ^{22}Na source). The cameras are operated in coincidence, so that (a) the x axes have to be in the same plane, (b) the gains at the X and Y inputs to the amplifier/summing circuit must be matched, and (c) the sample must be on the camera-camera axis. The resolution of such a system can be optimised by making adjustments until the width of the thermalised p-Ps

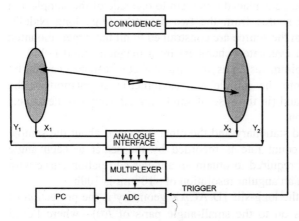

Figure 3.11 Schematic diagram of a 2D-ACAR system [36].

peak characteristic of annihilation in quartz is minimised. The resolution limit of a modern Anger camera system for 20m camera separation ~ 0.3 mrad.

The sample is mounted on a cold finger—preferably cooled by a closed-cycle helium refrig-erator to ~10^1K—which greatly reduces the thermal smearing of the electron-positron momentum density. The sample is again tilted so that each detector is irrad-iated fairly uniformly, irrespective of gamma emission angle.

The reader is referred to the articles by Berko [37] and West [38] for overviews of the analysis and interpretation of ACAR data and to Dugdale *et al* [39] on data reduction by the maximum entropy technique.

The advent of intense positron beams opens up new possibilities for ACAR measurements: small samples, surface and near-surface phenomena, and thin films are amongst the opportunities presented by the stationing the two-camera system at the end of a beam line [40].

3.9 Positron beams

Positron beams essentially separate the thermalisation of positrons implanted into a material from their eventual annihilation in another. While the field has been enlivened by a number of ingenious and exciting experiments—LEPD, PAES, etc. (reviews are to be found in 4, 35, 41, 42), in this section we shall concentrate on the basic elements of positron beam experimentation.

3.9.1 Positron moderators

A small fraction of positrons implanted into a moderator are thermalised close enough to the surface to be able to diffuse back to it in a time less than the annihilation lifetime. This fraction is clearly optimised if the moderating material is well-annealed—i.e., if it contains very few non-equilibrium defects. In the case of simple metals this means annealing, preferably *in situ*, to as high a temperature as possible (say ~$0.8T_m$) in as low an ambient pressure as possible (e.g. ~10^{-6} Pa or less). However, tungsten has been observed to operate as a moderator with useable efficiency after heating to a range of temperatures (often unmeasured), in a range of vacua (from 10^{-2} Pa to UHV), with slow (ramp) or rapid cooling, *in situ* or in a separate vacuum chamber followed by transport through air to the beam system. Whereas all appear to work reasonably well with tungsten because of its high positron work function, the same is not true, for example, with nickel, for which careful surface cleaning and maintenance is required. Even with tungsten, however, there may be a 30-40% difference in performance between a moderator prepared *in-* and *ex-situ*.

The fast-to-slow conversion efficiency of standard moderators can be increased by increasing their positron work functions (e.g. by diffusing sulphur to the surface of copper). Other methods for increasing slow positron yields have included depositing an epilayer of copper on a (high-Z) tungsten backing. Many moderator geometries have been used; venetian blinds made of annealed tungsten strips, tungsten meshes, thin ($\sim 10^2$nm) metallic films, backscattering geometries, conical geometry, etc. Although in perfect circumstances the moderation efficiency ε (= number of useable slow positrons per Bq) for some moderators can be $\sim 10^{-3}$, observed efficiencies are usually in the 10^{-4} range. [Note: in order for a mesh moderator to work efficiently it is essential that ~ 10V be applied between source and mesh.]

Choice of moderator depends on the application of the positron beam. For implantation defect spectroscopy the priority is to maximise moderator efficiency, whereas for electrostatic systems a well-collimated parallel beam requires a planar low-ϕ_+ surface cooled to minimise thermal smearing.

The highest moderation efficiencies recorded are for solid rare gases, which can be condensed directly on to the radioactive source capsule. These moderators rely on the fact that positrons do not fully thermalise and have a long effective diffusion length in the solid gas. Values of ε in the 10^{-3} range are common, with the value of 7×10^{-3} reported for solid Ne with a ^{22}Na source holding the record to date [43]. The energy distribution of positrons leaving such moderators naturally extend over a wider range; in some applications this is unimportant, and in others remoderation can still lead to an improvement on standard moderators.

Moderation efficiency could be greatly enhanced by drifting a larger fraction of thermalised positrons to the exit surface. Attempts to realise field-assisted moderation have to date largely foundered because of the interactions of positrons with the interfaces between the material across which an electric field is maintained and the conductive coatings to which the potentials are applied. One reported observation of the enhancement of positron emission by an electric field has been that from a solid gas moderator whose surface was charged by electron bombardment [44].

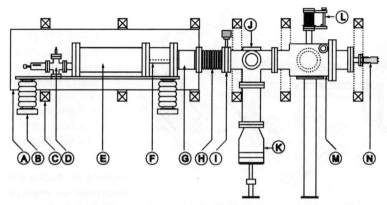

Figure 3.12 Magnetic-transport positron beam system. A-grounded shield: B-standoff insulators: C-coils for magnetic field: D-source/moderator: E-*ExB* plates: F-lead shielding: G-accelerator: H-bellows: I = aperture: J-guiding coils: K = turbopump: L = sample manipulator: M = sample chamber: N = CEMA/CCD camera.

3.9.2 Laboratory-based beams

A standard slow positron beam system for positron implantation spectroscopy is illustrated in Figure 3.12. The flight tube is evacuated; if UHV is required for positron-surface studies then the usual pumping and baking procedures must be followed and compatible materials used; additionally, sample cleaning and characterisation facilities such as an ion gun and a LEED/Auger head should be fitted. The system shown is surrounded by coils to provide a quasi-uniform axial transporting field. The sample is viewed through a thin foil window by a Ge detector for gamma photon counting or lineshape measurement. The positron energy (accelerating potential) is changed remotely by computer, and at each energy the computer sends small currents through the guiding coils in front of the target chamber to counter the small transverse shifts caused by *ExB* deflection in the accelerator region, where the electrostatic accelerating field may not be perfectly aligned with the guiding magnetic field.

In variants of the beam in Figure 3.12 the ***ExB*** filter may be substituted by a curved section of flight tube; positron acceleration may be achieved by floating the sample target at negative potentials (in which case any re-

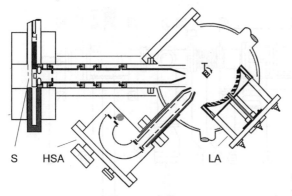

Figure 3.13 Electrostatic positron beam system: S = source assembly, T = target, HSA = hemispherical analyser LA = LEED/ Auger head.

emitted or backscattered positrons are returned to the sample), a further set of *ExB* plates may be held just in front of the sample to ensure that re-emitted positrons are not reflected at the accelerator and return to the sample. In an *ExB* filter the transverse deflection (perpendicular to *B* and *E*) is proportional to the time a positron spends between the plates; if a 100eV positron is deflected by 60mm, an unmoderated 250keV positron will only be deflected by ~ 1mm. The ceramic accelerator tube should have a series of identical rings or an equivalent system for ensuring that the accelerating field is as uniform as possible.

Positron intensities $\sim 10^5$ s^{-1} are achievable with standard positron beams of the type outlined above, using primary radioactive sources of ~ 4GBq.

Whereas most magnetic-transport beams are a few m long in order to shield the annihilation gamma detector from the source, this length is notrequired if we are interested in particle spectroscopies, especially if electrostatic transport and focusing is incorporated into the design. Canter [45] has published several papers on positron optics to which the reader is referred. Figure 3.13 shows an example from the author's laboratory.

Brightness enhancement is achieved in positron beams by repeated

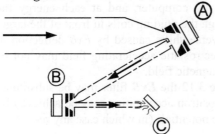

Figure 3.14 Two-stage reflection-geometry brightness enhancement optics: C = sample.

focusing and remoderation; the example in Figure 3.14 shows a brightness enhancement unit using backscattering geometry; positrons are focused on to remoderators at A and B, each time leaving the remoderator surface close to the normal. Although remoderation losses may be 70%, the beam area can decrease by a factor of ~ 50 at each stage; hence beam intensity per unit area can increase after n stages by a factor 15^n. The reader is referred to [45] for a full discussion.

3.9.3 Facility-based beams

The facility-based system is still the main hope for intense positron beam generation. In the case of LINACs the beam is pulsed; this property can be exploited in applications where timing is an advantage.

3.9.4 Beam bunching

LINAC beam pulses are of widths from several ns to µs; if subnanosecond timing resolution is required the positrons must be put through a buncher. Several buncher designs have been put into practice; the first, by Mills [46] used magnetic mirrors. Later bunchers use RF chopping and bunching techniques—for example, the system at Munich [47], which can now achieve a timing resolution below 200ps.

3.9.5 Polarized positron beams

Beta positrons are longitudinally spin polarised (they possess helicity) along the direction of emission. This property has been used in the study of magnetic phenomena in bulk positron measurements. It is of great interest and importance that positron beams retain a high degree (as high as 70%) of polarisation. A novel polarimeter can be used in which Ps is formed on a surface and its decay components recorded [48]. To achieve a high degree of polarisation a low-Z backing is needed behind the primary positron source, to minimise backscattering. This necessarily decreases the positron flux.

3.9.6 MeV positron beams

High-energy (~MeV) monoenergetic positron beams have been built and used in recent years. MV accelerators used in this work have been of Pelletron or Van de Graaff type. Source-free lifetime measurements in a wide range of materials, beam-based AMOC, and novel annihilation spectroscopies are possible with such systems. Work performed with these beams [see, e.g., 49] is described elsewhere in this volume.

3.9.7 Time-of-flight spectrometry

There have been a number of positron experiments over the past twenty-five years which have benefited from the advantages of the time-of-flight (TOF) method. In very low signal-to-background beam experiments monoenergetic positrons all have essentially the same TOF and appear as a peak on a time spectrum, whereas the very large number of unrelated, randomly-occurring 'events' are spread over all times.

In [50] beta positrons pass through a thin scintillator to produce start pulses, and the small fraction of fast positrons which are moderated after passing through the scintillator are guided to a CEM detector, which produces the stop pulse. The scintillator pulses are delayed so that a reverse time spectrum is obtained and dead time losses minimised. Another method for tagging positrons or Ps is to use secondary electrons ejected by the incident positron [51]; these are detected by a CEMA assembly, sometimes with a central hole through which positrons pass. Ps TOF spectroscopy takes its start signal from a pulsed positron beam, either bunched in the laboratory or in a LINAC burst. A new application of a LINAC beam is TOF-PAES, positron annihilation-induced Auger electron spectroscopy [52].

3.9.8 Positron microscopy

Two types of positron-beam-based microscopes have been developed and used over the past decade [35]. In the *positron re-emission microscope* [53, 54] a brightness-enhanced microbeam illuminates a surface (either in transmission or reflection geometry) and re-emitted positrons pass through micrscope optics to a position-sensitive detector. The unique contrast mechanism in the resulting intensity image is therefore based on the changes in effective positron re-emission probablities across the viewed surface. In the *scanning positron annihilation microscope* a ~ µm diameter positron beam is implanted into a sample target and its annihilation radiation detected. Optics from a scanning electron microscope have been adapted for this purpose [55, 56]. Position-sensitive Doppler broadening spectroscopy can be performed (additionally with some depth sensitivity) to yield information on near-surface defect structure. In the Munich microscope [57] the microbeam is time-bunched, so that lifetime spectroscopy is possible.

3.9.9 Plasma-generated positron beams

Surko and co-workers have pioneered the generation of very low-energy positron beams with extremely narrow (~20meV) energy spread by extracting positrons from a plasma of cold positrons stored in a Penning-Malmberg trap [58]. Positrons lose energy by collisions with gas molecules in a differentially-pump system until they are trapped in a potential well from which they can be ejected in a very controlled manner. This high-quality beam has already found many new applications in atomic physics [59] and holds great promise for wider exploitation in the future.

Problems

Problem 3.1 Standard lead blocks around positron experiments are usually 50mm thick. Why (apart from practical reasons associated with weight) is this? [The coefficient for absorption of 511keV annihilation radiation by lead is ≈ 0.16 mm^{-1}; for 1.28MeV radiation it is ≈ 0.072 mm^{-1}.]

Problem 3.2 What is the peak voltage of a pulse fed into 50Ω from a PM tube (gain 10^6) fitted with a fast plastic scintillator which has absorbed 511keV from a gamma ray?

Problem 3.3 What is the maximum energy that an incident annihilation (i.e., 0.511 MeV) gamma photon can deposit in a scintillator detector in a Compton collision?

Problem 3.4 Positrons in a 5mm-diameter beam are detected in two different ways: (a) directly by a channeltron (10mm-diameter mouth) placed at the end of an evacuated beam line, and (b) indirectly via annihilation radiation from the channeltron by a 75mm-diameter, 75mm-long NaI(Tl) crystal on a PM detector mounted outside the vacuum system directly to one side of the channeltron. The distance from the channeltron to the front face of the scintillator is 100mm, and the intrinsic detection efficiency of the crystal for 511keV gamma radiation is 40%. Estimate the ratio of the positron signal counts recorded in the same time by the two detectors.

Problem 3.5 The 'random' background underlying a positron lifetime spectrum—caused by unrelated pulses starting and stopping the TAC — appears flat. In fact it is approximately exponential in form. Explain this.

Problem 3.6 The angular resolution of a standard 1D angular correlation apparatus ~0.5 milliradians. To what angular resolution is the energy resolution of a Ge detector at 511keV (= 1.2keV) equivalent? [Use $p \approx m_0 c \theta$, where θ is the deviation of a gamma ray from the system axis.]

Problem 3.7 What is the angle between annihilation photons if the energy of the annihilating pair ($\approx$ the electron energy) associated with motion transverse to the emission direction is 5eV?

Problem 3.8 It is possible, with a positron beam, to measure the re-emitted positron fraction as a function of incident energy E, and hence to estimate quite precisely the value of y_0, the re-emitted positron fraction at zero E. From this we can estimate the primary moderation efficiency ε for a radioactive source from the expression, where L_+ is the positron diffusion length in the moderating material (~100nm) and λ the mean free path for β^+ in the moderator. Find ε for ^{22}Na positrons bombarding a piece of tungsten for which y_0 has been measured to be 0.35, if λ = 15 µm.

Problem 3.9 The magnitude of the transverse velocity of positrons passing through planar-geometry ExB plates is $|\mathbf{E}|/|\mathbf{B}|$. Calculate the transverse deflections of (a) 100 eV positrons and (b) 300 keV positrons passing through plates separated by 20mm and of length 300mm. The potentials applied to the plates are ±100V, and the axial magnetic field is 100 G.

Problem 3.10 Compute (a) the time of flight for 1eV and 19 keV positrons travelling along a flight path of length 1 m: and (b) the distance travelled in vacuum by o-Ps and p-Ps of kinetic energy 2eV in their mean lifetimes.

References

[1] R.N. West, *Positron Studies of Condensed Matter* (Taylor & Francis 1972)
[2] W Brandt & A Dupasquier: *Positron Solid-State Physics* (North-Holland 1983)
[3] A P Mills, Jr, W S Crane & K F Canter, *Positron Studies of Solids, Surfaces and Atoms* (World Scientific 1986)
[4] P J Schultz and KG Lynn, *Interaction of Positron Beams with Surfaces, Thin Films and Interfaces*, Rev. Mod. Phys. **60**, 701 (1988)
[5] A Dupasquier and A P Mills, Jr, eds, *Positron Spectroscopy of Solids* (IOS 1995)
[6] S DeBenedetti, CE Cowan, WR Konneker & H Primakoff, Phys.Rev. **77** 205 (1950)
[7] M Deutsch, Phys. Rev. **82**, 455 (1951)
[8] W Cherry, PhD dissertation (Princeton 1958)
[9] DW Gidley, A Rich, PW Zitzewitz & DAL Paul, Phys.Rev.Lett. **40**, 737 (1978)
[10] IK MacKenzie, in *Positron Solid State Physics*, ed A Dupasquier and W Brandt (North-Holland, Amsterdam, 1983) pp208-220
[11] HM Weng, YF Hu, CD Beling and S Fung, Appl. Surf. Sci. **116**, 98 (1997)
[12] DT Britton, M Härting, CM Comrie, S Mills, FM Nortier and TN van der Walt, Appl. Surf. Sci. **116**, 53 (1997)
[13] KG Lynn, M Weber, LO Roellig, AP Mills, Jr., and AR Moodenbaugh, in *Atomic Physics with Positrons*, eds JW Humberston and EAG Armour (Plenum: New York 1987) p.161.
[14] G Triftshäuser, G Kögel, W Triftshäuser, M Springer, T Hagner and K Schreckenbach, Mat. Sci. Forum **175-178** 221(1995).
[15] DG Costello, DE Groce, DF Herring & JW McGowan, Phys.Rev.B **5** 1433 (1972)
[16] KF Canter, Workshop on Intense Positron Beams in Europe, London, 1995
[17] M Hirose, T. Nakajyo and M Washio, Appl. Surf. Sci. **115**, 63 (1997)
[18] W Bauer, J Major, W Weiler, K Maier and HE Schaefer, in *Positron Annihilation*, eds PC Jain *et al* (World Scientific, Singapore: 1985) p 804
[19] R N West, J Mayers and PA Walters, J.Phys.E:Sci. Instrum. **14**, 478 (1981)
[20] See, for example, AP Jeavons, Nucl. Instrum. Methods **156**, 41 (1978) and R Sachot, AA Manuel and P Descouts in *Positron Annihilation*, eds PG Coleman *et al.* (North- Holland, Amsterdam: 1982), p. 892.
[21] PG Coleman, TC Griffith & GR Heyland, Proc.Roy.Soc..A **331**, 561 (1973)
[22] W Bauer *et al.*, in *Positron Annihilation*, eds. L. Dorikens-Vanpraet, M. Dorikens and D Segers (World Scientific, Singapore: 1989) p. 579
[23] GR Massoumi, N Hozhabri, WN Lennard & PJ Schultz, Phys.Rev.B **44**, 3486 (1991)
[24] S Pendyala, J Wm McGowan, PHR Orth and PW Zitzewitz, Rev.Sci.Instrum. **45**, 1347 (1974)
[25] CP Burrows and PG Coleman, Appl.Surf.Sci. **116**, 184 (1997)
[26] TN Horsky *et al.*, Phys. Rev. B **46**, 7011 (1992)

[27] DA Gedke and WJ McDonald, Nucl. Instrum. Methods **55**, 377 (1967)
[28] PG Coleman, TC Griffith and GR Heyland, J.Phys.E:Sci.Instrum. **5** 376 (1972)
[29] P Kirkegaard and M Eldrup, Comp.Phys.Commun.**3** 240 (1972) & **7** 401 (1974)
[30] DP Kerr, PD Fellows, DJ Sullivan and RN West, Phys. Letts.A **61**, 418 (1977).
[31] DT Britton and A van Veen, Nucl. Instrum. Methods A **275**, 387 (1989)
[32] S Mantl and W Triftshäuser, Phys. Rev. B **17**, 1645 (1978)
[33] KG Lynn and AN Goland, Sol. State Commun. **18**, 1549 (1976)
[34] see A van Veen, H Schut and P E Mijnarends, ref[35] p 191 (2000)
[35] *Positron beams and their Applications*, ed. P.G. Coleman (World Scientific: Singapore, 2000).
[36] Courtesy of University of Bristol positron group web page
[37] S Berko, in *Positron Solid State Physics*, eds A Dupasquier and W Brandt (North-Holland, Amsterdam: 1983) p 64
[38] RN West, in *Positron Spectroscopy of Solids,* eds. A Dupasquier and AP Mills, Jr (IOS, Amsterdam: 1995) p. 75
[39] SB Dugdale, MA Alam, HM Fretwell, M Biasini and DA Wilson, Mat. Sci. Forum **175-178**, 895 (1995)
[40] See for example KG Lynn *et al.*, Phys. Rev. Lett. **54**, 1702 (1985)
[41] AP Mills, Jr., in *Positron Solid State Physics*, eds. A Dupasquier and W Brandt (North-Holland, Amsterdam: 1983) p. 432
[42] AP Mills, Jr, in *Positron Spectroscopy of Solids*, eds. A Dupasquier and AP Mills, Jr (IOS, Amsterdam, 1995) p. 209
[43] AP Mills, Jr. and EM Gullikson, Appl. Phys. Lett. **49**, 1121 (1986)
[44] JP Merrison, M Charlton, BI Deutch and LV Jørgensen, J.Phys:Condens. Matter **4**, L207 (1992)
[45] KF Canter, in *Positron Spectroscopy of Solids*, eds. A Dupasquier and AP Mills, Jr (IOS, Amsterdam, 1995) p. 361
[46] AP Mills, Jr., Appl. Phys. **22**, 273 (1980)
[47] P Willutski *et al.*, Mat.Sci.Forum **175-178,** 237 (1995)
[48] J Van House and PW Zitzewitz, Phys. Rev. A **29**, 96 (1984)
[49] W Bauer *et al.*, Appl. Phys. A **43**, 261 (1987)
[50] KF Canter, PG Coleman, TC Griffith & G R Heyland, J.Phys. B **5**, L167 (1972)
[51] DW Gidley, AR Köymen andTW Capehart, Phys.Rev.Lett. **49**, 1779 (1982)
[52] R Suzuki, T Mikado, M Chiwaki, H Ohgaki and T Yamazaki, Appl. Surf. Sci. **85**, 87 (1995)
[53] G R Brandes, K F Canter and A P Mills, Jr, Phys. Rev. Lett. **61**, 492 (1988)
[54] A Goodyear and P G Coleman, Appl. Surf. Sci. **85**, 98 (1995)
[55] L J Siebel, R F J Neelissen, P Kruit, A van Veen and H Schut, Appl. Surf. Sci. **85**, 92 (1995)

[56] H Greif, M Haaks, U Holzworth, U Männig, M Tongbhoyai and K Maier, Appl. Phys. Lett. **71**, 2115 (1997)
[57] A David, G Kogel, P Sperr & W Triftshäuser, Phys.Rev.Lett. **8706** 7402 (2001)
[58] SJ Gilbert, C Kurz, RG Greaves & CM Surko, Appl. Phys. Lett. **70** 1944 (1997)
[59] J P Sullivan, J P Marler, S J Gilbert, S J Buckman and C M Surko, Phys. Rev. Lett. **8707**, 3201 (2001)

Answers to Problems

Problem 3.1. The fraction of gamma rays f penetrating a thickness d of lead = $\exp(-\mu d)$, where μ is the (energy-dependent) absorption coefficient. Thus $f = \exp(-0.16 \times 50) = 0.00034$ for 0.511MeV radiation and $\exp(-0.072 \times 50) = 0.028$ for 1.28MeV radiation. So a single block does a good job on annihilation radiation. 64mm lead would reduce the intensity of the more energetic gamma radiation to 1%; two blocks (100mm) to 0.075%.

Problem 3.2 $I = \dfrac{d}{dt}(gn)$, where g = gain, n = number of photoelectrons from photocathode: so $I = g\dfrac{dn}{dt} = 0.2g\dfrac{dN}{dt}$, where N = number of photons from phosphor:
hence $I = 0.2\lambda g N e^{-\lambda t}$ with a peak value of $0.2\lambda g N = 0.2gN/\tau = (0.2)(10^6)(2000)/(2 \times 10^{-9}) = 2 \times 10^{17}$ electrons s^{-1} = 32 mA. In 50Ω this is 1.6V; the PM pulse dies away with a decay constant of 2ns.

Problem 3.3 The Compton scattering formula states that the change in wavelength of a gamma ray scattered through an angle θ is $\lambda_2 - \lambda_1 = (h/mc)(1-\cos\theta)$. The maximum change in wavelength (and hence energy) thus occurs for $\theta = \pi$, when $\lambda_2 - \lambda_1 = 2h/mc$. To convert this to energy loss we use $\lambda = hc/E$ so that $hc(E_2^{-1} - E_1^{-1}) = 2h/mc$, or $(E_2^{-1} - E_1^{-1}) = 2/mc^2$. Now $E_1^{-1} = 1/mc^2$, as E_1 is the annihilation gamma energy mc^2, so that $E_1 = mc^2/3$. This means that a gamma photon can lose up to a maximum of 2/3 of its incident energy in a Compton interaction. Thus the answer to the problem is that the maximum energy deposited in the crystal (which determines the position of the 'Compton edge' on the pulse height spectrum) = (2/3)(0.511MeV) = 0.341 MeV.

Problem 3.4 Assuming 100% detection efficiency for the CEM, let the total positron signal count be N. The counts recorded by the scintillator will be lower because of solid angle and efficiency factors. To estimate solid angle use the following sketch:

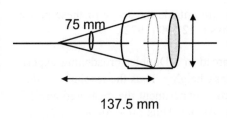

Fractional solid angle ≈ $\pi(37.5)^2/4\pi(137.5)^2 \approx 1.9\%$. This has to be doubled because there are two annihilation photons per positron → 3.8%. Then overall detection efficiency ≈ 0.4 x 3.8% ≈ 1.5%—i.e., total counts recorded by scintillator ≈ 0.015N. (So ratio ≈ 0.015.)

Problem 3.5 To simplify the situation somewhat, let us consider completely unrelated start and stop pulse trains (i.e., no signal). When a pulse starts the timing system it may be stopped (converted) by a stop pulse after a time interval t, in the period $t \to t+dt$, only if there is *no* pulse in the intervening period (i.e., $0 \to t$). Thus the probability $P(t)$ that both conditions are met is the product of the probabilities for both:
(a) probability that a random stop pulse occurs after time t (i.e., in $t \to t+dt$) = $n_2 dt$, where n_2 is the stop pulse rate);
(b) probability that *no* random stop pulse occurs between 0 and t =

$$1 - \int_0^t P(t)dt.$$

Thus $P(t)dt = n_2 dt \,[\,1 - \int_0^t P(t)dt\,]$.

One can deduce from this expression that $P(t) = n_2 \exp(-n_2 t)$. Now the background spectrum $B(t)dt$ is $n_1 n_2 dt_i$, where n_1 is the start pulse count rate and dt_i is the time width of channel i: hence $B(t)dt = n_1 n_2 \exp(-n_2 t)$. [Note: this does not take into account the finite widths of, or any dead times associated with, the start or stop pulses!] This exponential slope on the background is only significant (i.e., important to the analysis of the spectrum) when the stop pulse rate n_2 is not negligible in comparison to the inverse of the time range of the MCA spectrum. In most standard lifetime measurements in condensed matter the background can be, and is, treated as flat; but in fast-counting experiments (lifetime or time-of-flight) beware! One final point: when signal is present (as one would hope it is), both start and stop pulse rates n_1 and n_2 are subject to t-dependent corrections. Such details, and the dead-time problems referred to above, are outside the scope of these notes; but see for example P G Coleman, J.Phys.E **12** 590 (1972).

Problem 3.6 The measured energy spread in a Doppler broadening experiment is $W \sim pc/2$ (at half-maximum peak height). Thus the resolution $\Delta W \sim c\Delta p/2 = 1.2$ keV. In an angular correlation experiment the measured angle $\theta \sim p/mc$ and the angular resolution $\Delta \theta \sim \Delta p/mc$. Thus $\Delta \theta \sim \Delta p/mc = c\Delta p/mc^2 = (2.4)/(511) \approx 4.8$ mr. This is an order of magnitude worse than the angular resolution quoted for standard angular correlation experiments.

Problem 3.7 $\delta \theta \approx p_t/mc = 2v_t/c = 2[(mv_t^2/2)/(mc^2/2)]^{1/2} = 2[2E_t/mc^2]^{1/2}$, where E_t is the energy associated with p_t. Substituting $E_t = 5$eV and $mc^2 = 511$ keV we get $\delta \theta = 2[10/511000]^{1/2} = 8.8$ milliradians.

Problem 3.8 $\varepsilon \approx (0.35)(10^{-7})/(15 \times 10^{-6}) = 2.3 \times 10^{-3}$. Efficiencies up to an order of magnitude lower than this figure are commonly seen because of non-ideal conditions.

Problem 3.9 $|E| = 200/0.02 = 10^4$ Vm^{-1}. $|B| = 10^{-2}$ T. Thus $v_{\text{trans}} = 10^6$ ms^{-1}. Then (a) time t spent between the plates $= D(m/2E)^{1/2} = (D/c)(mc^2/2E)^{1/2} = (0.3/3 \times 10^8)(511 \times 10^3/200) = 5.06 \times 10^{-9}$s. Deflection $d = v_{\text{trans}} t = 50.6$mm. (b) $(d_{300\text{keV}})/(d_{100\text{eV}}) = (100/3 \times 10^5) = 1/54.7$, so $d_{300\text{keV}} = 0.93$mm.

Problem 3.10 (a) Time of flight $t = D/v = D(m/2E)^{1/2} = (D/c)(mc^2/2E)^{1/2} = (1/3 \times 10^8)(511 \times 10^3/2) = 1.69$ μs for 1eV positrons. For 10keV positrons, time of flight $= (1.69\mu\text{s})/100 = 16.9$ ns. (b) From (a), $D = tc(2E/2mc^2)^{1/2} = 140 \times 10^{-9} \times 3 \times 10^8 (2/511 \times 10^3)^{1/2} = 8.3$ cm. For p-Ps $t = (125 \times 10^{-12})/(140 \times 10^{-9})t_{\text{o-Ps}} = 0.074$ mm.

Chapter 4

Organic and Inorganic Chemistry of the Positron and Positronium

G. Duplâtre and I. Billard

Chimie Nucléaire, Institut de Recherches Subatomiques CNRS/IN2P3 et Université Louis Pasteur B.P. 28, 67037 Strasbourg Cedex 2, France

4.1 Positronium formation in condensed matter

The "spur" model, proven to be valid in condensed media, proposes that Ps formation would occur through the reaction of a (nearly) thermalized positron with one of the electrons released by ionization of the medium, at the end of the e^+ track, in a small region containing a number of reactive labile species (electrons, holes, excited molecules) [1].

Numerous experiments have been performed on the action of solutes on Ps formation in liquids. Parameters that have been examined are: nature and concentration of the solutes, temperature, nature of the solvent, presence of cosolutes, electric and magnetic field effects. Due to the ease of dissolution of a large variety of compounds, the most studied solvent is water, which allows larger possibilities of comparison with data from pulse radiolysis.

The main techniques used are positron annihilation lifetime spectroscopy (PALS) and the Doppler broadening (DB) or angular correlation (AC) techniques. The PALS parameters are the relative intensities (I_i) and the lifetimes τ_i (decay rate constants: $\lambda_i = 1/\tau_i$) of the decay components; subscripts i = 1, 2 and 3, in the order of increasing lifetimes, refer to p-Ps, free (solvated) e^+ and o-Ps [2]. When more components are present (e.g., e^+ or Ps bound-states with a solute) higher subscripts may be used.

The energy distribution spectra collected by DB can be resolved into a sum of Gaussians, each representative of a positron state [3], with intensities, I_i^D, and full widths at half maximum (fwhm), Γ_i. The subscripts are the same as for PALS. The DB intensities can be quantitatively correlated to those from PALS. Because p-Ps annihilates in an intrinsic mode, one has $\Gamma_1 \ll \Gamma_3 < \Gamma_2$. The DB results are conveniently presented in the form of the global fwhm of the energy distribution spectra. Similar treatments can be made, with a better resolution, in angular correlation (AC) experiments.

4.1.1 The spur model in polar solvents (Strasbourg Group) [2]

Based on a long series of experiments in polar solvents, the set of reactions shown below has been proposed (M and S stand for the medium and the solute, respectively). Some of the data that led to establish this model are presented in the following.

$$e^+ + M \rightarrow e^+ + M^+ + e^- \quad \text{ionization} \qquad (I)$$

$$e^+ + e^- \rightarrow Ps \qquad \text{Ps formation, fraction (1-f)} \quad (II)$$

$$e^- + M^+ \rightarrow M^* \qquad \text{recombination} \qquad (III)$$

$$e^+ \; (e^-) \rightarrow e^+_{loc} \; (e^-_{loc}) \qquad \text{localization} \qquad (IV)$$

$$e^+_{loc} + e^-_{loc} \rightarrow Ps \qquad \text{Ps formation, fraction f} \quad (V)$$

$$e^+_{loc} \; (e^-_{loc}) \rightarrow e^+_s \; (e^-_s) \qquad \text{solvation} \qquad (VI)$$

Main reactions with solutes:

$$e^- + S \rightarrow S^- \qquad \text{"total" inhibition} \qquad (VII)$$

$$e^+_{loc} \text{ (or } e^-_{loc}) + S \rightarrow \left[Se^+\right] \text{ (or } S^-\text{)} \quad \text{"partial" inhibition} \quad \text{(VIII)}$$

$$M^+ + S \rightarrow \text{products} \quad \text{enhancement} \quad \text{(IX)}$$

•Total inhibition. Solutes known to be efficient solvated electron (e^-_s) scavengers are found to inhibit Ps formation. Experimentally, the decrease in the o-Ps intensity, I_3, obeys a simple Stern-Volmer type equation:

$$I_3(C) = I_3(0)/(1 + kC) \quad (1)$$

where k is the solute total inhibition constant and C, its concentration. This is illustrated in Figure 4.1 for the halate ions [4] in water.

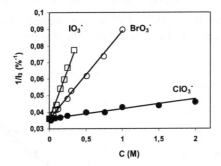

Figure 4.1: Variation of $1/I_3$ (%$^{-1}$) with concentration C (M) of ClO_3^-, BrO_3^- and IO_3^-. Water, 294 K.

The inhibition constants are strongly correlated with the reaction rate constants with solvated electrons, although a few solutes are outside the correlation line: either well above (e.g., $SeO_4^=$) or below (e.g., Ni^{2+}) this line [2]. The correlation is even better with the constants, $1/C_{37}$, related to the scavenging of the precursors of the solvated electrons (Figure 4.2): such ions as $SeO_4^=$ or Ni^{2+} are now well included. Note that such correlations also include implicitely a large number of solutes that do not react with electrons and do not inhibit Ps formation (e.g.: the halide ions and numerous organic molecules).

The total inhibition constants are little sensitive to temperature (T), usually slightly decreasing as T increases [5, 6]. This is consistent with the hypothesis that the processes involved, Ps formation and electron scavenging, occur on a short time scale, when the species are not yet in complete equilibrium with the medium. In water, electron (and presumably too, positron) solvation is very fast (0.24 ps) [7]. Once solvated, these particles are less mobile; although they could continue to produce Ps, their contribution must be very small. This is clearly shown in propanol, for instance, where, at low enough temperatures, the solvated positron lifetime is shorter, by up to 3 orders of magnitude, than the electron solvation time [8]. The implication of quasi-free particles in Ps formation is also sustained by the concentrations needed to suppress about half of Ps, typically in the range 0.05—0.2 M, which are similar to those used in fast pulse radiolysis experiments (see values of C_{37} in Figure 4.2), to suppress about half of the initial e^-_s yields [9]. By contrast with experiments on e^-_s scavenging, where much lower concentrations (10^{-4}—10^{-3} M) are needed, this denotes that the processes involved are very fast: large amounts of solutes must be present to be effective.

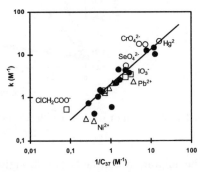

Figure 4.2: Variation of the Ps inhibition constant, k (M^{-1}), with the constant of capture of the precursors of e^-_{aq}, $1/C_{37}$ (M^{-1}), by solutes.

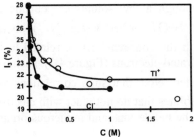

Figure 4.3: Variation of the o-Ps intensity, I_3 (%), in solutions of TlClO$_4$ and NaCl. Water, 294 K.

Proper care must be taken in the presence of chemical equilibria (e.g., ion pairing or dimerization) as the free and associated species may have different inhibiting powers [10-12].
- Partial inhibition. Some solutes inhibit only a fraction (f) of Ps at high concentrations. The variation of I_3 with C is conveniently described by [2]:

$$I_3(C) = I_3(0) \, [f/(1 + KC) + 1 - f] \qquad (2)$$

where K is the partial inhibition constant. This is illustrated in Figure 4.3 for Tl^+ and Cl^-, which show a similar inhibiting behavior although Cl^- is a hole (OH radicals) scavenger, not an electron scavenger, in contrast to Tl^+ [6]. However, complementary DB experiments show that these two solutes have in fact a very different action (see Figure 4.4).

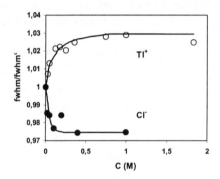

Figure 4.4: Variation of fwhm in solutions of $TlClO_4$ and NaCl. Water, 294 K.

In case of electron scavenging (and no Ps lifetime quenching, as is true for both Cl^- and Tl^+), no other positron states are present than free e^+ and Ps: then, the intensities from PALS and from DB are the same. The p-Ps and o-Ps intensities are expected to decrease so that the fwhm of the DB spectra should increase with solute concentration (the narrow components are suppressed). The variations of fwhm with C can be completely calculated, knowing the intensities I_i from PALS and the Γ_i previously established for a given solvent. This is illustrated by the solid line in Figure 4 for Tl^+: this ion, as expected from its high solvated electron scavenging rate constant, is thus shown to suppress Ps formation by electron capture.

By contrast, the variation of fwhm with $C(Cl^-)$ shows a decrease (Figure 4.4). This can only occur if a fourth positron state, not present in the case of

Tl^+, is involved. Thus, both Ps formation inhibition and the decrease in fwhm are very well explained by supposing that Cl^- scavenges the positrons. The solid line in Figure 4.4 is calculated by taking this reaction into account, with a specific fwhm, Γ_4, associated to the [PsCl] bound-state formed.

At present, all positron scavengers have been found to belong to the class of partial inhibitors, and only a small portion of the electron scavengers. The reason for partial inhibition is still debated.

For a given solvent, fraction f is similar for all solutes, whether positron or electron scavengers (see Figure 4.3). Table 4.1 collects typical values of f for a variety of polar solvents: f decreases from the more to the less protic solvents and is strongly correlated to the electron solvation time or, when this is not known, to the trap depth for e^-. Thus, although having viscosities largely exceeding those of the lighter alcohols, 1,3 propanol and glycol both show a larger f and a higher total Ps yield: as it seems, these α or β alcohols offer, through their neighbouring OH moieties, pre-organized sites that favor fast electron localization.

Table 4.1 Fraction f of partial inhibition in various polar solvents and its contribution to the total Ps yield. The stars denote non protic solvents.

Solvent	F	fI_3 (%)	$(1-f)I_3$ (%)
Water	0.28	7.8	20.2
Methanol	0.10	2.2	19.8
Ethanol	0.08	1.8	20.6
Propanediol	0.14	3.1	19.6
Glycol	0.17	3.8	18.5
Dimethylsulfoxide*	0.09	1.3	13.0
hexamethyl-phosphorotriamide*	0.05	1.4	26.1
N-ethylacetamide*	0.03	0.7	22.5

Table 4.1 shows that f is also correlated with the total Ps yield. Thus, in the series including water and the alcohols, fraction $(1-f)I_3$ is about constant and the difference in the total yields is essentially due to fI_3. This picture is strengthened when examining the variations with temperature of the Ps yields, for all systems that have been studied. Thus, for Cl^- in glycol [13], the

initial total Ps yield increases greatly with T, and this is entirely due to the increase in fraction fI_3, while fraction $(1-f)I_3$ is not affected by T, in agreement with its ascription to very early processes [14].

Although limited, the decrease in I_3 due to partial inhibition occurs at much lower solute concentrations (about 0.01—0.02 M) than that promoted by total inhibitors. This is reflected in the range of values for either type of inhibition constants: in water, the highest k values are at about 20 M^{-1}, while typical K values are in the range 40—140 M^{-1} [2]. By contrast with k, the partial inhibition constants are usually strongly temperature-dependent, with an activation energy close to that of the viscosity of the solvent [13].

• Enhancement of Ps formation. As expected from the spur model, all solutes that are efficient hole scavengers, thus somehow preventing the recombination process and increasing the electron availability (see reactions I—IX) enhance Ps formation. In water, strong positive ion scavengers are essentially the halide and pseudo-halide ions, together with amines. A convenient empirical equation to describe the Ps intensity variation is as follows [2]:

$$I_3(C) = I_3(0) (1 + \alpha C)/(1 + \beta C) \qquad (3)$$

where α/β (>1) is the maximum fraction of enhancement and β, the solute enhancement constant. A typical example is provided by allylamine in water. As was the case for Tl^+ (Figure 4.4), the variation with C of fwhm from DB measurements is completely calculated (not fitted) on the basis of the PALS results [15]. Comparison with pulse radiolysis data can be made using the reaction rate constants (k_{OH}) for OH radical scavenging (OH recombines with e^- to yield OH^-): a good correlation is found between β and k_{OH} [2].

Hole capture by a solute (S) is probably not sufficient to explain the Ps yield enhancement as recombination might well occur as easily between the electron and either M^+ (the hole) or S^+ (the trapped hole). An explanation to the phenomenon is probably that e^+ cannot react with the electron once recombined with M^+ whereas it can pick up an electron shallowly trapped by S^+. The weakness of recombined S^+/e^- pairs has been shown in some instances in pulse radiolysis experiments, where a delayed formation of e^-_s has been observed from such a state [16]. The concentration range necessary for efficient enhancement of Ps formation is similar to that related to total inhibition, indicating that the processes involved occur on a very short timescale. The absence of temperature effects on the enhancement constant, β, as found for the total inhibition constant, k, points to the same conclusion.

- Combined Ps partial inhibition and enhancement. A solute that scavenges e^+ is likely to be also a hole scavenger. This is verified in a number of cases, such as I^- in water [17]: on increasing the concentration of this ion there is first a rapid decrease in I_3 due to e^+ capture, followed by an increase due to hole scavenging. As the temperature effects on β are very weak, while those on both K and fI_3 are strong, the shape of the variations of I_3 with C can change from an apparently pure enhancement at low T, to pure partial inhibition, at high T, as shown in Figure 4.5 for the acetate ion in methanol [18].

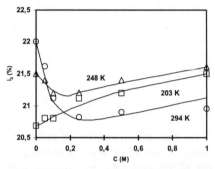

Figure 4.5: Variation of the o-Ps intensity, I_3 (%), with sodium acetate concentration, C (M), at various temperatures in methanol.

- Consecutive e^- and e^+ scavenging reactions. The capture of e^- by a solute can lead to a product which is itself a positron scavenger, leading to consecutive e^- and e^+ capture reactions. This is true for halogenated compounds, such as $HgCl_2$ or CCl_4: both are efficient e^- scavengers; the product may be the radical anion, $SCl^{\bullet-}$, or there may occur halide detachment, $S^\bullet + Cl^-$. In either case, the anion formed can capture the positron in the spur.

Specific experiments have been carried out to demonstrate the existence of such reactions. A typical example is provided by CH_3I in methanol [19]. As expected from its strong e^- scavenging ability, this solute is an efficient total inhibitor in PALS experiments. However, the DB data show that the fwhm vs C variation cannot be recovered on the hypothesis that only the e^+

and Ps components are present (broken line in Figure 4.6): an excellent fit is obtained when supposing the formation of a positron bound-state involving the solute (solid line in Figure 4.6): the reaction scheme must imply the consecutive captures of e$^-$, then of e$^+$. By comparing the Γ_4 parameters of such e$^+$ bound-states to those of the corresponding positron/halide bound-states, it is possible to know if, on the time scale of the lifetime of the bound-states (*ca.* 400 ps), halogen detachment occurs [19].

To further assess the reality of these reactions, a series of experiments was carried out on mixtures of solutes. It is expected, and experimentally verified, that adding an e$^-$ scavenger to a halogenated compound solution should result in a decrease of the e$^+$ bound-state yield, while adding an e$^+$ scavenger should increase this yield [19, 20].

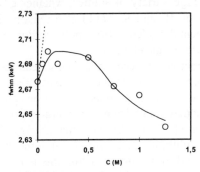

Figure 4.6: Variation of fwhm with CH$_3$I concentration (M). Methanol. For the broken and solid lines, see text.

- Anti-inhibition. As evoked in the case of enhancement, the capture of e$^-$ in a shallow trap can allow further reaction with e$^+$ to form Ps. This can give rise to the anti-inhibition effect [21, 22]: adding a solute known as a Ps inhibitor (shallow trap) to a solution of another Ps inhibitor (deep trap) can result in an increase in the Ps yield.
- Comments to the model proposed through reactions (I)—(IX). Although this model can explain most of the experimental facts, some doubts remain on its details; furthermore, other possibilities are open and have been proposed, although they never proved to explain as many data as this model.

Thus, the existence of fraction f even in some frozen solutions is unexplained. This is the case for Cl$^-$ in sulfolan, where partial inhibition of

Ps formation appears both in the liquid and solid (plastic) phases [23]; the data confirm however that the increase in the total Ps yield with T arises from the increase in fraction fI_3^0. The experiments are scarce and should be developed; the hypothesis of fast localization processes in solids would need support from pulse radiolysis experimental data.

Partial inhibition has received other explanations, one being based on the anti-inhibition effect: a solute scavenges e^- in a shallow trap, so that e^+ can pick-up this trapped electron to form Ps, resulting in a balance between Ps inhibition and enhancement. The possibility that the shallowly trapped e^- (or e^+) delocalizes over several solute molecules at high enough concentration, thereby gaining mobility and reactivity towards e^+ (or e^-) has also been evoked [24, 25]. On this basis however, it is difficult to explain why f should be the same for a large variety of solutes, whether e^+ or e^- scavengers; the temperature effects on fI_3 and K would also remain obscure. Furthermore, there is no experimental evidence from pulse radiolysis on the existence of such delocalized states of the electron.

In spite of numerous experiments, many features of the e^+ spur (or "blob") are still unknown and debated. In particular, discordant views are commonly expressed about the number of electrons participating effectively to Ps formation: this may range from rather large (~ 10—30) [26, 27] to only about one (in average) [19, 20, 28]. Note that in case of successive e^- and e^+ captures, the total inhibition constant, k, and the bound-state formation constant, K_4, are found to be similar, while this is *a priori* expected only if the number of electrons involved is close to 1 [19, 20].

4.1.2 The spur model in nonpolar solvents

In spite of the wealth of data accumulated in the past decades, Ps formation in nonpolar solvents is less understood than in polar solvents. A reason for this is that data from pulse radiolysis that could be used for comparison are less numerous than for polar solvents. Globally, the spur processes described above are also present: the main differences between the polar and nonpolar solvents arise from the absence of solvation in the latter case.

Due to the lesser mobility of the solvated e^-_s and e^+_s, as compared to that of the quasi-free particles, solvation sets a time limit to Ps formation in polar solvents. The absence of solvation in nonpolar solvents therefore usually results in a much higher Ps yield. Furthermore, the Onsager radius, the distance at which the attractive potential between charged particles (here, e^+

and e⁻) is overcome by their thermal energies, is also much higher in nonpolar than in polar solvents, providing higher chances for Ps formation.

As in polar solvents, correlations are found between the ability of solutes to inhibit Ps formation and their reaction rate constants with the electrons. However, with a few exceptions (e.g.: CCl_4), the variation of I_3 with C does not follow a simple equation as eq.(1). In many cases, the following empirical equation has been used [29]:

$$I_3(C) = I_3(0)/[1 + (\delta C)^\gamma] \qquad (4)$$

where δ is the inhibition constant; γ has been found to be close to 1/2 in many instances [29], as expected on the basis of a model from pulse radiolysis (WAS equation) [30]. However, the WAS equation has been strongly criticized and Eq. (4) may not be considered as based on reliable physical grounds. In particular, it is not understood why some solutes would result in $\gamma = 1$ and others in $\gamma = 1/2$. The simple Stern-Volmer equation (Eq.1) that applies to total inhibitors in polar solvents probably reflects the fact that only a few effective processes are responsible for Ps formation and inhibition: it is hardly conceivable that the combination of numerous reactions in inhomogeneous conditions would lead to such a simple equation for solutes whose inhibiting ability ranges from weak to strong. In nonpolar solvents therefore, the breakdown of Eq. (1) may simply result from a more intricate situation than in polar solvents. The solutes studied in non polar solvents are neutral molecules: electron capture results therefore in the formation of radical anions which may in turn react with e^+ either to form a bound-state, if the electron happens to be tightly bound, or to form Ps if the electron is shallowly trapped (anti-inhibition). Also, the addition of non negligible amounts of such solutes may increase the Ps formation probability by mere increase of the electron density of the medium (i.e., increase in the electron yield). Although these possibilities have been evoked in the literature, no quantitative attempts have appear to predict their effect on the I_3 vs C variations.

The absence of a time limit for Ps formation in nonpolar solvents (no solvation) leads to some difficulties when modeling this process. Having still some Ps being formed on the time scale of the PALS spectra would probably be detected upon analysis (bad variance) as the fitting programs are based on purely decaying exponentials. Therefore, most of Ps formation must have come to an end at short time (about 10 ps at most).

4.1.3 Quantitative approaches and modeling of Ps formation

Three levels of quantification of the spur processes are possible, from the most rigorous to the most approximate.

- *Ab initio* and Monte-Carlo calculations. Attempts have appeared in pulse radiolysis to describe the dynamics of free electron production, recombination and solvation on a microscopic scale [31-34]. This requires the knowledge of a number of physical parameters: solvated electron and free ion yields, electron and hole mobilities, slowing-down cross-sections, localization and solvation times, etc. The movement and fate of each reactant is examined step by step in a probabilistic way and final results are obtained by averaging a number of calculated individual scenarios.

Most published work refers to the treatment of model pure solvents, with a limited number of species: the electron and the hole, and their daughter species. Interference of a solute is more conveniently treated using, at least in part, deterministic equations.

Extending such calculations to the positron spur introduces additional difficulties, for two main reasons: (i) even excepting solutes, one has to deal with a multicomponent inhomogeneous system (at least, e^+, e^- and hole); (ii) much less is known on the properties of e^+ than of e^-. At present, only one paper on Monte Carlo calculations on Ps *formation* has appeared [28].

- Deterministic calculations. On the basis of general physical laws, equations can be set to describe the processes globally, as a function of time. Two main sources of difficulty must be overcome: (i) the processes relate to inhomogenous kinetics and (ii), the reactants may not be in equilibrium with the medium (diffusion may not be established).

In pulse radiolysis, models have been proposed that may be helpful in the case of the positron spur. One popular approach is the "prescribed diffusion model", where, in essence, the difficulty of solving the general diffusion equations is circumvented by supposing that the inital distribution of a reactant with reference to another is of a Gaussian shape, and remains such as time evolves [35]. It is interesting to note that using this model for the set of reactions (I)—(IX) with reasonable values for the various physical parameters leads to good results as concerns the shape of the various variations of I_3 with solutes (total and partial inhibitions, enhancement and mixed effects); however obtaining a linear variation of $1/I_3$ *vs* C for total inhibition (eq. 1) is only possible within a narrow window of values of the parameters (initial distributions of the reactants, diffusion coefficients, etc).

Furthermore, having initially too many electrons per spur leads to difficulties in obtaining a reasonable Ps yield (e.g., in water, $I_3 = 28$ % only).

An interesting idea has been proposed where probabilistic considerations are largely taken into account [36-38]. However, none of these models has found (published) applications in the field of Ps formation.

• Statistical and semi-empirical calculations. Although not rigorous on physical grounds, a convenient and useful way of describing the various positron spur processes is based on simple first order (homogeneous) kinetics. The trick is to focus on the processes of interest, reducing all reactions to first order processes: (i) each of the "i" reactions is quantified by a time and space-averaged probability, σ_i, with the solute concentration explicited where necessary; (ii) the total Ps intensity (I_{Ps}) is given by the ratio of those channels leading to Ps formation over all other channels.

Taking reactions (I)—(IX) as an example, and focussing on Ps inhibition through e^- scavenging, reactions II, III, IV and VII may be treated as follows, where only the electrons are considered:

In the absence of solute : $I_{Ps}(0) = \sigma_{II}/(\sigma_{II} + \sigma_{III} + \sigma_{IV})$ (5)

At concentration C of solute: $I_{Ps}(C) = \sigma_{II}/(\sigma_{II} + \sigma_{III} + \sigma_{IV} + \sigma_{VII}C)$ (6)

Eq. (6) can be rewritten: $I_{Ps}(C) = I_{Ps}(0)/(1 + kC)$ (7)

Eq.(7) thus identifies with eq. (1), with $k = \sigma_{VII}/(\sigma_{II} + \sigma_{III} + \sigma_{IV})$. All empirical equations given before can be obtained in this way and extension to more complex cases (e.g., successive and competing spur reactions) is easily done (19). Developments of this approach are possible, such as by taking into account the statistical distribution of the solute molecules in the spurs. Although their physical grounds are of course questionable, the equations derived provide a quantitative approach to the data and the parameters (k, K, f, etc) prove to be very useful for comparisons (between solutes, as a function of temperature, of solvent; to correlate with data from other fields, etc). The usefulness of such simple treatments may well reflect a genuinely simple situation for the actors of the spur processes where probabilistic factors may have a more important impact than detailed dynamics.

4.1.4 Positronium formation in solids

The spur processes described for the liquid state should apply to solids. As compared to the scheme of reactions (I)—(IX), solvation is to be excluded. Localization in the sense that the particle would preserve some non negligible mobility as such, is also to be discarded in most cases.

Whereas Ps can be formed anywhere in liquids, considered as a very soft state of matter, this is no more true in most solids. Thus, besides the chemical concepts of the spur model, structural considerations must be taken into account, since Ps, as a physical particle, cannot form and survive in the absence of sufficient free space in the matrix (except if delocalized).

• Chemical effects: the spur model. The clearest example supporting the validity of the model in solids comes from solid solutions of Me (Me = Co, Cr, or Fe) and Al trisacetylacetonates (acac) [39]. The latter compound, which is chemically inert, behaves like an alkane, with a high o-Ps yield (I_3 = 43 %): this is ascribable to the large Onsager radius (low dielectric constant), the absence of localization/solvation processes and the high electron density; the latter also results in a short o-Ps lifetime (1.23 ns). The solute is a Me(III) salt, and therefore an efficient electron scavenger: very little Ps is formed in pure Me(acac)$_3$ (I_3 < 2 %) and its addition to the Al(acac)$_3$ matrix results in a strong inhibition, which is perfectly described by eq. (1), with a very high inhibition constant (64 M^{-1} for Me = Co as compared to 24 M^{-1} in benzene at 294 K) [39, 40].

Most generally, spur aspects appear whenever a chemically very active compound (essentially, electron scavengers) is present in the solid matrix, either as a normal constituent or as an additive. This has been demonstrated recently in a variety of irradiated polymers, where radiolysis of the matrix induces the production of either free radicals, inhibiting Ps formation by electron scavenging, or of trapped electrons, enhancing the Ps formation probability, depending on the temperature [41].

• Structural effects: the free volume model. Demonstrative examples of the role of free spaces on Ps formation in solids are provided by solids in which no Ps is formed when pure, and where the Ps yield increases as some dopant impurity is added. This is the case for p-terphenyl, in which the Ps yield increases as either chrysene or anthracene are added. Both dopant molecules, when introduced in the p-terphenyl matrix, promote the formation of extrinsic defects having roughly the size of a naphthalene molecule [42]. Similarly, doping the ionic KIO$_4$ matrix by IO_3^- ions induces the formation of oxygen vacancies which promote the formation of Ps [43].

The free spaces where Ps can form and o-Ps can have a reasonably long lifetime may be extrinsic defects, as just illustrated, or intrinsic defects, such as created when heating a pure solid compound. More generally, they may correspond to the natural voids present in any solid matrix (e.g., "free volume" in polymers, treated elsewhere in this book). Ps can be formed not only in molecular solids, including frozen liquids, but also in a number of ionic solids, even when the open spaces are rather small. For example, Ps is formed in such a highly packed lattice as KCl [44, 45] where the largest space available corresponds to the tetrahedral sites circumscribed by 4 Cl⁻ anions, with a radius of only 0.0845 nm, resulting in an o-Ps lifetime of about 0.65 ns.

4.2 Positron chemistry

Positron chemistry is a specific field which aims at determining which solutes react with e^+ to form a bound-state, comparing the related constants and studying the effects of temperature [3, 5, 6, 13, 18, 25, 46-48]. The results are scarce, because the bound-states can only be characterized through AC or DB experiments, which are less used than PALS: as it seems, the lifetimes of all e^+ bound-states known are very close to those of the free positrons, so that PALS cannot sufficiently distinguish these two states and is therefore unable to provide useful information.

At present, essentially the halide, pseudo-halide and sulfide anions have been shown to react with e^+, either directly [3, 6, 25, 46, 48] or after previous e^- capture by hologenated compounds (see Figs. 4.4 and 4.6) [19, 20]. However, it is not excluded that most anions are able to react with the positron.

As expressed by reaction (VIII), all positron scavengers characterized in polar solvents lead to partial inhibition and therefore are supposed to react specifically with the localized particles. The reasons for this are not well established but, in the same way as for those solutes that are very poor quasi-free electron scavengers although reacting effectively with the solvated electron (e.g., H^+), the explanation may lie on thermodynamics. Too much energy may be released upon reaction with the quasi-free particles, either e^- or e^+, so that the bound-state is unstable: localization or solvation would reduce the energetics of the process, allowing the reaction to occur. Note that most of the partial inhibitors, whether electron (e.g. H^+, Tl^+) or positron (Cl⁻,

I⁻, S⁼) scavengers, are monatomic species for which the release of extra energy is difficult because of the lack of vibrational modes.

It has also been suggested that positron scavenging would involve the hydrated positron, e^+_{aq} [25, 46]. Although the possibility cannot be completely discarded, the arguments given are not valid for several reasons [49]. The e^+_{aq} capture reaction is <u>supposed</u> to occur in competition with e^+_{aq} annihilation, and this sets arbitrarily the time scale for both processes. (i) All authors recognize that e^+_{aq} cannot contribute to a significant extent to Ps formation; therefore, the e^+_{aq} capture could not result into Ps inhibition. (ii) From the above basic hypothesis, the rate constant of e^+_{aq} capture (k_{BS}) must have, by definition, a value as expected for a diffusion-controlled reaction with a solvated particle (around 10^{10} M⁻¹s⁻¹): thus, finding such values by fitting the bound-state intensity (I_4) vs C experimental plot [50] cannot be taken as a proof of the validity of the hypothesis. Taking any other process than e^+_{aq} decay (e.g., e^+ localization, or solvation, etc.) as the competing process will result in completely different values for k_{BS}. Furthermore, there are cases where partial inhibition is so efficient that the involvement of e^+_{aq} becomes highly improbable: with the ferrocyanide ion for instance [48], inhibition is almost completed at about 2 10^{-3} M only, which would correspond to k_{BS} higher than 10^{12}M⁻¹s⁻¹.

Nevertheless, some contribution to the increase in I_4 with solute concentration due to e^+_{aq} capture (i.e., capture of solvated positrons that would not have formed Ps) cannot be discarded [5].

The possibility of e^+ bound-state formation has attracted the attention of theoreticians [51-55], as exposed elsewhere in this book. However, discrepancies may appear when comparing (kinetic) data from AC and DB and (thermodynamic) theoretical predictions. For the halides for instance, theory predicts an order of stability for the positron bound-states as F⁻ > Cl⁻ > Br⁻ > I⁻, while the experimental bound-state formation constants are in the reverse order. An evident reason for this arises from the solvation energies of the ions, not included in the theoretical evaluations, which decrease importantly the trap depth of F⁻ towards e^+.

4.3 Positronium states in condensed matter

The thermodynamic properties of Ps can be calculated to a high precision *in vacuo*. In matter, Ps interacts with the surrounding molecules or ions, so that its wave function is altered. On physical grounds, these interactions result essentially in processes that may be coined Ps trapping, in a broad sense. In liquids, considered as very soft matter, Ps is supposed to dig its own trap ("self-trapping"). In solids on the contrary, the rigid arrangement of the molecules or ions imposes geometrically well defined potential wells wherein Ps may or not become trapped. The state of Ps in a given medium can be inferred from its lifetime. However, applying an external magnetic field can give more precise information, through the determination of the contact density parameter, η.

4.3.1 Ps trapping in liquids: the bubble model

Due to its high zero-point kinetic energy, Ps is supposed to dig a cavity, or "bubble" in liquids [56, 57]. Various levels of approximation are possible for a quantum mechanical approach to the problem: the potential well (of depth U) constituting the bubble may be considered or not as infinite and/or rigid [58-60]. Some typical values of (rigid) well depth and radius are given in Table 4.2 [61]: the bubble radius, R_b, remains in a rather narrow range, about 0.3—0.45 nm, independently of the solvent or temperature.

From the bubble model, it is expected that the o-Ps pick-off decay rate constant should increase with the surface tension. Thus, the following empirical equation [62], later theoretically explicited [63], has been proposed, with $\beta = 0.5$ for liquid hydrocarbons:

$$\lambda_3^0 = \alpha \, \gamma^\beta \qquad (8)$$

Eq. (8) has been quantitatively verified for a large number of solvents [61]. However, at least 3 classes of solvents should be considered [61]: more systematic examination of experimental data is required to better understand the reasons for this and settle the applicability of the model.

For a rigid bubble, the zero-point kinetic energy of Ps is correlated to R_b^{-2} which is proportional to $\gamma^{-1/4}$. From the expression of the Ps energy level in the bubble, a constant ratio of 0.329 $eV^{1/2}$ $cm^{1/4}$ $dyn^{-1/4}$ is effectively found between the Ps average momentum and $\gamma^{-1/4}$ for a variety of liquids [61].

Table 4. 2 Ps bubble parameters for some liquids: radius (R_b) and well depth (U).

Solvent	T (K)	τ_3 (ns)	R_b (nm)	U (eV)
Water	294	1.81	0.291	
N-dimethylacetamide	294	2.66	0.396	1.069
	393	3.19	0.452	0.881
Sulfolan	303	2.03	0.381	1.017
Benzene	294	3.10	0.408	0.968
	353	3.58	0.446	0.856

Further support to the model should be found by examining the values of Γ_1, as deduced from AC or DB measurements: since p-Ps annihilates in an intrinsic mode, this parameter should reflect directly the average Ps kinetic momentum. However, the data on Γ_1 are usually poorly defined [61, 64], so that the correlation with γ is more conveniently sought using experimental values of Γ_3, provided that the momentum distribution of the valence electrons participating in the o-Ps pick-off annihilation is reasonably solvent independent. Such a correlation has been effectively found for a variety of solvents at various temperatures [61], leading to :

$$\Gamma_3 \text{ (keV)} = 0.83 \, \gamma^{1/4} \, \text{(dyn/cm)}^{1/4} \quad (9)$$

Although less expected than that between λ_3 and γ, because pick-off is an extrinsic process, the correlation found between Γ_3 and $\gamma^{1/4}$ appears more demonstrative as it is valid for a very large number of solvents having quite different properties.

4.3.2 Ps trapping in solids: the free volume model

The sites where Ps can become trapped and decay in solids correspond to pre-existing voids, whether of intrinsic or extrinsic origin. A quantitative equation has been established, which has proven to be reliable in many cases, correlating λ_3 with the radius (R) of the supposedly spherical trap:

$$\lambda_3 (\text{ns}^{-1}) = 2\left[1 - R/R_0 + (1/2\pi)\sin(2\pi R/R_0)\right] \quad (10)$$

where R_0 (nm) = R + 0.166 [65, 66]. Equation (10) has been verified in a large number of cases, including both ionic and molecular solids. For example, various dopants introduced in a p-terphenyl matrix lead to expect the formation of defects having roughly the size of a naphthalene molecule [67]. Their radius should range between 0.252 nm (largest circle from the centre of the naphthalene molecule) and 0.185 nm (second-largest circle), which, from Eq. (10), should result in o-Ps lifetimes between 1.66 and 1.12 ns, respectively. This expectation is experimentally confirmed, the various shapes of the dopant molecules and the way they insert in the p-terphenyl lattice giving rise to o-Ps components with lifetimes between 1.62 ns (phenanthrene) and 1.12 ns (chrysene, 2,3-benzofluorene).

One of the most successful and popular applications of Eq.(10) relates to the determination of free volumes in polymers [68].

Extensions of Eq. (10) to traps having a large size [69] or a cylindrical shape [70, 71] are available.

4.3.3 Ps states in condensed matter: the contact density parameter

Applying an external magnetic field, B, results in a mixing of the o-Ps (m = 0) magnetic substate with the p-Ps state and, subsequently, in an increase in the p-Ps lifetime together with a decrease in the o-Ps (m = 0) lifetime. A PALS spectrum typically containing only 3 components at B = 0, p-Ps (I_1, τ_1), e^+ (I_2, τ_2) and o-Ps (I_3, τ_3), should thus imply 4 components at B > 0: p-Ps [I_1, $\tau_1(B)$], e^+ (I_2, τ_2), o-Ps (m = 0) [$I_3/3$, $\tau_4(B)$] and o-Ps (m = ± 1) (2 $I_3/3$, τ_3). The expression for $\lambda_4(B) = 1/\tau_4(B)$ comes [72]:

$$\lambda_4(B) = (x^2\lambda_3^0 + \lambda_1^0)/(1+x^2) \qquad (11)$$

$$x = \left[(1+y^2)-1\right]/y \text{ and } y = 0.02756 \text{ B(T)}/\eta \qquad (12)$$

$$\lambda_1^0 = \eta^2\lambda_s + \lambda_p \quad \text{and} \quad \lambda_3^0 = \eta'\lambda_t + \lambda \qquad (13)$$

where λ_s and λ_t are the singlet (1/125 ps^{-1}) and triplet (1/142 ns^{-1}) intrinsic decay rate constants, respectively, and λ_p is the pick-off decay rate constant. Two parameters are involved (η and η'), which merge into a single parameter, η, the contact density parameter, for a non excited spherical Ps wave-function; η is the ratio of the hyperfine splitting energies in the studied medium and in vacuum. By definition, $\eta = 1$ for Ps in vacuum. For $\eta < 1$, the overlap of the e^+ and e^- wave functions in Ps is less than it is in vacuum, and the reverse is true for $\eta > 1$.

The PALS spectra can be analyzed in various ways. (i) Through a four-component analysis of the decay spectra. However, for low B values, $\tau_4(B)$ and τ_3 may not be resolved. (ii) By seeking for an average o-Ps lifetime, $\tau_{3av}(B)$, in a three-component analysis of the spectra. (iii) By evaluating the spectra through parameter R(B), which is the normalized ratio of two integrals of counts, f(B), taken between time t_a and t_b: $R(B) = f(B)/f(0)$. Times t_a and t_b are chosen such that: (i) the short-lived components (p-Ps and free e^+) contribute negligibly to the integral; (ii) the counts at t_b would be still significantly above the background. By definition, one has $R(0) = 1$. At very high field, $\tau_4(B)$ may become so short that the o-Ps (m = 0) component, representing 1/3 of the total long-lived o-Ps component, decays outside of the time window; thus, $R(\infty) = 2/3$.

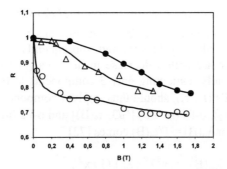

Figure 4.7: Variation of parameter R as a function of the magnetic field, B(T). (●): isooctane; (Δ): polyethylene terephthalate; (○): nitrobenzene in n-hexane.

Fitting the variations of either $\tau_4(B)$, $\tau_{3av}(B)$ or R(B) with B enables one to derive the value of η. Fig 4.7 shows the variations of R(B) for a nonpolar solvent, isooctane ($\eta = 0.88$) [73] and a polymer, polyethylene terephthalate: in the latter case there is a rapid decrease of R(B), denoting a low value of η (0.23).

In all liquids studied until now, except some amides, Ps is found to have a spherical non-excited wave-function, with η varying between about 0.6 and 0.9, going from the most to the least polar solvents, as shown in Table

4.3. Attempts have been made to correlate η with some macroscopic properties of the liquids, without great success [73, 74].

Table 4.3 Positronium states in condensed matter: contact density parameters. When more than one site is present, the amount of the first site is also indicated (%).

Medium	τ_3 (ns)	# of sites	Site 1	Site 2
Isooctane	3.88	1	0.88	
Benzene	3.15	1	0.84	
n-hexane	3.75	1	0.82	
Methanol	3.32	1	0.68	
Water	1.81	1	0.65	
Naphthalene	3.03	1	0.61	
Formamide	2.29	2	0.51 (68%)	$\eta = 0.1$
				$\eta' = 0.76$
Teflon	4.14	1	1.00	
p-terphenyl + 1% anthracene	1.44	1	0.80	
Naphthalene	1.03	1	$\eta = 0.25$	
			$\eta' = 0.13$	
Mylar	1.82	2	1.00 (77%)	$\eta = 1$ (fixed)
				$\eta' = 0.14$
KCl	0.65	1	0.59	
$Th_3(PO_4)_3$	0.70	1	$\eta = 0.22$	
			$\eta' = 0.36$	

In amides, the variations of R(B) with B are peculiar, showing a rapid initial decrease followed by a much slower one [75]. These data are fitted on the assumption that two Ps states are present: the first state is abundant (68 % in formamide and about 80 % in methylformamides) and "normal", although with a low η value; the second state is anomalous, requiring η ≠ η' (Table 3). These two states are provisionally ascribed to two solvation states for Ps, either on the oxygen or on the nitrogen of the amides [75].

As seen by applying a magnetic field, the Ps states are more varied in solids than in liquids [44, 76, 77]. Positronium appears to be in a "normal" state, with η= η', or there are two states, either normal or anomalous (Table 4.3). No clear correlation has been found yet between these various possibilities and properties of the solids. The existence of two different Ps states in a solid would correspond to the presence of two different sites, or classes of sites, where Ps annihilates. However, complementary experiments should be carried out to substantiate this hypothesis.

The origin of the anomalous states of Ps is not clear. Theory predicts that η should be different from η' only when Ps is in an excited state [72]. However, it seems unlikely that Ps would be excited in some solids and not in others otherwise very similar (e.g.: Ps is normal in Teflon and liquid naphthalene, not in solid naphthalene). Therefore, finding η ≠ η' cannot be indicative of the presence of an excited state of Ps. Most probably, the use of Eqs. (11-13) is incorrect. The number of systems that have been studied is still too limited to derive firm conclusions. However, it has been proposed that those cases where η ≠ η' may correspond to some distorted states of Ps (78), which would arise from specific geometric conditions imposed over Ps by the matrices: for instance, either when the traps have a shape which is far from spherical (e.g., rod-like in the case of solid naphthalene) or if they are surrounded by ions of quite different charges resulting in interactions of different strength on e^+ and e^- in Ps (e.g., in thorium phosphate).

4.4 Positronium chemistry in liquids

In spite of numerous experiments, this broad and rich field still possesses many unexplored or badly understood aspects. Its importance is twofold: (i) development of basic knowledge on Ps reaction thermodynamics, mechanisms and kinetics and (ii) improvement of our understanding of Ps behavior in matter towards more fruitful applications of this probe.

4.4.1 Positronium reactions

An interesting aspect of Ps chemistry on fundamental grounds is the variety of possible reactions, although only some of these have been effectively characterized. Thus, Ps addition, substitution, tunneling and hot reactions have been suspected, but not proven to occur; only Ps oxidation, bound-state formation and spin conversion reactions have been firmly proven.

For decades, Ps chemistry has been quantitatively treated on the basis of a constant reaction rate coefficient (k'). It is only recently that the possibility of a time-dependent k' has been examined. Therefore, the various known Ps reactions are presented in the following, before some fundamental aspects of Ps kinetics are discussed.

• Ps oxidation. Only very strong oxidizers react effectively with Ps [2]. The nature of the reaction can be characterized by coupling PALS and DB (or AC) measurements. The equations that have been commonly used in case of Ps oxidation, for the o-Ps PALS and DB parameters, are as follows:

$$\text{PALS lifetime: } 1/\tau_3(C) = \lambda_3(C) = \lambda_3(0) + k'C \qquad (14)$$

$$\text{PALS measured intensity: } I_{3m}(C) = I_3(C)\left[\lambda_2^0 - \lambda_3(0)\right]/\left[\lambda_2^0 - \lambda_3(C)\right] \qquad (15)$$

$$\text{DB measured intensity: } I_3^D = I_3(C)\,\lambda_3(0)/\lambda_3(C) \qquad (16)$$

$I_3(C)$ represents the true, physical o-Ps intensity and λ_2^0 is the free e^+ decay rate constant. Similar equations hold for p-Ps, with subscript 1 instead of 3.

The linear variation of $1/\tau_3$ with C expected from Eq. (14) has been verified in a very large number of cases. Furthermore, Eqs. 15 (I_{3m}) and 16 (I_3^D) show that the DB intensities can be completely calculated on the basis of the PALS experimental results: having PALS data, it is thus possible to calculate the variation of fwhm with C with no fitting parameters. Figure 4.8 illustrates such a calculation [79, 80]: the agreement with the experimental data is excellent, strongly confirming that the reaction is Ps oxidation by the solutes. Note that when k' is small (e.g., k' < 0.1 $M^{-1}s^{-1}$) [2], this parameter may not reflect the effective presence of a (weak) reaction with Ps, but rather an increase in the average electron density of the medium due to the solute, resulting in an increase in the o-Ps pick-off rate.

In case of Ps oxidation (more generally: reaction), the intensity (I_{3m}) of the longest-lived component measured in PALS, improperly called the o-Ps component, does not represent the true o-Ps yield, I_3. Thus, to derive information on the inhibiting properties of a solute that also reacts with Ps, I_3 must be calculated by using Eq. (15); its variations with C can thereafter be examined in the light of such equations as given in I.1. Papers have appeared that ignored this necessary correction on the experimental I_{3m} value, and thus produced misleading conclusions.

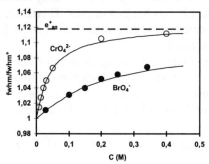

Figure 4.8: Oxidation reactions: DB results. Water, 294 K. The solid lines are completely calculated from the PALS results. The broken line denotes the asymptote for pure solvated positron.

The use of DB or AC to characterize the reaction observed in PALS is compulsory as far as quantitative comparison of the experimental k' with theoretical expressions is sought.

• Ps bound-state formation. Although a close examination of published data [2, 81] may reveal a larger potential variety, Ps bound-state formation reactions have only been characterized with nitrocompounds. The equations for the PALS and DB (AC) parameters are similar to those given in Eqs. (14)—(16), with λ_c, the bound-state decay rate constant, instead of λ_2^0. The appearance of a new component in the PALS spectra, of lifetime $\tau_c = 1/\lambda_c$, is however not the source of identification of the Ps bound-state reactions: as it seems, and is the case for the positron bound-states, the Ps bound-state lifetimes are not resolved from the free positron lifetime, τ_2^0, in the PALS spectrum analysis. Curiously, no published data are available where such reactions would be characterized through DB experiments.

The nature of the bound-state formation reactions has been disclosed through the variations of k' with T. Whereas, in the case of oxidation, k' varies linearly with T/η (η: viscosity of the medium) [82, 83], the variations of the apparent reaction rate constant for nitrocompounds show a maximum, as illustrated in Figure 4.9 for nitrobenzene in toluene [84]: at low temperature, the activation energy of the reaction rate constant is similar to that of the viscosity, denoting a diffusion controlled process; the important decrease in k' at high T reveals that the reaction is reversible:

$$2\gamma \xleftarrow{\lambda_3^0} \text{o-Ps+S} \underset{k_b}{\overset{k_f C}{\rightleftarrows}} [\text{PsS}] \xrightarrow{\lambda_c} 2\gamma \qquad (X)$$

Although more complex models have been proposed to describe the process [57, 85, 86], involving the Ps bubble state and its shrinking upon reaction, the equations based on a reversible reaction with a forward and reverse reaction rate constants as in scheme (X) enables the fitting of the data perfectly, as shown by the solid line in Figure 4.9. The kinetic equations corresponding to such a scheme are tedious to derive, particularly as concerns the intensities (still more when a magnetic field is applied). However, they do not present insuperable mathematical difficulties and should be used instead of the approximate expressions that have appeared casually (e.g., "steady state" treatment of the reversible reaction). From scheme (X), it is not expected that the variation of λ_3 with C be linear, but the departure from linearity may be rather small, so that the shape of the λ_3 vs C plots may not be taken as a criterion to ascribe the nature of the reaction.

98 *Principles and Applications of Positron and PositroniumChemistry*

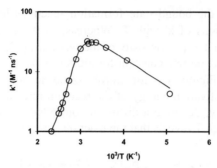

Figure 4.9: Variation of the apparent reaction rate constant, k' (M^{-1} ns^{-1}), with 1/T (K^{-1}). Ps bound state formation: nitrobenzene in toluene.

Further information on the Ps bound-states has been sought by applying a magnetic field [87-91], leading to so-called "anomalous" variations of either $\tau_{3m}(B)$ or R(B) with B. In fact, these are shown to be perfectly predictable [90, 91], by considering the following statements: (i) since the solutes are diamagnetic, the reaction with either the triplet or singlet state of Ps results in the formation of an equivalent triplet or singlet Ps bound-state. (ii) As for Ps, the m = 0 states of the Ps complex are liable to magnetic field effects; the corresponding equations are similar to eqs. (11)—(13), where all parameters relating to p-Ps and o-Ps (including the hyperfine splitting), must be replaced by the corresponding parameters of the singlet and triplet bound-state.

Such an "anomalous" variation of R(B) with B [90] is shown in Figure 7: the solid line is the fit to the data on the basis of the above-mentioned statements. Qualitatively, the rapid variation of R(B) with B is easily understood: as B increases, the (m=0) bound-state decay rate constant, $\lambda_c(B)$, also increases, in accordance with eq.(11). This increase is very rapid, because of the high sensitivity of $\lambda_c(B)$ to B, due to the low hyperfine splitting of the bound-state [89-91], which corresponds to having an efficient leak towards the right of reaction (X). Thus, the overall reaction rate constant is greatly increased, resulting in the rapid decrease of R(B) with B [90].

A number of papers have appeared, aiming at describing the properties of the bound-state as a function of temperature [89, 91] or seeking for the properties of the solvents that are responsible for its stability [57, 92, 93], including those that would affect the bubble states of Ps [86].

• Ps spin conversion reactions. Ps can suffer a spin conversion reaction with paramagnetic solutes without necessity of any change in the spin state of the latter [94]. Due to the spin statistics, after interacting with the odd electron of a solute molecule or ion, Ps (whether singlet or triplet) has normally 3 chances in 4 to become o-Ps, and 1 in 4 to be found as p-Ps; due to the much shorter lifetime of p-Ps as compared to that of o-Ps, this corresponds *in fine* to transform a large portion of o-Ps into p-Ps:

$$\text{Ps} + \text{S} \xrightarrow{k'C} 1/4\text{p-Ps} + 3/4\text{o-Ps} + \text{S} \qquad \text{(XI)}$$

Such reactions have now been verified in numerous cases with either transition metals ions [15, 95] or organic free radicals [61, 73, 96]. They are easily characterized: (i) in PALS, the variation of λ_3 with C are not linear:

$$\lambda_3 = \left[\lambda_1^0 + \lambda_3^0 + k'C - (\lambda_1^0 - \lambda_3^0)^{1/2} (\lambda_1^0 - \lambda_3^0 + k'C)^{1/2} \right]/2 \qquad (17)$$

(ii) In DB, the change of o-Ps into p-Ps results in a decrease in fwhm since $\Gamma_1 \ll \Gamma_3$. Figure 4.10 illustrates the difference between an oxidation and a spin conversion reaction for Cr(acac)$_3$ in benzene [40]: the solid and broken lines are entirely *calculated* on the basis of the PALS data.

In the early treatments of these reactions [94], and in many works since, reaction (XI) is written as follows:

$$\text{o-Ps} + \text{S} \underset{3k'C}{\overset{k'C}{\rightleftarrows}} \text{S} + \text{p-Ps} \qquad \text{(XII)}$$

This is obviously incorrect as, on chemical grounds, the o-Ps and p-Ps reaction rate constants cannot be different: the statistical spin substate factor should not appear in the rate at which the reaction occurs but rather, as in reaction XI, in the yield of the products of the reaction. Formally, reaction schemes XI and XII lead to exactly the same type of kinetic equations to describe the PALS parameters, particularly, λ_3. However, if one wishes to compare the experimentally determined k' with some theoretical expression such as the diffusion-controlled reaction rate constant, reaction XII will lead to a value of k' which is 4 times lower than that yielded by reaction XI: if o-

Ps encounters a solute molecule on a diffusion-controlled rate, this results in an effective reaction (o-Ps becomes p-Ps) only once in 4 trials.

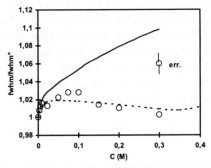

Figure 4.10: Oxidation (solid line) vs spin conversion (broken line). Cr(III) acetylacetonate in benzene, 353 K.

Although several calculated fwhm vs C variations coincide closely with the experimental data, as shown in Figure 4.10, a larger number of experimental data show a much more pronounced decrease in fwhm than theoretically expected. Various unsatisfactory explanations have been proposed to this discrepancy [73]. Close examination of the data shows that the DB data would imply larger p-Ps amounts at a given C, than predicted by the equations, whereas the PALS variations of τ_3 with C are perfectly described by those same equations. This will be commented farther on.

Regarding the mechanism of Ps spin conversion reactions, a study on the free radical HTMPO in methanol has shown that k' increases linearly with T/η at low T, as expected for a diffusion-controlled reaction, then slackens out [96]. This is explained on the basis of spin interaction dynamics, which allow a good fitting of the data. The efficiency of the reaction between two paramagnetic species depends both on the strength and duration of the interaction. Upon heating, the diffusion of the species is enhanced, resulting in a higher rate of encounter, but, at the same time, the duration of the encounter is reduced. The combination of these two opposite effects is expected to lead to a plateauing in the measured reaction rate constant.

4.4.2 Fundamentals of positronium kinetics

• Positronium diffusion: free state vs bubble state. The basic equation that has been widely used in Ps chemistry literature for the Ps diffusion-controlled reaction rate constant (k_D) arises from a combination of the Smoluchowski and Stokes-Einstein equations, yielding [97]:

$$k_D = 2N_{Av}k_BT(2 + R_{Ps}/R_S + R_S/R_{Ps})/(3000\eta) \qquad (18)$$

where N_{Av} is the Avogadro number, k_B, the Boltzmann constant, R_{Ps} and R_S the Ps and solute radii, respectively, and η, the viscosity of the solution. Several modifications to eq. (18) are possible: an interesting approach has been proposed, involving the dielectric constant of the liquids [82, 83].

When the solute is spherical, or close to be so, its radius is easily obtained; otherwise, estimations can be made on the basis of the geometry and arrangement of the constituting atoms or ions. For solutes having a complex stucture (e.g., micelles), a distinction should be made between the hydrodynamic radius (which appears in the Stokes-Einstein equation of the diffusion coefficient) and the reaction radius [98]. For Ps, R_{Ps} should represent the bubble radius. However, as shown in Table 4.4, the experimental data are systematically very well recovered by using the free Ps radius, $R_{Ps} = 0.053$ nm: using the bubble radius results in a calculated value of k_D (noted k_{Db}) that is too small by a factor of 2 or 3. Table 4.4 does not include such cases where k' << k_D, as these do not correspond to purely diffusion-controlled reactions.

All three kinds of well-characterized Ps reactions appear to be diffusion-controlled, at least in specific solvents and in given temperature ranges.

For small bubbles, a modification of the Stokes-Einstein equation has been proposed, leading to a 50% increase in the contribution from the Ps bubble part in the equation of the reaction rate constant [99]:

$$k_{Db} = 2N_{Av}k_BT(2.5 + R_b/R_S + 1.5R_S/R_b)/(3000\eta) \qquad (19)$$

However, this does not change the conclusions drawn above. At 298 K for $CrO_4^=$ in water, for instance, k_{Db} increases from 7.44 to 9.38 $M^{-1}ns^{-1}$, which is still significantly less than the experimental k' = 15.7 $M^{-1}ns^{-1}$ (Table 4.4).

Table 4.4 Comparison between experimental Ps reaction rate constants (k') and diffusion-controlled reaction rate constants calculated from eq. (18) by using either the bubble (k_{Dd}, R_b) or the free Ps (k_D, R_{Ps} = 0.053 nm) radius. (a), unpublished results; (b), [82]; (c) [61]; (d), [84]. ox = oxidation; sp = spin conversion; bs = bound-state formation. The rate constants are in $M^{-1}ns^{-1}$. NDMA: N-dimethylacetamide; Φ-NO_2: nitrobenzene.

Solute	Solvent	Reaction	T (K)	R_S (nm)	k'	k_D	K_{Db}
UO_2^{2+}	water	ox (a)	294	0.212	9.7	10.4	6.80
$CrO_4^=$	water	ox (b)	274	0.316	7.65	7.15	3.53
			298		15.7	15.1	7.44
			308		19.9	19.3	9.51
HTMPO	benzene	sp (c)	294	0.35	22.7	21.9	10.1
			323		37.5	41.2	19.1
	NDMA	sp	294		15.0	14.7	6.75
			353		30.7	35.5	16.4
	m-cresol	ox	294		0.88	0.7	0.3
Φ-NO_2	toluene	bs (d)	197	0.42	4.31	4.95	
			278		25.6	23.7	

• Establishment of Ps diffusion. In conventional chemistry, most of the reactions occur on a somewhat long time scale (>> 1 μs). For light particles possessing high diffusion constants however, such as e^-_{aq} or Ps, the time scale of reaction is much shorter and the time required for diffusion to be established should be considered e.g., [35, 100]. In the case of Ps, the time (t) dependence of the reaction rate constant is as follows [98, 100-102]:

$$k_D(t) = k_D \left[1 + (R_{Ps} + R_s)^{1/2} (R_{Ps} R_s \frac{6\eta}{k_B T t})^{1/2} \right] \quad (20)$$

where k_D is given by Eq. (18). From Eq. (20), it is possible to estimate how long it takes for diffusion to be established. For a conventional solute of radius 0.35 nm, a typical time for o-Ps to reach the diffusion-controlled rate would be about 1 ns. This goes up to 23 ns for a solute with a very large radius, such as a micellar aggregate [98]: during its time of life, Ps never reaches the diffusion-controlled rate.

Because of the time-dependence, the PALS spectrum is no longer constituted of a sum of decaying exponentials; its mathematical expression depends on whether the lifetime of the product of the reaction is longer (e.g., Ps trapping in micelles) [98] or shorter (e.g., Ps oxidation) [103] than that of o-Ps. In the former case for instance, neglecting the time-dependence effect on the p-Ps trapping rate constant, the continuous PALS spectrum, S(t), comes [98]:

$$S(t) = \sum_{i=1,2,4} \lambda_i^0 I_i \exp(-\lambda_i^0 t) + S_3(t) \quad (21)$$

$$S_3(t) = T_1 + T_2 \quad (22)$$

$$T_1 = \alpha \frac{\lambda_3^0 - \lambda_4^0}{\alpha - \lambda_4^0} \exp(-\alpha t - \beta t^{1/2}) + \lambda_4^0 \frac{\alpha - \lambda_3^0}{\alpha - \lambda_4^0} \exp(-\lambda_4^0 t)$$

$$T_2 = \lambda_4^0 \beta \frac{\lambda_3^0 - \lambda_4^0}{(\alpha - \lambda_4^0)^{3/2}} \frac{\pi^{1/2}}{2} \exp\left[\beta^2/4(\alpha - \lambda_4^0)\right] \exp(-\lambda_4^0 t)(\text{erf1} - \text{erf2})$$

$$\text{erf1} = \text{erf}\left[(\alpha - \lambda_4^0)^{1/2} t^{1/2} + \beta/2(\alpha - \lambda_4^0)^{1/2}\right], \text{erf2} = \text{erf}\left[\beta/2(\alpha - \lambda_4^0)^{1/2}\right]$$

$$\alpha = \lambda_3^0 + k_D C, \beta = 2k_D C (R_{Ps} + R_S)^{1/2} (6\eta R_{Ps} R_S/k_B T)^{1/2}$$

Subscripts 1, 2, 3 and 4 relate to p-Ps, free e^+ and o-Ps in the aqueous and organic (trapped Ps) subphases, respectively; erf is the error function. It has been shown, in the case of Ps trapping in micelles, that a physically satisfactory analysis of the PALS spectra is obtained by using Eq. (21) with the Ps bubble radius, not with the free Ps radius [98]. For usual, smaller solutes reacting with Ps, the difference between k_D and $k_D(t)$ should not be very large, as may be inferred by averaging Eq. (20) over time.

A complete set of equations has also been published for the case of Ps spin conversion reactions, but these are not amenable to any explicit description of the PALS spectra [104]. However, the time-dependence of the reaction rate coefficient may represent a hint as to the difficulty often encountered to calculate the DB data (fwhm *vs* C) from the PALS results (λ_3 *vs* C) in the case of spin conversion reactions. For a solute of R_S = 0.35 nm, o-Ps will have an effective reaction rate coefficient higher than expected from Eq. (19), until about 1 ns. But p-Ps will have a much higher average rate coefficient over its whole life. Numerical calculations show that taking account of the time-dependence may lead to specified values of τ_3 with I_1^D significantly higher than obtained with a constant reaction rate coefficient (see 4.4.1).

4.5 Some chemical applications of (PAT) in liquids and solids

In principle, e^+ and Ps could be used as probes to obtain information on a very large number of physical and chemical properties of matter. In practice however, application of these particles has been restricted to some specific areas, particularly those where conventional techniques are unable to provide any, or, at least, sufficient information.

Information on properties or characteristics of matter is usually sought when these are studied as a function of some parameter (temperature, concentration, irradiation dose, etc). Besides some weaknesses on the theoretical side to predict absolute values of the measured PALS parameters (τ_2, I_3 and even τ_3) the physico-chemical causes that may lead to specific values of these are usually not unequivocal: PAT should not be considered as potential one-shot analytical techniques.

4.5.1 Stability constants

The stability constant (K) of an equilibrium as shown in scheme (XIII) can be determined from the changes in τ_3 with solute concentrations.

$$A+B \xrightleftharpoons{K} AB \tag{XIII}$$

The condition for this is that Ps should react with, at least, one of the compounds involved: A, B or AB. For instance, in case of Ps oxidation, Eq. (14) can be extended as:

$$1/\tau_3 = 1/\tau_3^0 + \sum_i k'_i C_i \qquad (23)$$

Each compound has its own reaction rate constant, k'_i, and effective concentration, C_i. The latter depend on the nominal (selected) concentrations of A and B, and on K. The unknown parameters, k'_i and K, can thus be determined by studying mixtures of A and B of various nominal concentrations; k'_A and k'_B are conveniently measured separately, by using pure solutions of either A or B, respectively.

This approach has been used successfully to derive the stability constants of a number of ionic complexes such as $NiSO_4$ [105] and metal cation polyhalides [106, 107], as well as of molecular associations [108, 109]. Figure 4.11 shows the difference in the λ_3 vs C variations for Ni^{2+} perchlorate (no ion association) and $NiSO_4$ (ion association; the sulfate anion has no action on τ_3): due to the lesser availability of the odd electrons of the metal ions, the Ps spin conversion reaction is less efficient when $NiSO_4$ ion pairs are present.

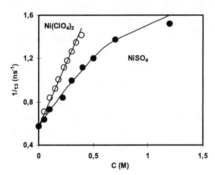

Figure 4.11: Variation of $1/\tau_3$ (ns^{-1}) in aqueous solutions of $Ni(ClO_4)_2$ and $NiSO_4$.

Equation (23) shows that the measured parameter, λ_3, is a linear function of the concentrations of the solutes. This is similar to what one has with, e.g., spectrophotometry. In PALS however there is no such means as to change the excitation wavelength to enrich information: deriving reliable stability constants through PAT requires the study of a large number of solute concentrations. Quantitative information is difficult to obtain when too many equilibria are present [110].

One advantage of PALS is that, as far as τ_3 is concerned, the medium appears transparent to Ps even at high concentrations of some solutes (e.g. ClO_4^-, Cl^-, alkali cations). A promising application would be then to measure stability constants in concentrated solutions, to assess the validity of equations for the activity coefficients at high ionic strengths [111].

The necessity of somewhat high solute concentrations for significant changes in τ_3 results in that essentially equilibria having a rather low stability constant can be studied. This provides the advantage of characterizing weak ion association complexes, such as were found in $AgClO_4$ solutions [112, 113].

The use of τ_3 to derive information on chemical equilibria is commendable as rigorous equations can be set to describe the various chemical reactions. However, I_3 may be used, inasmuch as Eq.(1) is valid, as shown in the case of the dimerization of a hydroxylamine derivative in methanol [12].

4.5.2 Polyelectrolyte solutions

Polyelectrolytes present a long chain of charged ions. When a metal cation is added to a solution of a polyanion, the high charge density along the polymeric chain will first result in the formation of inner sphere complexes between the anionic groups and the cation. When more metal ions are added, the charge density decreases until a specific point, where these cations are no more tightly bound to the anions. This behavior is illustrated in Figure 4.12 [114].

Up to a certain ratio, r, of (twice) the Co^{2+} and polyphosphate (PP) anions concentrations, τ_3 remains unaltered. As the reaction of Ps with Co^{2+} is spin conversion [15], this is due to the formation of the inner sphere PP/Co^{2+} complexes, which make the odd electrons of Co^{2+} unavailable to Ps. At a given value $r = R_{Co}$, tight binding ceases and the Co^{2+} ions recover the reactivity towards Ps they would have in pure water; R_{Co} is found to coincide closely with the theoretically expected value [15].

An additional information relates to the microviscosity of the PP solutions. Their macroscopic viscosity is very high. However, the Ps reaction rate constant measured above R_{Co} is only slightly lower than what it is in water: on a microscopic scale, the diffusive properties of Ps inside the net created in the PP solution by the polymeric chains are similar to those in neat water

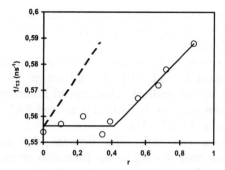

Figure 4.12: Variation of $1/\tau_3$ (ns^{-1}) with $R=2C(Co^{2+})/C(PP)$ in aqueous solutions of sodium polyphosphate (PP) at 294 K. Broken line: Co^{2+} in the absence of PP.

4.5.3 PAT applications to primary radiolysis processes

Determining the total inhibition constant, k, of a solute allows one to predict the values of its reaction rate constant with solvated electrons as well as its C_{37}, through the correlations established (Figure 4.2). The only alternative to obtain such information, also in an indirect way, is through the use of costly pulsed electron accelerators (Linac) [9, 115, 116], so that PAT appear quite interesting on this respect. Femtosecond lasers enable one to follow directly such processes as e⁻ localization, solvation and recombination as a function of time, even in water [117-119]. However, the energy of the electrons released upon laser irradiation is much lower than that of electrons ionized by impinging high energy electrons (Linac) or positrons (PAT), so that the information derived in either case may be somewhat different.

From the correlations depicted in Sec. 4.1, information is also obtained on hole scavenging, temperature effects, e⁻ solvation, etc. Among many other applications of PAT in fast radiolysis, the aggregation of alcohol molecules in an alkane and the solvation of e⁻ in such a system is notheworthy. The ethanol molecules remain dispersed in the alkane at low alcohol concentration (C), as reflected by the small change in I_3. With increasing C, the alcohol molecules aggregate and trap the electrons which can then solvate: I_3 decreases very rapidly. The variations of I_3 also parallel those of the electron yield [120, 121].

In contrast with a previous claim [122], no clear correlation has been found recently between the Ps yields in liquids and either the e^+ or e^- mobilities [27, 28].

By comparison with conventional pulse radiolysis techniques based on absorption spectra, PAT offer the advantage of not depending on the transparency of the medium. Thus, useful information on, e.g., quasi-free electron scavenging, can be obtained through PAT in non glassy solids or frozen solutions [14, 23, 123).

Correlations can also exist between the processes governing Ps formation and those leading to the formation of excited fluorescent states: these may arise from the recombination of cations and electrons produced by radiation at short times, in a similar way as Ps is formed [124]. Valuable information is still to be obtained from such comparisons, which have remained scarce.

4.5.4 PAT applications to phase transitions

All PALS and DB (AC) parameters can be sensitive to phase transitions and PAT have been commonly used for such studies [125, 126]. A typical example is shown in Figure 4.13, for I_3 in the case of sulfolan; similar changes are observed for τ_3 and fwhm [23]. With a roundish molecular shape, this compound forms a plastic phase below the melting temperature and before forming a brittle solid phase. In the plastic phase, the molecules can still rotate around their axis, but not move in a translational mode.

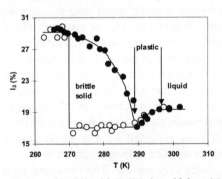

Figure 4.13: Variation of I_3 (%) with T (K), in sulfolan: (●) heating, (○) cooling. Supercooling of plastic phase down to 269 K; transition at 288 K.

The changes in the o-Ps lifetimes should be explainable on the basis of eq. (10) and its connection with the free volume. It is interesting to note that in sulfolan, the latter does not change at the liquid/plastic phase transition. The changes in I_3 cannot yet be quantified. Changes in the dielectric constant (in the Onsager radius) should be one of the main factors to consider.

PAT should be used for such studies only when other techniques fail to reveal clearly a given phase transition. Thus, in Figure 4.13, it is interesting to note that the liquid/plastic transition is not detected through the PALS and DB parameters, although it is easily seen by direct observation. On the opposite, the plastic/solid transition is well marked, although it is not easily seen by using conventional techniques.

4.5.5 PAT applications to defect formation in solids

In metals, most semi-conductors and ionic solids, no Ps is formed: the useful probe is thus e^+. Being charged, this probe is sensitive to negatively charged defects, such as cation vacancies, in which it can be trapped. In most chemical applications, and particularly in molecular solids, the useful probe is Ps: the trap is then expected to be neutral (e.g.: molecular vacancy).

• Intrinsic and extrinsic defects in ionic solids. Both types of defects have been studied in a mixed-valence compound, Tl_4Cl_6 [127, 128]. In the untreated material, the e^+ lifetime, τ, varies in a sigmoidal way with T upon heating, denoting the appearance of cation vacancies of low formation energy (0.31 eV), in agreement with conductivity measurements. At low T, the solid is little defected: e^+ decays in the bulk with some specific lifetime, τ_b. On heating, defects are generated (e.g., Schottky pairs: anion + cation vacancies). Some of the e^+ injected into the lattice now get trapped into the attractive cation vacancies and decay therein with a longer lifetime, τ_t. At high enough temperature, the defect concentration is high enough for all e^+ to end up and decay in the traps. By measuring the average e^+ lifetime, τ, between τ_b (low T) and τ_t (high T) as a function of T, quantitative information is obtained on the defect, such as its formation activation energy (E_f):

$$\tau = \tau_b(1 + k_t\tau_t)/(1 + k_t\tau_b) \qquad (24)$$

The trapping rate constant obeys the Arrhenius equation, as:

$$k_t = A\exp(-E_f/k_B T) \tag{25}$$

where the pre-exponential factor, A, includes the vacancy concentration.

In crushed crystals, τ varies in a completely different way during the first heating run and in a similar way as in the untreated material in subsequent heating and cooling cycles: the extrinsic defects created upon crushing are thus shown to be annealed out at about 403 K.

• Intrinsic defects in molecular solids. Figure 4.14 shows the variations of τ_3 with T in succinonitrile [125, 129]. As sulfolan, this compound exhibits a brittle (monoclinic) phase up to 233 K, then a plastic body-centered cubic phase. The phase transition is well evidenced, with some hysteresis. However, above the phase transition point, τ_3 increases further in a sigmoidal way characteristic of defect formation. By using Eq. (24), the activation energy of the intrinsic molecular defect is obtained (0.36 eV). Analyzing the PALS spectra in four components confirms that τ_3 does represent the average between an o-Ps bulk lifetime, which decreases with T, and the constant lifetime of o-Ps trapped in the vacancies whose intensity increases with T.

The occurrence of well defined sigmoidal variations of the e^+ or o-Ps lifetimes with T may fail to be observed. This is the case in AgI, a compound well known to present a transition to an ionic superconducting phase above 423 K, due to the production of large amounts of cation vacancies. However, none of the PALS or DB parameters show evidence of a change at the phase transition temperatures [130]. As it seems, the lifetime of the cation vacancy is too short to allow e^+ trapping, due to the intense movement of the Ag^+ ions from vacancy to vacancy through the lattice.

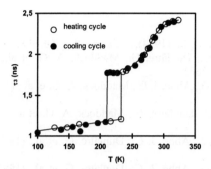

Figure 4.14: Phase transition and intrinsic defect formation in molecular solids: variation of τ_3 (ns) with T (K) in succinonitrile.

Extending PAT applications to the study of solid state chemistry, defect formation and annealing, nature of defects, ionic conductivity, etc, is a challenging field yet almost unexplored.

4.5.6 Spin cross-over in solid metal complexes

DB or AC measurements are of course very valuable wherever spin interactions can be involved. A typical example is for an iron complex which suffers a change from high to low spin state when cooling below about 100 K. As Ps suffers a spin conversion reaction with the former state, measuring the narrow component (p-Ps) intensity of AC spectra gives direct information on the temperature at which the spin cross-over occurs [131].

Although some papers have appeared on the subject [132-135] applications of PAT in this field are still scarce and should be developed.

References

[1] Mogensen, O. E., (1974), *J. Chem. Phys.* **60**, 998.
[2] Abbé, J. C., Duplâtre, G., Maddock, A. G. et al., (1981), *J. inorg. nucl. Chem.* **43**, 2603.
[3] Duplâtre, G., Abbé, J. C., Maddock, A. G. et al., (1980), *J. Chem. Phys.* **72**, 89.
[4] Abbé, J. C., Duplâtre, G., Maddock, A. G. et al., (1980), *Chem. Phys.* **49**, 165.
[5] Talamoni, J., Abbé, J. C., Duplâtre, G. et al., (1983), *Radiat. Phys. Chem.* **21**, 431.
[6] Talamoni, J., Abbé, J. C., Duplâtre, G. et al., (1982), *Radiat. Phys. Chem.* **20**, 275.
[7] Gauduel, Y., Pommeret, S., Migus, A. et al., (1991), *J. Phys. Chem.* **95**, 533.
[8] Byakov, V. M., Grafutin, V. I., Koldaeva, O. V. et al., (1977), *Chem. Phys.* **24**, 91.
[9] Duplâtre, G. and Jonah, C. D., (1985), *Radiat. Phys. Chem.* **24**, 557.
[10] Duplâtre, G., Abbé, J. C., Maddock, A. G. et al., (1978), *Radiat. Phys. Chem.* **11**, 249.
[11] Duplâtre, G. and Jonah, C. D., (1991), *J. Phys. Chem.* **95**, 897.
[12] Machado, J. C., Magalhaes, W. F., Marques Netto, A. et al., (1989), *Chem. Phys. Lett.* **163**, 140.
[13] Talamoni, J., Duplâtre, G., Abbé, J. C. et al., (1984), *Chem. Phys.* **83**, 471.
[14] Duplâtre, G., Abbé, J. C. and Talamoni, J. ; Jain, P. C., Singru, M., Gopinathan, K. P., Eds.; World Scientific: 1985, p 233.
[15] Duplâtre, G., Haessler, A. and Abbé, J. C., (1985), *J. Phys. Chem.* **89**, 1756.
[16] Delaire, J. A., Delcourt, M. O. and Belloni, J., (1980), *Radiat. Phys. Chem.* **15**, 255.
[17] Abbé, J. C., Duplâtre, G., Haessler, A. et al., (1986), *Radiat. Phys. Chem.* **28**, 19.
[18] Talamoni, J., Abbé, J. C., Duplâtre, G. et al., (1982), *Chem. Phys. Lett.* **90**, 242.
[19] Didierjean, F., Abbé, J. C. and Duplâtre, G., (1992), *J. Phys. Chem.* **96**, 8074.
[20] Didierjean, F., Duplâtre, G. and Abbé, J. C., (1991), *Radiat. Phys. Chem.* **38**, 413.
[21] Anisimov, O. A. and Molin, Y. N., (1976), *High Energ. Chem.* **9**, 471.
[22] Wikander, G., (1982), *Chem. Phys.* **66**, 227.
[23] Abbé, J. C., Duplâtre, G. and Machado, J. C. ; Jain, P. C., Singru, R. M., Gopinathan, K. P., Eds.; World Scientific: Singapore, 1985, p 620.
[24] Lévay, B. and Mogensen, O. E., (1980), *Chem. Phys.* **53**, 131.
[25] Mogensen, O. E., (1979), *Chem. Phys.* **37**, 139.
[26] Stepanov, S. V., Byakov, V. M., Wang, C. L. et al., (2001), *Mat. Sc. Forum* **363/365**, 392.

[27] Wang, C. L., Kobayashi, Y. and Hirata, K., (2000), *Radiat. Phys. Chem.* **58**, 451.
[28] Billard, I., Goulet, T., Jay-Gérin, J. P. et al., (1998), *J. Chem. Phys.* **108**, 2408.
[29] Wikander, G., Mogensen, O. E. and Pedersen, N. J., (1983), *J. Phys. Chem.* **77**, 159.
[30] Warman, J. M., Asmus, K. D. and Schuler, R. M., (1969), *J. Phys. Chem.* **73**, 931.
[31] Fischetti, M. V. and DiMaria, D. J., (1985), *Phys. Rev. Lett.* **55**, 2475.
[32] Fischetti, M. V., (1991), *IEEE Trans. Electron Devices* **38**, 634.
[33] Goulet, T. and Jay-Gérin, J. P., (1986), *Radiat. Phys. Chem.* **27**, 229.
[34] Goulet, T., Jung, J. M., Michaud, M. et al., (1994), *Phys. Rev. B* **50**, 5101.
[35] Schwartz, H. A., (1969), *J. Phys. Chem.* **73**, 1928.
[36] Green, N. J. B. and Pimblott, S. M., (1991), *Mol. Phys.* **74**, 795.
[37] Green, N. J. B. and Pimblott, S. M., (1991), *Mol. Phys.* **74**, 811.
[38] Pimblott, S. M., Mozumder, A. and Green, N. J. B., (1989), *J. Chem. Phys.* **90**, 6595.
[39] Machado, J. C., Carvalho, C. F., Magalhaes, W. F. et al., (1993), *Chem. Phys.* **170**, 257.
[40] Magalhaes, W. F., Abbé, J. C. and Duplâtre, G., (1989), *Chem. Phys.* **136**, 141.
[41] Suzuki, T., Kondo, K., Hamada, E. et al., (2001), *Radiat. Phys. Chem.* **60**, 535.
[42] Goworek, T., Rybka, C. and Wawryszczuk, J., (1978), *phys. stat. sol. B* **89**, 253.
[43] Takriti, S. and Duplâtre, G., (1988), *Radiochim. Acta* **43**, 45.
[44] Consolati, G. and Quasso, F., (1991), *Appl. Phys.* **52**, 295.
[45] Bussolati, C., Dupasquier, A. and Zappa, L., (1967), *Nuov. Cim. B* **52**, 529.
[46] Mogensen, O. E., Pedersen, N. J. and Andersen, J. R., (1982), *Chem. Phys. Lett.* **93**, 115.
[47] Talamoni, J., Abbé, J. C., Duplâtre, G. et al., (1981), *Chem. Phys.* **58**, 13.
[48] Duplâtre, G., Talamoni, J., Abbé, J. C. et al., (1984), *Radiat. Phys. Chem.* **23**, 531.
[49] Beling, C. D. and Smith, F. A., (1983), *Chem. Phys.* **81**, 243.
[50] Mogensen, O. E. and Shantarovich, V. P., (1974), *Chem. Phys.* **6**, 100.
[51] Kao, C. M. and Cade, P., (1984), *J. Chem. Phys.* **80**, 3234.
[52] Cade, P. E. and Farazdel, A., (1977), *J. Chem. Phys.* **66**, 2598.
[53] Patrick, A. J. and Cade, P. E., (1981), *J. Chem. Phys.* **75**, 1893.
[54] Schrader, D. M., (1979), *Phys. Rev. A* **20**, 933.
[55] Schiller, R. and Nyiokos, L., (1980), *J. Chem. Phys.* **72**, 2245.
[56] Ferrel, R. A., (1957), *Phys. Rev.* **108**, 167.
[57] Buchikhin, A. P., Goldanskii, V. I. and Shantarovich, V. P., (1973), *Dokl. Phys. Chem.* **212**, 879.
[58] Byakov, V. M. and Stepanov, S. V., (2000), *Radiat. Phys. Chem.* **58**, 687.

[59] Mukherjee, T., Das, S. K., Ganguly, B. N. et al., (1998), *Phys. Rev. B* **57**, 13363.
[60] Mukherjee, T., Gangopadhyay, D., Das, S. K. et al., (1999), *J. Chem. Phys.* **110**, 6844.
[61] Magalhaes, W. F., Abbé, J. C. and Duplâtre, G., (1991), *Struct. Chem.* **2**, 399.
[62] Tao, S. J., (1972), *J. Chem. Phys.* **56**, 5499.
[63] Levay, B. and Vertes, A., (1976), *J. Phys. Chem.* **80**, 37.
[64] Mogensen, O. E. and Jacobsen, F. M., (1982), *Chem. Phys.* **73**, 223.
[65] Eldrup, M., Lightbody, D. and Sherwood, J. N., (1981), *Chem. Phys.* **63**, 51.
[66] Schrader, D. M. and Jean, Y. C., Positron and Positronium Chemistry; Elsevier: Amsterdam, 1988.
[67] Goworek, T., Rybka, C. and Wawryszczuk, J., (1977), *Phys. stat. sol., B* **84**, K49.
[68] Jean, Y. C., (1990), *Microchem. J.* **42**, 72.
[69] Ito, K., Nakanishi, H. and Ujihira, Y., (1999), *J. Phys. Chem. B* **103**, 4555.
[70] Gidley, D. W., Frieze, W. E., Dull, T. L. et al., (1999), *Phys. Rev. B* **60**, R5157.
[71] Jasinka, B., Dawidowicz, A. L., Goworek, T. et al., (2000), *Phys. Chem. Chem. Phys.* **2**, 3269.
[72] Mills, A. P., Jr., (1975), *J. Chem. Phys.* **62**, 2646.
[73] Didierjean, F., Billard, I., Magalhaes, W. F. et al., (1993), *Chem. Phys.* **174**, 331.
[74] Bertolaccini, M., Bisi, A., Gambarini, G. et al., (1974), *J. Phys.* **C7**, 3827.
[75] Billard, I., Abbé, J. C. and Duplâtre, G., (1994), *Chem. Phys.* **184**, 365.
[76] Bisi, A., Consolati, G., Gambarini, G. et al., (1985), *Nuov. Cim., D* **6**, 183.
[77] Bisi, A., Consolati, G. and Zappa, L., (1987), *Hyperf. Inter.* **36**, 29.
[78] Goworek, T., Badia, A. and Duplâtre, G., (1994), *J. Chem. Soc. Faraday Trans.* **90**, 1501.
[79] Abbé, J. C., Duplâtre, G., Maddock, A. G. et al., (1980), *J. Radioanal. Chem.* **55**, 25.
[80] Duplâtre, G., Abbé, J. C., Talamoni, J. et al., (1981), *Chem. Phys.* **57**, 175.
[81] Fantola Lazzarini, A. L. and Lazzarini, E., (1987), *Chem. Phys. Lett.* **141**, 459.
[82] Fantola Lazzarini, A. L. and Lazzarini, E., (1985), *Chem. Phys. Lett.* **113**, 380.
[83] Fantola Lazzarini, A. L. and Lazzarini, E., (1986), *J. Phys. Chem. B* **20**, 183.
[84] Madia, W. J., Nichols, A. L. and Ache, H. J., (1975), *J. Am. Chem. Soc.* **97**, 5041.
[85] Shantarovich, V. P. and Jansen, P., (1978), *Chem. Phys.* **34**, 39.
[86] Gangopadhyay, D. and Ganguly, B. N., (1999), *J. Chem. Phys.* **111**, 9687.
[87] Rochanakij, S. and Schrader, D. M. ; Jain, P. C., Singru, R. M., Gopinathan, K. P., Eds.; World Scientific: 1985, p 193.
[88] Rochanakij, S. and Schrader, D. M., (1988), *Radiat. Phys. Chem.* **32**, 557.

[89] Duplâtre, G., Billard, I. and Abbé, J. C., (1994), *Chem. Phys.* **184**, 371.
[90] Billard, I., Abbé, J. C. and Duplâtre, G., (1989), *J. Chem. Phys.* **91**, 1579.
[91] Billard, I., Abbé, J. C. and Duplâtre, G., (1992), *J. Chem. Phys.* **97**, 1548.
[92] Kobayashi, Y. and Ujihira, Y., (1981), *J. Phys. Chem.* **85**, 2455.
[93] Kobayashi, Y., (1991), *J. Chem. Soc. Faraday Trans.* **87**, 3641.
[94] Ferrel, R. A., (1958), *Phys. Rev.* **110**, 1355.
[95] Duplâtre, G., Haessler, A., Abbé, J. C. et al. ; Coleman, P. G., Sharma, S. C., Diana, L. M., Eds.; North-Holland: 1982, p 849.
[96] Abbé, J. C., Duplâtre, G., Haessler, A. et al., (1984), *J. Phys. Chem.* **88**, 2071.
[97] Tao, S. J., (1970), *J. Chem. Phys.* **52**, 752.
[98] Bockstahl, F. and Duplâtre, G., (1999), *Phys. Chem. Chem. Phys.* **1**, 2767.
[99] Byakov, V. M. and Petuchov, V. R., (1984), *J. Radioanal. Nucl. Chem.* **85**, 67.
[100] Jonah, C. D., Miller, J. R., Hart, E. J. et al., (1975), *J. Phys. Chem.* **79**, 2705.
[101] Waite, T. R., (1957), *Phys. Rev.* **107**, 471.
[102] Peak, D. and Corbett, J. W., (1972), *Phys. Rev. B* **5**, 1226.
[103] Bockstahl, F. and Duplâtre, G., (2000), *Phys. Chem. Chem. Phys.* **2**, 2401.
[104] Würschum, R. and Seeger, A., (1995), *Z. für Phys. Chem.* **192**, 47.
[105] Duplâtre, G., Haessler, A. and Marques Netto, A., (1984), *J. Radioanal. Chem.* **82**, 219.
[106] Duplâtre, G., Al-Shukri, L. M. and Maddock, A. G., (1980), *J. Radioanal. Chem.* **60**, 159.
[107] Duplâtre, G., Al-Shukri, L. M. and Haessler, A., (1980), *J. Radioanal. Chem.* **55**, 199.
[108] Bartal, L. J. and Ache, H. J., (1973), *J. Phys. Chem.* **77**, 2060.
[109] Jean, Y. C. and Ache, H. J., (1977), *J. Phys. Chem.* **81**, 2093.
[110] Eldrup, M., Shantarovich, V. P. and Mogensen, O. E., (1975), *Chem. Phys.* **11**, 129.
[111] Pitzer, K. S. In *Activity coefficients in electrolyte solutions*; Pytkowicz, R. M., Ed.; CRC Press: Boca Raton, 1979; Vol. 1, p 157.
[112] Duplâtre, G. In *Fundamentals of Radiochemistry*; Adloff, J. P., Guillaumont, R., Eds.; CRC Press: Boca Raton, 1993, p 228.
[113] Maddock, A. G., Abbé, J. C., Duplâtre, G. et al., (1977), *Chem. Phys.* **26**, 163.
[114] Zana, R., Millan, S., Abbé, J. C. et al., (1982), *J. Phys. Chem.* **86**, 1457.
[115] Jonah, C. D., Miller, J. R. and Matheson, M. S., (1977), *J. Phys. Chem.* **81**, 1618.
[116] Jonah, C. D., (1975), *Rev. Sci. Instrum.* **46**, 62.
[117] Gauduel, Y., Migus, A. and Antonetti, A. In *Chemical Reactivity in Liquids: fundamental aspects*; Moreau, M., Turq, P., Eds.; Plenum Press: New York, 1988, p 15.
[118] Lu, H., Long, F. H. and Eisenthal, K. B., (1990), *J. Opt. Soc. Am. B* **7**, 1511.

[119] Pommeret, S., Antonetti, A. and Gauduel, Y., (1991), *J. Am. Chem. Soc.* **113**, 9105.
[120] Byakov, V. M. and Grafutin, V. I., (1980), *High Energ. Chem.* **14**, 201.
[121] Byakov, V. M., Bugaenko, V. L., Grafutin, V. I. et al., (1978), *High Energ. Chem.* **5**, 346.
[122] Jacobsen, F. M., Mogensen, O. E. and Trumpy, G., (1982), *Chem. Phys.* **69**, 71.
[123] Bockstahl, F., Billard, I., Duplâtre, G. et al., (1998), *Chem. Phys.* **236**, 393.
[124] Abbé, J. C., Duplâtre, G., Tabata, Y. et al., (1988), *J. Chim. Phys.* **85**, 29.
[125] Eldrup, M., Lightbody, D. and Sherwood, J. N., (1980), *Faraday Disc.* **69**, 175.
[126] Sharma, M., Kaur, C., Kumar, J. et al., (2001), *J. Phys.: Cond. Matter* **13**, 7249.
[127] Fernandez Valverde, S. and Duplâtre, G., (1986), *J. Radioanal. Nucl. Chem. Lett.* **107**, 307.
[128] Fernandez Valverde, S. and Duplâtre, G., (1986), *J. Chem. Soc. Faraday Trans. I* **82**, 2825.
[129] Eldrup, M., Lightbody, D. and Sherwood, J. N., (1979), *Phys. Rev. Lett.* **43**, 1407.
[130] Pérez Diaz, M. V., Abbé, J. C. and Duplâtre, G., (1988), *Phys. stat. sol., A* **109**, 337.
[131] Saito, H., Nagai, Y., Hyodo, T. et al., (1995), *Mat. Sci. Forum* **175/178**, 765.
[132] Adam, A., Cser, L., Kajcsos, Z. et al., (1972), *phys. stat. sol., B* **49**, K79.
[133] Kajcsos, Z., Vertes, A., Szeles, C. et al.; Jain, P. C., Singru, R. M., Gopinathan, K. P., Eds.; World Scientific: 1985, p 195.
[134] Vertes, A. and Ranogajec-Komor, M., (1980), *Radiochem. Radioanal. Lett.* **44**, 331.
[135] Vertes, A., Kajcos, Z., Marczis, L. et al., (1984), *J. Phys. Chem.* **88**, 3969.

Chapter 5

Physical and Radiation Chemistry of the Positron and Positronium

Sergey V. Stepanov and Vsevolod M. Byakov
Institute of Theoretical and Experimental Physics, Moscow 117218, Russia
sergey.stepanov@itep.ru

5.1 Introduction

Numerous applications of positron annihilation spectroscopy (PAS) in investigations of the physico–chemical properties of matter require a precise understanding of the process of Ps formation. Usually it proceeds on a picosecond time scale and is strongly influenced by early (pico- and femtosecond) processes initiated by ionizations in the track of a fast positron. These early intratrack processes initiate all subsequent chemical transformations and, consequently, take a key position in radiation chemistry.

In this chapter we will briefly discuss mechanisms of the positron slowing down, the spatial structure of the end part of the fast positron track, and Ps formation in a liquid phase. Our discussion of the energetics of Ps formation will lead us to conclude that (1) the Ore mechanism is inefficient in the condensed phase, and (2) intratrack electrons created in ionization acts are precursors of Ps. This model, known as the recombination mechanism of Ps formation, is formulated in the framework of the blob model. Finally, as a particular example we consider Ps formation in aqueous solutions containing different types of scavengers.

5.2 Energy deposition and track structure of the fast positron

5.2.1 *Ionization slowing down*

Usually positrons are born in nuclear β^+-decay with typical initial energies about several hundreds of keV. Upon implantation into a medium they lose energy via ionization of molecules. Within 10^{-11} s the positron energy drops to the ionization threshold. Further subionizing positrons reach thermal equilibrium by exciting primarily intra- and intermolecular vibrations. This usually takes an additional 10^{-13} s [1, 2].

Roughly half of the positron kinetic energy is lost in rare head-on collisions, resulting in knocking out δ-electrons with kinetic energies of about several keV. Tracks of these electrons form branches around the positron trail (Fig. 5.1) [3]. The other half of the energy is spent in numerous glancing collisions with molecules. The average energy loss in such a collision is 30–50 eV (at maximum it is 100 eV). A secondary electron knocked out in a glancing collision produces, by turns, a few ion–electron pairs inside a spherical microvolume, called a "spur" in radiation chemistry. Its radius, a_{sp}, is determined by the thermalization length of the knocked-out electrons in the presence of the Coulombic attraction of parent ions (Problem 1).

While positron energy W is greater than $W_{cyl} \sim 3$ keV, the mean distance l_i between adjacent ionizations produced by the positron is greater

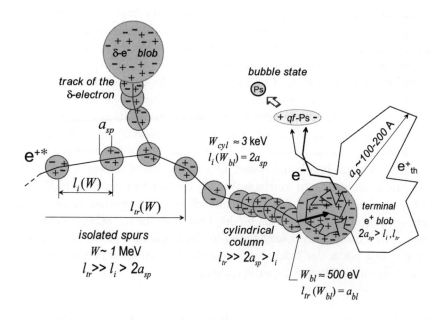

Fig. 5.1 Scheme of the end part of the e^{+*} track and Ps formation.

than spur size $2a_{sp}$. This means that at high positron energies, spurs are well separated from each other. The trajectory of the fast positron is a quasi-straight line because l_i is less than the positron transport path l_{tr}. The latter is the mean distance passed by the positron before it changes the initial direction of its motion by 90°.

When $l_i < 2a_{sp} < l_{tr}$ or $W_{bl} < W < W_{cyl}$ spurs overlap, forming something like a cylindrical ionization column. When the e^+ energy becomes less than the blob formation energy, W_{bl} ($\sim$ 500 eV, see below), $2a_{sp}$ becomes the largest parameter: $2a_{sp} > l_i, l_{tr}$. This means that the positron is about to create a blob. It contains a few tens of ion–electron pairs ($n_0 \approx W_{bl}/W_{iep} \approx 30$) because the average energy W_{iep} required for formation of one ion–electron pair is 16–22 eV [4]. The diffusion motion of the positron in the blob becomes more pronounced with decreasing energy. The positron frequently changes the direction of its momentum due to elastic scattering and the ionization of surrounding molecules. Roughly speaking, all intrablob ionizations are confined within a sphere of radius a_{bl}. The positron finally becomes subionizing and its energy loss rate drops by almost 2 orders of magnitude [6].

Typical dependencies of $l_i(W)$ and $l_{tr}(W)$ versus the energy of the positron are shown in Fig. 5.2. The calculation of $l_i(W) = W_{iep}/\text{LET}(W)$ is based on LET (Linear Energy Transfer) data of $e^\pm$. The estimation of the transport path has been done in the framework of the Born approximation (the wavelength of the positron with energy $\gtrsim$ 100 eV is small in comparison with the size of molecules), where the Born amplitude was calculated simulating a molecule of the liquid by an iso-electronic atom.

At low energies positron scattering becomes more and more efficient and we must regard its motion as diffusion-like with the energy-dependent mean free path (equal to l_{tr}) between successive "collisions", which completely randomize the direction of the velocity of the particle. If the probability of passing distance r without such a collision is $\exp(-r/l_{tr})$, the mean square distance $\overline{r^2} = 2l_{tr}^2$. After n collisions the mean square displacement becomes $n\overline{r^2}$. Calculation of the same quantity assuming that particle propagation is governed by the usual diffusion equation gives $n\overline{r^2} = 6D_p t$. Thus, we find that the diffusion coefficient is $D_p(W) = l_{tr}^2/3\tau = l_{tr}v_p/3$, where $\tau = t/n = \bar{r}/v_p = l_{tr}/v_p$ is the average time between successive collisions and v_p is the positron velocity. Integrating the relationship $d(r^2) = 6D_p dt = 2l_{tr}v_p dt = 2l_{tr}dx = 2l_{tr}dW/|dW/dx|_{ion}$ over the energy, we obtain the mean square displacement R_{ion} of e^{+*} during its ionization slowing down within this energy interval $W_i > W > W_f$:

$$R_{ion}(W_i, W_f) = \left(2 \int_{W_f}^{W_i} l_{tr}(W) \frac{dW}{|-dW/dx|_{ion}}\right)^{1/2}. \quad (1)$$

Now we can define W_{bl} and a_{bl}. These quantities are obtained from the following equations:

$$l_{tr}(W_{bl}) = a_{bl}, \qquad a_{bl} = R_{ion}(W_{bl}, Ry) - a_{bl}. \quad (2)$$

Here Ry stands for a typical ionization potential. Equation (2) indicates that the terminal positron blob is a spherical nanovolume, which confines the end part of its trajectory. This is where ionization slowing down is the most efficient (the thermalization stage of the subionizing positron is not included here). The mathematical formulation of this statement is twofold. Just after the first blob formation "step", which is $l_{tr}(W_{bl})$ (the thick arrow in Fig. 5.1), the positron reaches the center of the blob. After that, the end part of the ionization slowing-down trajectory is embraced by the blob; i.e., the slowing-down displacement of the positron, $R_{ion}(W_{bl}, Ry) - a_{bl}$ is equal to the "radius" of the blob, a_{bl}.

The solution of these equations in the case of liquid water is shown in Fig. 5.2. It gives $W_{bl} \approx 500$ eV and $a_{bl} \approx 40$ Å. One may assume that the values of a_{bl} and W_{bl} do not differ significantly from one liquid to another because the ionization slowing-down parameters depend mostly

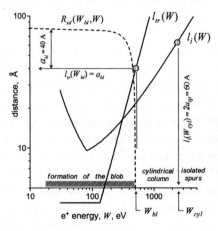

Fig. 5.2 Dependencies of $l_i(W)$ and $l_{tr}(W)$ vs the energy W of the positron in liquid water. This figure illustrates the solution of equations for the blob parameters W_{bl} and a_{bl}: $l_{tr}(W_{bl}) = a_{bl}$, $2a_{bl} = R_{ion}(W_{bl}, Ry)$.

on the ionization potential and the average electron density, parameters that are more or less the same in all molecular media. At first sight it seems strange that the obtained blob size $a_{bl} \approx 40$ Å is smaller than the dimensions of ion–electron pairs (60–200 Å), well known from radiation chemistry [2]. This is because the dimension of the ion–electron pair is determined by the thermalization length of the subionizing electron (in the field of the parent cation), while the positron blob radius is determined by the ionization slowing-down of the energetic positron when it loses the last ~ 500 eV of its energy.

5.2.2 Thermalization stage

At the end of the slowing-down by ionization and electronic excitation, the spatial distribution of e^{+*} coincides with the distribution of the blob species (i.e., $\sim \exp(-r^2/a_{bl}^2)$). Such a subionizing positron having some eV of excess kinetic energy may easily escape from its blob because there is no Coulombic interaction between the blob and the e^{+*} (the blob is electrically neutral). It is expected that by the end of thermalization, the e$^+$ distribution becomes broader with the dispersion:

$$a_p^2 \approx a_{bl}^2 + \langle R_{vib}(T, W_0)\rangle_{W_0}. \quad (3)$$

Here $R_{vib}(T, W_0)$ is determined by Eq. (1) where $|dW/dx|_{ion}$ should be replaced by $|dW/dx|_{vib}$, i.e., the stopping power of subionizing e^{+*} related to the excitation of vibrations (T is the temperature in energy units). The estimation of a_p requires quantitative data on $|dW/dx|_{vib}$, scattering properties of subionizing e^{+*}, and the spectrum of its initial energies W_0 after the last ionization event. In Eq. (3) $\langle\ldots\rangle_{W_0}$ denotes the average over W_0. In contrast to the parameters related to ionization slowing down, a_p strongly depends on the properties of each particular liquid and may reach hundreds of Å especially in non-polar media [5, 6].

Between positively charged ions and knocked-out intrablob electrons there exists strong Coulombic attraction. Thus out-diffusion of the electrons (even during their thermalization) is almost completely suppressed and the distribution of ions is close enough to that of electrons (Problem 2). This case is known as ambipolar diffusion when ions and electrons expand with the same diffusion coefficient equal to the duplicated diffusion coefficient of the ions (Problem 3). Thus blob expansion proceeds very slowly and may be practically neglected in the problem of Ps formation.

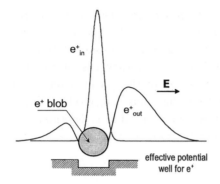

Fig. 5.3 The spatial distribution of the positron density at the Ps formation stage when an external electric field **E** is imposed. e^+_{in} is that part of the positron density that is bound within the blob and not biased by the external electric field. e^+_{out} is the part perturbed by the field. The depth of the trapping potential is about several tenths of eV.

5.2.3 Interaction between the positron and its blob

The interaction between the thermalized e^+ and the terminal blob has not been discussed before to our knowledge. One may think that this interaction is negligible. If we adopt the structure of the blob considered above (in which the distribution of intrablob electrons is only slightly broader than that of ions) and calculate electrostatic potential $\varphi(r)$ of the blob, we find that $\varphi(r)$ is everywhere repulsive to the positron. The maximum value of this interaction is about $3T$ (Problem 4).

However, there is obviously an opposite effect [7]. Residing within the blob, the thermal e^+ rearranges the intrablob electrons, so total energy of the system decreases because of the Debye screening. The corresponding energy drop may be estimated using the Debye-Huckel theory [8]. One finds the energy drop $\sim \frac{e^2}{\varepsilon(r_D + a_{bl}/n_0^{1/3})}$. However, the Debye radius $r_D \approx (4\pi r_c c_{iep})^{-1/2} \approx 4$ Å is quite small in comparison with the second term, $a_{bl}/n_0^{1/3}$, the average distance between intrablob electrons.* Because r_D is much smaller than $a_{bl}/n_0^{1/3}$, the screening energy becomes on the order of $T n_0^{1/3} r_c / a_{bl}$, which is some tenths of eV.

Thus, one may think that the redistribution of the blob charges induced by the e^+ may dominate over the repulsive interaction mentioned above, and could lead to e^+ trapping within the blob. Moreover, the estimated value of the energy drop indicates that not only the thermalized positron

---

*$r_c = e^2/\varepsilon_\infty T$ is the Onsager radius and $c_{iep} \approx n_0 / \frac{4}{3}\pi a_{bl}^3$ is the concentration of ion-electron pairs within the blob.

but also the epithermal one at the end of its thermalization process could be trapped within the blob. The trapping of the thermal and epithermal positrons by the blob is important for calculation of the Ps formation probability and its behavior in the external electric field [7].

5.3 Positronium formation in condensed media

5.3.1 *The Ore model*

Historically, the first Ps formation mechanism was suggested by Ore for the purpose of interpretation of experiments on e^+ annihilation in gases [9]. It implies that the "hot" positron, e^{+*}, having some excess kinetic energy, pulls out an electron from molecule M, thereby forming a Ps atom and leaving behind a positively charged radical–cation $M^{+\cdot}$:

$$e^{+*} + M \rightarrow Ps + M^{+\cdot}. \qquad (4)$$

This process is most effective when energy W of the positron lies within the interval named the "Ore gap":

$$I_G - Ry/2 < W < W_{ex} \quad (\text{or } I_G). \qquad (5)$$

Here I_G is the first ionization potential of the molecule, W_{ex} is its electronic excitation threshold and $Ry/2 = 6.8$ eV is the Ps binding energy in vacuum. It is accepted that a positron with energy lower than $I_G - Ry/2$ cannot pick up an electron from a molecule. When $W > W_{ex}$ (I_G) electronic excitations and ionizations dominate and Ps formation becomes less effective.

5.3.2 *Quasi-free Ps state*

The availability of a large free space in gases allows one to neglect the zero-point kinetic energy of Ps caused by the presence of surrounding molecules. So in this case, Ps binding energy is simply $-Ry/2$. In the condensed phase because of the high concentration of molecules, the final state of reaction (4) differs from that in vacuum. First, Ps zero-point kinetic energy arising due to Ps repulsion from the molecules is essentially increased (both e^- and e^+ are repelled from the cores of surrounding atoms because of exchange and Coulombic repulsions, respectively). This effect may be referred to as "confinement" of the Ps. Secondly, molecular electrons screen the e^+–e^-

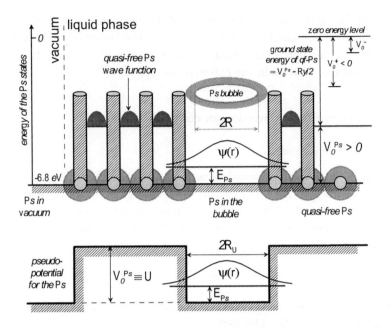

Fig. 5.4 Energy diagram of the Ps states. Ps energy in vacuum is -6.8 eV. V_0^{Ps} is the Ps work function, i.e., the energy needed for Ps to enter the liquid without any rearrangement of molecules. qf-Ps senses molecules (atoms) of the liquid as an external potential with repulsive cores (for convenience of representation in the figure this potential is shifted down to -6.8 eV). Being averaged, it may be represented as a potential step of height $V_0^{Ps} \equiv U$. Usually within some picoseconds qf-Ps transfers to the bubble state with the free-volume radius, R, the center-of-mass wave function of the Ps, $\Psi(r)$, and the ground-state energy $E_{Ps} - Ry/2$. Note that at 2γ-annihilation of p-Ps total energy of two γ-quanta is equal to $2mc^2 + E_{Ps} - Ry/2$. R_U is the radius of the potential well, which localizes Ps.

Coulombic attraction and increase average e^+–e^- separation in the positronium. Both these factors decrease binding energy between the positron and electron constituting the positronium and therefore reduce the width of the Ore gap.

Elementary estimation of the Ps confinement effect is the following. The ground-state energy of a point particle with mass $2m$ in an infinite potential well of radius R is $\frac{\pi^2 \hbar^2}{4mR^2} = \frac{Ry}{2}\left(\frac{\pi a_B}{R}\right)^2$, where m is the electron mass and

$a_B = 0.53$ Å is the Bohr radius. For $R = \pi a_B \approx 1.66$ Å this energy becomes equal to the binding energy, $-Ry/2$, of the free Ps atom taken with an opposite sign (for the H atom it happens at $R = 1.835 a_B = 0.97$ Å). This gives a hint that Ps might become unstable with respect to a breakup of e^+ and e^-, and means that in a dense medium the width of the Ore gap becomes narrower and may even completely disappear.

Of course, simulation of the Ps repulsion from molecules by an infinitely deep potential well is a crude approximation. A more realistic approach, based on the finite well Ps bubble model [10, 11], gives information about the depth U of the well, which has a meaning of the Ps work function, V_0^{Ps}, i.e., the energy needed for Ps to enter the liquid without any rearrangement of molecules and stay there in the delocalized quasi-free state, qf-Ps. This state has no preferential location in the bulk. The qf-Ps state corresponds to the bottom of the lower-energy band available to the interacting e^+–e^- pair. Obviously, this state precedes the formation of the Ps bubble. The same state may be obtained from the Ps bubble state if the free-volume radius R of the bubble is tending to zero, Fig. 5.4.

V_0^{Ps} is an analog of the work function V_0^- of the excess quasi-free electron, e_{qf}^-, but with one important distinction. V_0^- is at the same time the ground-state energy of e_{qf}^-, because the energy of the electron at rest after having been removed from the liquid to infinity is zero by definition [13]. Ps work function $V_0^{Ps} > 0$ differs from the qf-Ps ground-state energy, $\approx V_0^{Ps} - Ry/2 < 0$, by a constant shift, because its energy far outside the liquid is not zero, but $-Ry/2$. However, if Ps is dissociated in vacuum, the energies of the largely separated e^+ and e^- would be equal to zero. Energetics of the qf-Ps state may be illustrated with the help of the Born–Haber cycle:

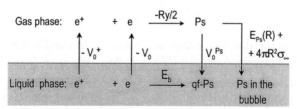

It consists of the following steps: (1) removal of the quasi-free electron and positron from the liquid to vacuum accompanied by some expenditure (or release) of energy, namely the sum of the respective work functions, $-V_0^- - V_0^+$, taken with negative sign; (2) formation of the Ps in vacuum with the binding energy $-Ry/2$; 3) return of the Ps atom to the liquid phase supplied with energy, V_0^{Ps}. From the Born–Haber cycle we are able to estimate binding energy, E_b, of the qf-Ps state:

$$E_b = -V_0^- - V_0^+ + V_0^{Ps} - Ry/2 < 0. \qquad (6)$$

One should expect that V_0^+ is negative because of the prevalence of the polarization interaction of e_{qf}^+ with the medium molecules over repulsion from the nuclei [12]. So we may roughly equate V_0^+ to the energy P_- of polarization interaction, e_{qf}^-, which can be calculated by decomposition of V_0^- into the sum of kinetic K_- and polarization P_- terms [13]:

$$V_0^- = K_- + P_-, \quad P_- = -\frac{e^2}{2R_{WS}} \frac{\varepsilon_\infty - 1}{\varepsilon_\infty + 2} \left[\frac{24}{7} + \frac{\varepsilon_\infty + 2}{\varepsilon_\infty} \right], \qquad (7)$$

where R_{WS} is the Wigner–Seitz radius of the molecule ($4\pi R_{WS}^3/3 = 1/n$, n being the number density of molecules) and ε_∞ the high-frequency dielectric permitivity. For example, in liquid water ($R_{WS} = 1.93$ Å, $\varepsilon_\infty = 1.78$) $P_- \approx -4.3$ eV, in hydrocarbons $P_- \approx -(2 \text{ to } 3)$ eV.

Based on the Ps bubble model, [10, 11] experimental data on o-Ps lifetimes, and the widths of the "narrow" component of the ACAR spectrum, it was found that in practically all investigated liquids of various chemical nature, values of $V_0^{Ps} \equiv U$ are close to $+3$ eV. Thus, from Eq. (6) we obtain $E_b \approx -(0.5 \text{ to } 1)$ eV. This means that qf-Ps is a loosely bound structure although the long-range Coulombic attraction between e^+ and e^- always provides for the existence of the bound state. Because e^+–e^- separation, r_{ep}, in qf-Ps is expected to be large, we may rewrite E_b in the following form:

$$E_b \approx -\frac{Ry}{2} \cdot \frac{1}{\varepsilon_\infty^2} \cdot \frac{3a_B}{r_{ep}}, \qquad (8)$$

where $3a_B$ is the distance between e^+ and e^- in the Ps atom in vacuum; i.e., the Ps diameter $\langle \varphi | r_{ep} | \varphi \rangle / \langle \varphi | \varphi \rangle = 3a_B$, where $\varphi \propto \exp(-r_{ep}/2a_B)$. The scaling $E_b \propto \varepsilon_\infty^{-2}$ directly follows from the Schrödinger equation for Ps and is confirmed experimentally [14]. Using the above-mentioned numerical assessment for E_b we conclude that r_{ep} is about 5 Å. In such a starched Ps state the density of the "native" electron on the positron is rather small and e^+ will primarily annihilate with electrons of the medium, with the average lifetime of about 0.5 ns contributing to the free e^+ components of the LT and ACAR spectra. Thus, the qf-Ps component could hardly be resolved experimentally, but it should renormalize the contribution ascribed to the "free" e^+ annihilation.

One more useful relationship between the numbers involved can be obtained from the criterion of the energetic stability of the Ps bubble state. The only modification of the above Born–Haber cycle is on the right: Ps should be returned to the liquid not in the quasi-free state, but in the bubble state. It requires $E_{\mathrm{Ps}} + 4\pi R^2 \sigma_\infty$ instead of V_0^{Ps}. In most liquids E_{Ps} is usually $0.4 - 0.5$ eV, reaching the maximal in water, 0.73 eV, and the minimal, 0.05 eV, in liquid He [10]. Similarly, practically everywhere $4\pi R^2 \sigma_\infty$ is equal to 0.3-0.4 eV (in water 0.58 eV, in liquid He 0.023 eV). It is worth mentioning that here R is the free-volume radius, but not the radius of the surface tension, and σ_∞ is the surface tension of the plane surface. As shown in [10, 11] this expression implicitly takes curvature correction on the surface tension into account. PV-term (P is the external pressure and $V = 4\pi R^3/3$) comes into play only in the case of liquid He, where the radius of the Ps bubble is very large, $R = 17.5$ Å, and the PV-term is equal to 0.014 eV. In all other cases it is negligible. The Ps bubble state is stable with respect to dissociation into e_{qf}^+ and e_{qf}^- if

$$-V_0^- - V_0^+ + E_{\mathrm{Ps}} + 4\pi R^2 \sigma_\infty - Ry/2 < 0. \qquad (9)$$

The most uncertain quantity here is V_0^+, for which one obtains the lowest threshold: $V_0^+ > -V_0^- + E_{\mathrm{Ps}} + 4\pi R^2 \sigma_\infty - Ry/2$. For water it gives $V_0^+ > -4.3$ eV, which occasionally coincides with P_-. It implies that the above-mentioned relationship $V_0^+ \approx P_-$ underestimates V_0^+ due to neglect of the e^+ zero-point kinetic energy contribution.

5.3.3 Disappearance of the Ore gap

Now let us estimate the low boundary, W_{low}, of the Ore gap in molecular liquids. Because the Ore process is just an electron-transfer reaction, we assume that no rearrangement of molecules occurs and, therefore, the final positronium state will be quasi-free (formation of the bubble requires much longer time). The corresponding Born–Haber cycle is the following:

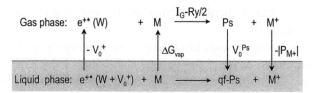

If translational kinetic energy of the obtained qf-Ps is nearly thermal, the excess kinetic energy, W, of the projectile positron in this case should coincide with the low boundary of the Ore gap: $W \approx W_{low}$. In parentheses we displayed the total energy of e^+ in the liquid and gas phases. ΔG_{vap} is the vaporization energy of the molecule (a few tenths of eV), and P_{M+} (negative) is the energy of polarization interaction of the radical–cation $M^{+\cdot}$ with the environment. Total energy balance permits Ps formation if $W + V_0^+ - \Delta G_{vap} - (I_G - Ry/2) - V_0^{Ps} + |P_{M+}| > 0$. Using the approximate relationship between liquid and gas phase ionization potentials, $I_L \approx I_G - |P_{M+}| + V_0^-$, we obtain:

$$W_{low} \approx I_L - V_0^- - V_0^+ + \Delta G_{vap} + V_0^{Ps} - Ry/2. \tag{10}$$

For example, in liquid water $I_L = 8.8$ eV, $V_0^- \approx -1.2$ eV. This relationship gives $W_{low} \approx 8-9$ eV. It exceeds the lowest energy threshold $W_{ex} \approx 6.7$ eV of electronic excitations, so in water, apparently, the Ore gap completely disappears. This takes place in the majority of molecular liquids, which explains the inefficiency of the Ore mechanism there.

In [15] it was suggested that in polymers, because of the presence of large free-volume space, the final Ps state in the last Born–Haber cycle may not be quasi-free. Ps forms immediately in one of the preexisting voids. If it did, the total energy gain would be larger and the Ore gap could exist. However, because the final Ps state has obviously zero translational momentum while the projectile positron possesses a lot, this reaction must be suppressed owing to the conservation of momentum law. e^{+*} cannot promptly transfer its momentum to surrounding molecules because of the large difference in masses. Absorption of the released energy during formation of the localized Ps in a cavity cannot be immediate and requires some time. Probably we should expect qf-Ps formation in the bulk at first. Then it will move somewhere around the void, lose energy and momentum, and its center-of-mass will gradually approach this cavity after awhile.

5.4 Recombination mechanism of Ps formation

The preceding discussion reserves the unique possibility of Ps formation in molecular media. It is the recombination mechanism, which postulates that Ps is formed via combination of the deenergized positron with one of the electrons formed at the end part of its track. This process is not energetically forbidden in the sense discussed above.

Over the last 30 years the recombination mechanism has become extremely widespread [16, 17]. It has been used to interpret extensive data on Ps chemistry, and explain variations of Ps yields from 0 to 0.7 in very different chemical substances where parameters of the Ore gap are practically the same. Variations of Ps formation probability under phase transitions have also received natural explanation. Experimentally observable monotonic inhibition of Ps yields (practically down to zero) in solutions of electron acceptors contradicts the Ore model, but is well incorporated in the recombination mechanism. It explains the anti-inhibition effect, including experiments on Ps formation in moderate electric fields in pure liquids and mixtures.

There are two models which utilize this mechanism, the spur model [18, 16] and the blob model (diffusion-recombination model) [19, 20]. In spite of the fact that both models answer the question about the Ps precursor in the same way, they differ as to what constitutes the terminal part of the e^+ track and how to calculate the probability of the Ps formation.

5.4.1 The spur model

Quantitative formulation of the spur model was given by Tao [21]. It is based on the following assumptions:

(1) the positron and secondary electrons knocked out during the last several acts of ionizations thermalize within the same volume, and positive ions also reside there;

(2) secondary electrons recombine with the same probability either with the positive ions or with the positron. Therefore, the probability that the positron gets an electron for subsequent Ps formation is equal to $n_0/(1+n_0)$, where n_0 is the "initial" number of ion-electron pairs in the terminal spur;

(3) probability of the "elementary act" of e^+–e^- recombination is taken in the Onsager form, $1 - e^{-r_c/b}$, where b is the "initial" distance between e^+ and e^- by the end of their thermalization.

Thus, in the framework of the spur model, Ps formation probability is written as:

$$P_{Ps} = \frac{n_0}{1+n_0} \cdot \left(1 - e^{-r_c/b}\right) \cdot e^{-t_{Ps}/\tau_2}. \tag{11}$$

The last exponential factor takes into account a possibility of the free-positron annihilation occuring during the Ps formation time, t_{Ps} (on the order of some picoseconds) with the annihilation rate $1/\tau_2$ ($\lesssim 2$ ns^{-1}). Obviously, the contribution from this factor is negligible. In nonpolar molecular media at room temperature r_c is ≈ 300 Å. Typical thermalization lengths b of electrons are $\lesssim 100$ Å. Thus, the Onsager factor, $1 - e^{-r_c/b}$, is also very close to unity. Therefore, to explain observable values of Ps yields, which never exceed 0.7, we must conclude according to Eq. (11) that the terminal positron spur contains on average 2 to 3 ion–electron pairs.

However, Eq. (11) apparently contradicts the correlation between experimentally measured P_{Ps} and the free-ion yields, G_{fi}, at least in hydrocarbons (the higher the P_{Ps} the higher the G_{fi}) [22].

Besides, the spur model is inconsistent with the structure of the positron track (Fig. 5.1). The problem is that the end of the positron track is not a spur, containing 2 to 3 ion–electron pairs, but the blob, containing about 30 overlapped ion–electron pairs [2, 23, 24]. In essence this statement relies on the following two well known facts related to behavior of the energized electron (positron) in the blob formation stage $W \lesssim W_{bl}$, which we discussed above: (1) Ionization energy losses of such a positron (electron) are about 1–4 eV/Å. It means, that e$^+$ very effectively ionizes molecules at the end of its track; (2) Motion of the positron in the blob is diffusion-like. Efficient e$^+$ scattering results in the blob having an approximately spherical shape.

Therefore, it is difficult to justify the approach [25] to the Ps formation process where the end part of the e$^+$ track is simulated by a straight-line sequence of isolated ion–electron pairs separated by more than 100 Å. However, having considered the data for more than 50 liquids, the authors [25] came to the conclusion that multiparticle effects in the terminal positron spur are important for correct description of the intratrack processes.

It is also hard to believe that ion–electron recombination (IER) in the end part of the e$^+$ track, which in [24] is assumed to be a cylindrical column of ionizations, precedes Ps formation. IER is related to much larger energy release (about W_{bl} amount of energy has to be eventually transferred to molecular vibrations) than energy release at Ps formation. Therefore it should require longer time.

The statement that the terminal part of the e^+ track is a blob but not a spur, is not just a question of terminology. Processes (IER, Ps formation) related to energy dissipation and screening of local electric fields, proceed there in a different way. At present, experimental data clearly indicate that e^+ behavior in the blob is quite different from that of intrablob electrons and ions. e^+ is rather mobile and easily escapes from the blob during its thermalization (see experiments on Ps formation in electric fields [26, 27] and measurements of e^+ mobility [28]). Particularly, it implies that the multiparticle nature of the terminal part of the e^+ track cannot be correctly taken into account via the factor $n_0/(n_0 + 1)$ in Eq. (11).

5.4.2 The blob model

Let us summarize the most important properties of the blob, i.e., the end part of the positron track [3].

Size and energy release. The terminal blob consists of about $n_0 \approx 30$ overlapped ion–electron pairs. Spatial distribution of these species may be described by Gaussian function $n_0 \frac{\exp(-r^2/a_{bl}^2)}{\pi^{3/2} a_{bl}^3}$, where the "blob radius" $a_{bl} \approx 40$ Å. Total energy deposition in the blob is $W_{bl} \sim 500$ eV.

Quasi-neutrality condition. Because of strong Coulombic interaction, intrablob electrons adjust their motion to the distribution of the primary positive ions and screen them. As a result, the width of the spatial distribution of the electrons is only slightly larger than that of ions.

Expansion of the blob. Because of attraction between ions and electrons, expansion of the blob is governed by the law of ambipolar diffusion. As a result out-diffusion of electrons is almost completely suppressed, but the diffusion coefficient of ions is increased by a factor of two. Thus, blob expansion proceeds very slowly and may usually be neglected in the problem of Ps formation.

Elongated thermalization and recombination. After the last ionization event in the terminal blob, the positron and intrablob electrons become sub-ionizing. Having no possibility of exciting electronic transitions in molecules, they lose energy by exciting intra- and intermolecular vibrations. Molecular excitation is caused by an alternating electric field of the moving electrons. If the Fourier spectrum of this field contains harmonics with frequencies close to that of molecular vibration, the excitation gets very probable. In the blob, because of the high density of ion–electron pairs and efficient screening, fluctuations of the local electric field are significantly suppressed. Thus, in spite of the fast motion of intrablob electrons, they cannot easily transfer kinetic energy to excitations of vibrations. This effect lengthens thermalization of electrons and ion-electron recombination.

5.4.3 "Elementary act" of the Ps formation

If the positron is thermalized outside the blob, the only way for it to form Ps is to diffuse back and pick up one of the intrablob electrons. Otherwise e^+ annihilates as a free positron. To pick up an electron the "initial" separation between e^+ and e^- must be comparable to the average distance $(4\pi a_{bl}^3/3n_0)^{1/3} \approx 20$ Å between intrablob species. According to Eq. (8) the binding energy of $e^+ \cdots e^-$ pair with $r_{ep} \approx 20$ Å is about 0.1 eV. Translational kinetic energies of each particle composing the pair must be less than the binding energy, otherwise the pair will break up. Therefore just before the Ps formation these e^+ and e^- should not possess excess kinetic energy, in other words, be almost thermalized. In other words, it means that the Ps formation cross section has a maximum at thermal energies.

In adiabatic approximation (positions of molecules are fixed) such a pair is not on the bottom of its energy spectrum; it continues to release energy via excitation of molecular vibrations and finally reaches an equilibrium qf-Ps state. In comparison with the ground-state energy of qf-Ps, the $(e^+ \cdots e^-)$ pair has about 1 eV of excess energy, which is accumulated in the form of a potential energy of the mutual Coulombic attraction. What happens with this excess energy during $(e^+ \cdots e^-) \to$ qf-Ps transformation?

Let total momentum of the $(e^+ \cdots e^-)$ pair be zero at the beginning. As particles get closer, kinetic energy of both the positron and electron increases. If then one of the particles creates a phonon (vibration excitation) its energy drops down, while the momentum changes predominantly to the opposite direction accompanied by some diminishing of its absolute value. As a result the $(e^+ \cdots e^-)$ pair acquires non-zero translational momentum and its center-of-mass starts to move. Adopting ≈ 0.01 eV/Å as a typical value of the $e^-(e^+)$ LET due to excitation of vibrations, we may estimate total path E_b/LET ≈ 50–100 Å of the $(e^+ \cdots e^-)$ pair before it becomes qf-Ps. As e^+ and e^- get closer, their polarization interaction with the medium decreases because the larger number of surrounding molecules begin to "feel" the $(e^+ \cdots e^-)$ pair as an electrically neutral object. So the LET of such a pair may decrease and its total path may even increase.

Losing all excess potential energy, the $(e^+ \cdots e^-)$ pair becomes the qf-Ps, which is in thermal equilibrium with the environment. In a liquid, qf-Ps immediately finds the nearest place with lower density, stays there, and over a longer time scale continuously evolves into the bubble state. Further gain of the Coulombic attraction energy and decrease of the positronium zero-point kinetic energy are the driving forces of rearrangement of molecules and appearance of additional free space around Ps. Schematically, the consecutive stages of Ps formation may be displayed as follows:

$$e^+_{qf} + e^-_{blob} \rightarrow (e^+ \cdots e^-) \rightarrow \text{qf-Ps} \rightarrow \text{Ps in the bubble.} \quad (12)$$

Crude estimation of the Ps bubble formation time based on the solution of the Euler equation for incompressible fluid, taking into account the Laplace pressure and the quantum-mechanical "pressure" exerted by the Ps atom on the wall of the bubble, gives several picoseconds in a majority of liquids at normal conditions.

5.4.4 Role of localized states of electrons and positron

From the viewpoint of the energy balance, solvated (localized) electrons, e^-_s, and positrons, e^+_s, might participate in Ps formation. Actually e^-_s already resides in a small void and its binding energy there is about 1 eV. However, consideration of the data in Fig. 5.5 shows that in some liquids, electron localization occurs over a longer time than needed for Ps formation; e^-_s does not yet exist during the Ps formation process [30].

In water e^- solvation time is much shorter ($\tau^s_e = 0.3$ ps at room temperature [31]) and one may think that e^-_s may contribute to Ps formation. At present there is no clear theoretical judgment on such a possibility. However, there is some experimental evidence (see the next section) against participation of e^-_s in Ps formation.

Theoretical arguments are twofold. On one hand, one may expect that e^+ also gets solvated over a time comparable with τ^s_e. Mobility of solvated particles drastically drops and they simply do not have enough time to meet each other during the free-positron lifetime (~ 0.5 ns). Really, corresponding diffusion displacement of e^+_s is smaller than e^+ thermalization

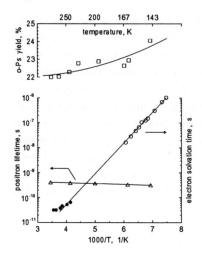

Fig. 5.5 Temperature dependencies for Ps formation probability, electron solvation time and positron lifetime in n-propanol. With changing temperature from 150 K to 300 K the electron solvation time τ_e^s in n-propanol varies within 5 orders of magnitude. At the highest studied temperature τ_e^s approaches 10 ps. At low temperatures it exceeds e$^+$ lifetime τ_2 more than 1000 times. If e_s^- had really contributed to Ps formation, Ps yield would have to decrease. However, it even slightly increases. This fact favors the presolvated electron as the main precursor of Ps formation.

displacement ($a_p \lesssim 100$Å) [32].

On the other hand, it is clear that the solvated state of the positron is very different from that of the electron. Because of spin-exchange repulsion, an excess electron, being solvated, forms a small void and preferentially resides inside it. On the contrary, the positron is not involved in spin exchange and prefers to reside in the bulk of a liquid because of the prevalence of polarization interaction. e$^+$ may be trapped by positive density fluctuations (aggregation of molecules) [12]. However, density of a liquid state is too high to permit clustering around e$^+$. Therefore e$^+$ solvation may be suppressed and this may explain the very high experimental values of e$^+$ mobility [28].

Further studies are needed in this direction. Below, we do not include localized particles in equations describing Ps formation keeping, e$^+$ solvation time as an adjustable parameter.

5.5 Quantitative formulation of the blob model

5.5.1 *Reactions of hot electrons and positrons*

Despite the inefficiency of the Ore mechanism in the condensed phase discussed above, reactions of sub-ionizing electrons and the positron during the slowing-down process cannot be ignored because of the possibility of resonant capture by solute molecules, $e^{\pm *} + S \to S^{\pm *}$. The term "resonant" means that the scavenger, S, selectively reacts with $e^{-*}(e^{+*})$ having a certain energy. If so, the capture cross section, $\sigma_L^{\pm}(W)$, of the solute in the liquid has a bell-like shape, which is determined by the structure of the electronic excited states, $S^{\pm *}$, and their energy relaxation in the liquid [33].

If e^- is weakly bound in S^{-*}, an additional reaction leading to Ps formation becomes possible:

$$S^{-*} + e^+ \to Ps + S,$$

but ion–electron recombination is suppressed. In the opposite case of the strongly S^{-*} state, both reactions are inhibited.

The fraction, φ_e, of the electrons escaping resonance capture by S (c_s the concentration of the solute) depends on the energy spectrum, $f(W_0)$, of sub-ionizing electrons and stopping power, $|dW/dx|_{vib}$ (Problem 5) [29]:

$$\varphi_e(c_S) = \left\langle \exp\left(-c_S \int_0^{W_0} \frac{\sigma_L(W)dW}{|dW/dx|_{vib}}\right) \right\rangle_{W_0}. \tag{13}$$

Here, as in Eq. (3) $\langle\ldots\rangle_{W_0}$ denotes an average over $f(W_0)$. For a number of scavengers the data on the e^- cross section $\sigma_G(W)$ in the gas phase are known. Following [34, 35] we adopt that $\sigma_L(W)$ has the same shape as $\sigma_G(W)$, but its energy position is different. It depends on the ground-state energy V_0^- of e_{qf}^- and energy P_{S^-} of polarization interaction of S^{-*} with the environment (P_{S^-} includes only electronic polarization). If in the gas phase $\sigma_G(W)$ has a maximum at W_G, in the liquid-phase energy position,

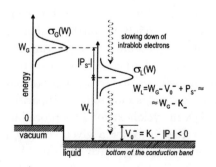

Fig. 5.6 Resonance trapping of electrons (positrons) having excess kinetic energy in the gas and liquid phases.

W_L, of the corresponding maximum of $\sigma_L(W)$ will be at (Fig. 5.6)

$$W_L = W_G - V_0^- - |P_{S-}|. \qquad (14)$$

When P_{S-} and the polarization part of V_0^- approximately cancel each other, Eq. (14) is reduced to $W_L \approx W_G - K_-$. Thus, the difference $W_G - W_L$ is determined by zero-point kinetic energy K_- of electrons in the medium. For example, when S has a maximum of $\sigma_G(W)$ close to the thermal region, in the liquid phase it even may lose trapping ability, i.e., $W_G - K_- < 0$, because of a lack of electrons with small energies (K_- is the lowest possible kinetic energy of the quasi-free electrons in a liquid).

If $\sigma_G(E)$ has the simplest form $\sigma_0 \delta(W - W_G)$, from Eq. (13) we easily obtain:

$$\varphi_e(c_S) = \begin{cases} 1 - F_- + F_- \exp(-c_S/c_{37}^-), & W_G > K_-, \\ 1, & W_G < K_-, \end{cases} \qquad (15)$$

$$F_-(W_L \equiv W_G - K_-) = \int_{W_G - K_-}^{\infty} f(W_0) dW_0, \quad \frac{1}{c_{37}^-} = \frac{\sigma_0}{\text{LET}(W_G - K_-)}.$$

Here $\text{LET}(W_G - K_-) = |dW/dx|_{vib}$ at $W = W_G - K_-$. The value $1 - F_-$ represents the fraction of electrons knocked out of the molecules with the energy below E_L and therefore escaping resonance capture. In a given liquid the value of $1 - F_-$ rises with increasing W_G. This effect is illustrated in Fig. 5.7 [36].

Actually, in the gas phase energy dependence of the electron capture cross section of haloid–benzenes has resonance-like maxima. Their positions for C_6H_5I, C_6H_5Br and C_6H_5Cl increase with e^- energy 0.1, 0.7 and 1 eV, respectively [33]. The values of $1 - F_-$ (i.e., the limiting intensities of the o-Ps component) obtained from the fit (Fig. 5.7) rise in the same proportion, 0.03, 0.19 and 0.33, respectively. 65 It is worth comparing the obtained values of c_{37}^- with the values of parameter α entering the Warman-Asmus-Schuler empirical equation. α characterizes the electron-scavenging efficiency of a solute. Unfortunately, we do not know particular numbers of α for C_6H_5X compounds in benzene, so for rough comparison we may use available values of α for CH_3X in cyclohexane. This procedure seems possible first of all because the values of α for a given scavenger do not differ significantly in different solvents having nearly the same V_0^-. Secondly, in a given solvent, values of α for C_6H_5X compounds are rather close to that of CH_3X. According to [37] in cyclohexane, $\alpha(CH_3I) = 22$

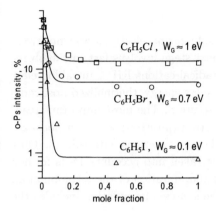

Fig. 5.7 o-Ps formation probability in benzene solutions of C_6H_5Cl (□) and C_6H_5Br (○), C_6H_5I (△). The data demonstrate a correlation between partial inhibition fraction F_- corresponding to W_G values. Solid lines are calculated according to Eq. (15).

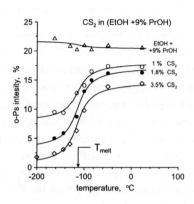

Fig. 5.8 Temperature dependencies of the o-Ps intensity in (EtOH + 9% PrOH) mixture containing different amounts of CS_2 (numbers in the figure are volume percents of CS_2). The lines are guided by eye.

M^{-1}, $\alpha(CH_3Br) = 16$ M^{-1}, and $\alpha(CH_3Cl) = 5\text{--}5.6$ M^{-1}. Their ratios practically coincide with the c_{37}^- values obtained from our fit (Fig. 5.7). Both correlations indicate that only intratrack electrons are Ps precursors.

Another example of the same effect is given in Fig. 5.8 [38]. In ethanol o-Ps intensity does not change with temperature around the melting point, -114 °C. However, it decreases appreciably in the solid phase with the increase of CS_2 concentration. It is seen that the electron-scavenging efficiency of CS_2 strongly depends on the phase state of ethanol. In the liquid state V_0^- is between -0.4 eV [39] and -0.65 eV [2]. Below the melting point V_0^- becomes positive, 0.34 eV [2] or even 1.1 eV [40]. Obviously this growth is due to increasing e^- zero-point kinetic energy $K_- = V_0^- + |P_-|$. Its value is ≈ 3 eV (substitute $R_{WS} = 2.84$ Å and $\varepsilon_\infty = 1.85$ into Eq. (7)). Thus, in the solid state the threshold energy moves closer to thermal energies, and CS_2 becomes a more efficient electron acceptor.

Sub-ionizing hot positrons can also react with a solute in a similar way. Value $\varphi_p = 1 - F_+ + F_+ \exp(-c_S/c_{37}^+)$ represents the fraction of positrons escaping resonance capture by the solute.

5.5.2 Diffusion-recombination stage

In molecular liquids slowing-down of subionizing electrons and positrons usually takes less than 10^{-13} s. The next stage involves fast intrablob reactions with participation of primary radical-cations $RH^{+\cdot}$, intrablob electrons which lost more or less all their excess energy, thermalized positrons, and possibly presented molecules of a solute S. The most important reactions which launch on this stage are electron (positron) solvation (localization), ion–electron recombination (IER), and reactions with solutes. Out-diffusion of the positron should also be taken into account. Ps is formed just in this stage. Depending on the liquid, duration of the recombination stage varies from 10^{-12} to 10^{-10} s. The simplest formulation includes the following competing reactions:

electron (positron) solvation: $\quad e^- \xrightarrow{\tau_e^s} e_s^-, \quad e_{qf}^+ \xrightarrow{\tau_p^s} e_s^+,$ (16a)

ion–electron recombination: $\quad e^- + RH^{+\cdot} \xrightarrow{k_{ie}} RH^* \to \text{Products},$ (16b)

ion–molecule reaction: $\quad RH^{+\cdot} + RH \xrightarrow{\tau_{imr}} R^{\cdot} + RH_2^+,$ (16c)

capture by S: $\quad e^- + S \xrightarrow{k_{eS}} S^-, \quad e_{qf}^+ + S \xrightarrow{k_{pS}} e^+S,$

$\quad RH^{+\cdot} + S \xrightarrow{k_{iS}} \text{Products},$ (16d)

qf-Ps formation: $\quad e_{qf}^+ + e^- \xrightarrow{k_{ep}} \text{qf-Ps}.$ (16e)

Here e^- denotes the intrablob electron, and τ_e^s and τ_p^s are solvation times of e^- and e_{qf}^+, respectively. τ_{imr}^{-1} is the ion–molecule reaction rate of the radical–cation $RH^{+\cdot}$. k_{ie}, k_{ep}, k_{eS}, k_{pS} and k_{iS} are recombination and capture rate constants.

In some cases when electron trapping leads to formation of the weakly bound state, S^{-*} (it may also takes place in reaction with hot electrons), reaction $e^+ + S^{-*} \to$ Ps becomes possible and should be taken into account. On the contrary, in the case of a strongly bound electron and sufficient affinity of S^- to the positron, this reaction may lead to e^+S^- complex formation and, therefore, decreased Ps yield.

The blob model describes reactions (16) in terms of nonhomogeneous kinetics via Eqs. (17–18) on concentrations of the particles. It is an adequate approach to the problem because the number of particles involved is large and local motion of the intrablob electrons and the positron is fast enough:

$$\frac{\partial c_j(r,t)}{\partial t} = D_j \Delta c_j - \sum_{q(\neq j)} k_{qj} c_q c_j - c_j/\tau_j, \qquad j,q = \{i,e,p\} \quad (17)$$

$$c_i(r,0) = c_e(r,0) = n_0 \cdot \frac{\exp(-r^2/a_{bl}^2)}{\pi^{3/2} a_{bl}^3}, \qquad c_p(r,0) = \frac{\exp(-r^2/a_p^2)}{\pi^{3/2} a_p^3}. \quad (18)$$

Here c_i, c_e and c_p are the concentrations of the positive ions, electrons and the positron probability density at a point **r** measured from the center of the blob at time t. D_p is the diffusion coefficient of the positron, $D_i \equiv D_e \equiv D_{amb} \approx 0$ is the ambipolar diffusion coefficient of the blob, $a_i^2 \approx a_e^2 \approx a_{bl}^2$ is the dispersion of the intrablob species, and a_p^2 is the dispersion of the positron space distribution by the end of its thermalization. Decay rate $\tau_e^{-1} = 1/\tau_e^s + k_{eS} c_S$ is the sum of the electron solvation rate and possible capture by solute molecules; $\tau_p^{-1} = 1/\tau_2 + 1/\tau_p^s + k_{pS} c_S$ accounts for the free e^+ annihilation, solvation and reaction with S. Similarly, $\tau_i^{-1} = 1/\tau_{imr} + k_{iS} c_S$, where τ_{imr}^{-1} is the rate of the ion–molecule reaction.

To calculate qf-Ps formation probability, P_{Ps}, we must integrate the term $k_{ep} c_e c_p$ over whole space and time:

$$P_{\text{Ps}} = k_{ep} \int_0^\infty dt \int c_e(r,t) c_p(r,t) d^3 r. \quad (19)$$

As we discussed above, in liquids this state transforms to the bubble state rather quickly (within some ps). So in this case it can be compared with experimentally observed intensity $I_3 = 3 P_{\text{Ps}}/4$ of the LT spectrum and intensity $Y_1 = P_{\text{Ps}}/4$ of the narrow component of the ACAR spectrum.

Considering the diffusion–recombination stage below, we neglect an interaction between the thermalized positron and its blob. This approximation, as we discussed above, assumes that the appearance of a positive potential in the blob, caused by outdiffusion of electrons, is nearly cancelled by the negative potential caused by e^+ screening inside the blob. In this case we can apply the prescribed diffusion method to obtain the solution of Eq. (17). Let us write $c_j(r,t)$ in the following form:

$$c_j(r,t) = n_j(t) G_j(r,t), \qquad \int G_j(r,t) d^3 r = 1, \qquad j = \{i,e,p\}, \quad (20)$$

where $n_e(t)$ and $n_i(t)$ are numbers of electrons and ions surviving to time t, and $n_p(t)$ is the free-positron survival probability. $G_j(r,t)$ is the green function of diffusion equation:

$$G_j(r,t) = \frac{\exp[-r^2/(4D_j t + a_j^2)]}{[\pi(4D_j t + a_j^2)]^{3/2}}, \qquad (21)$$

where, as before, $D_i \equiv D_e \equiv D_{amb} \ll D_p$ and $a_i \approx a_e \approx a_{bl}$. Substituting Eqs. (20) and (21) into Eq. (17) and integrating over whole space, we obtain a much simpler system of ordinary differential equations on $n_j(t)$:

$$\begin{aligned}
\dot{n}_i &= -k_{ie} n_e n_i / V_{ie} - n_i/\tau_i, & n_i(0) &= n_0, \\
\dot{n}_e &= -k_{ie} n_e n_i / V_{ie} - k_{ep} n_e n_p / V_{ep} - n_e/\tau_e, & n_e(0) &= \varphi_e n_0, \quad (22) \\
\dot{n}_p &= -k_{ep} n_e n_p / V_{ep} - n_p/\tau_p, & n_p(0) &= \varphi_p,
\end{aligned}$$

where

$$\frac{1}{V_{jq}(t)} = \int G_j G_q d^3 r = \frac{1}{V_{jq}^0 (1 + t/\tau_{jq})^{3/2}} \qquad (23)$$

$$V_{jq}^0 = [\pi(a_j^2 + a_q^2)]^{3/2}, \qquad \tau_{jq} = \frac{a_j^2 + a_q^2}{4(D_j + D_q)}, \qquad j,q = \{i,e,p\}.$$

Obviously, it is possible to omit the term $k_{ep} n_e n_p / V_{ep}$ in Eq. (22), because it has a negligible effect on the disappearance of the electrons.

From Eqs. (19–21) it follows that

$$P_{Ps} = \int_0^\infty \frac{k_{ep} n_e n_p dt}{V_{ep}^0 (1 + t/\tau_{ep})^{3/2}}, \qquad (24)$$

$$V_{ep}^0 = [\pi(a_{bl}^2 + a_p^2)]^{3/2}, \qquad \tau_{ep} = \frac{a_{bl}^2 + a_p^2}{4(D_{amb} + D_p)}.$$

In the next section, some examples of application of this approach to aqueous solutions will be given.

5.6 Application of the blob model to aqueous solutions

5.6.1 Calculation of **Ps**, **H$_2$** and hydrated electron yields

In spite of some important distinctions in the structure of the terminal positron blob and spurs, ionization columns and blobs of δ-electrons, where reactions of radiolytic hydrogen (H$_2$) formation take place, these processes have much in common and it is natural to describe them in the framework of the unified approach. According to [20, 48], H_2-formation proceeds on a picosecond timescale via reaction

$$e^- + H_2O^+ \xrightarrow{k_{eh}} H_2O^* \xrightarrow{+H_2O} \begin{cases} H_2 + 2OH \\ H + OH + H_2O \\ 2H_2O \end{cases} \quad (25)$$

which is similar to Eq. (12). Joint consideration of these processes, as well as formation of the yield of hydrated electrons (e^-_{aq}), allow us to decrease the uncertainty of adjustable parameters involved and make more reliable conclusions.

During the last 50 years a great deal of experimental data has been accumulated on the yields of Ps, H_2 and e^-_{aq} in various aqueous solutions. Here, we apply the blob model to calculate these yields in solutions of NO_3^-, H_2O_2, $HClO_4$, Cl^-, Br^-, I^- and F^-. They were selected to embrace the greatest possible variety of solute properties with respect to intratrack reactive species.

Formation probabilities of H_2, e^-_{aq} and Ps are determined as integrals from the corresponding terms in Eq. (22):

$$P_{H_2O^*} = \int_0^\infty \frac{k_{ei} n_i n_e dt}{n_0 V_{ie}}, \quad P_{e^-_{aq}} = \int_0^\infty \frac{n_e dt}{n_0 \tau_e^s}, \quad (26)$$

$$P_{Ps} = \int_0^\infty \frac{k_{ep} n_e n_p dt}{V_{ep}}.$$

In the calculations we used an approximate relationship, $P_{H_2O^*} \approx 3 P_{H_2}$, between the formation probability of excited water molecules and molecular hydrogen [4, 48].

Joint fit of the experimental yields of Ps, H_2 and e^-_{aq} is shown in Figs. 5.9 and 5.10. Original experimental data are taken from [41, 42]; other papers are quoted in [5]. Electron scavengers NO_3^-, H_2O_2 and $HClO_4$ are characterized by rate constants, k_{eS}, with respect to presolvated intrablob electrons, while the halogen anions, being the scavengers of the hot positron as well as radical-cations $H_2O^{+\cdot}$, are described by four parameters: c_{37}^+, F_+, k_{pS} and k_{iS}. In calculations we do not use particular numbers for n_0, a_{bl}, a_p and D_p. Obtained values are shown in Table 5.1. We also calculated that $k_{ie} n_0 \tau_e^s / V_{ie} = 0.14$ and $k_{ep} n_0 \tau_e^s / V_{ep} = 0.49$. Roughly, statistical accuracy of all the obtained numbers is 10–20%. We also fixed the rate of

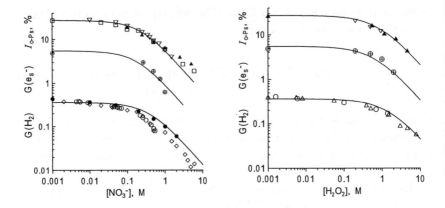

Fig. 5.9 Concentration dependencies of the o-Ps component I_3 in the lifetime spectra, yields of e_s^-, $G_{e_{aq}^-} = G_{iep} \cdot P_{e_{aq}^-}$, and radiolytic hydrogen, $G_{H_2} = G_{iep} \cdot P_{H_2}/3$, in aqueous solutions containing NO_3^- ions and hydrogen peroxide, H_2O_2. Here, $G_{iep} \approx 6.0$ (fitted value) is the initial yield of ion–electron pairs (the ionization yield). Solid lines are calculated using Eqs. (22-24, 26), with parameters presented in Table 5.1.

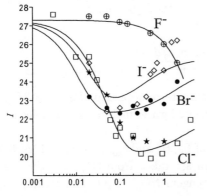

Fig. 5.10 Concentration dependencies of the o-Ps intensity in aqueous solutions containing halogen anions: $\oplus$ – F^-, $\diamond$ – I^-, ● – Br^-, $\star$ – Cl^-, $\square$ – Cl^-. Solid lines are theoretical calculations.

ion-molecule reaction to the electron hydration rate ($\tau_{imr} = \tau_e^s$). Positron solvation time is found to be longer than τ_e^s.

5.6.2 Discussion of the results

(1) There is a quantitative agreement between the obtained relative rate constants of halogen anions with $H_2O^{+\cdot}$ $k(H_2O^{+\cdot} + Cl^-) : k(H_2O^{+\cdot} + Br^-) : k(H_2O^{+\cdot} + I^-) \approx 1 : 2.4 : 5.1$ and analogous ratios $(0.3 - 0.5) : (0.9 - 1.1) : (1.8 - 2.2)$ determined in glassy solutions frozen at 77 K [44].

(2) Ratios of k_{eS} for NO_3^-, H_2O_2 and $HClO_4$ (5.0 : 1.5 : 0.27; Table 5.1, line 1) drastically differ from the ratios of the rate constants of the same solutes with e_{aq}^- (20 : 13 : 24 $M^{-1}ns^{-1}$ [43])[†], but correlate much better than corresponding ratios of experimental values of c_{37}^- (2.4 : 0.7 : ($\lesssim$ 0.1) [43]). It supports the concept that just the presolvated intratrack electron, not a solvated one, is a common precursor of Ps and H_2.

(3) Using $\tau_e^s = 0.3$ ps from the products $\tau_e^s k_{eS}$ we straightforwardly calculate absolute rate constants of NO_3^-, H_2O_2 and $HClO_4$ with quasi-free electrons, namely $2 \cdot 10^{13}$, $5 \cdot 10^{12}$ and 10^{12} $M^{-1}s^{-1}$, respectively. These values are comparable with the corresponding rate constants in those hydrocarbons where excess electrons are in a quasi-free state only [2].

(4) If we roughly adopt $a_p \approx 2a_{bl}$, from the values $k_{ie}n_0\tau_e^s/V_{ie} = 0.14$ and $k_{ep}n_0\tau_e^s/V_{ep} = 0.49$ we obtain the ratio k_{ei}/k_{ep} (small, about 0.1), which is in qualitative agreement with the estimation of the same ratio in hydrocarbons [3].

Correlation between Ps inhibition constants and $1/c_{37}^-$ stressed earlier in [20] is now proved by a large number of data [45, 46, 47, 48]. This presents strong experimental evidence that just a presolvated electron is the main precursor of the Ps.

5.7 Conclusion

In this chapter an attempt has been made to compare two mechanisms of Ps formation, the Ore mechanism and the recombination one. They have a clear-cut difference. The first method suggests that a hot positron picks up an electron from a molecule and immediately forms Ps just with that same electron. According to the recombination mechanism, a fast positron loses all its energy on ionizations and vibrational excitations and, getting thermalized, picks up one of the knocked-out intratrack electrons. Obviously,

---

[†]Really the last value $2.4 \cdot 10^{10}$ $M^{-1}s^{-1}$ is the rate constant of the reaction $e_{aq}^- + H_{aq}^+ \rightarrow H$. Note that $k(e_{aq}^- + ClO_4^-) < 10^5$ $M^{-1}s^{-1}$.

Table 5.1 Reaction parameters of intratrack primary species obtained from application of the blob model to description of Ps, e_{aq}^- and radiolytic hydrogen yields in aqueous solutions.

	NO_3^-	H_2O_2	$HClO_4$	Cl^-	Br^-	I^-	F^-
$\tau_e^s k_{eS}$, M^{-1}	5.0	1.5	0.27	-	-	-	-
$\tau_e^s k_{pS}$, M^{-1}	-	-	-	-	-	-	0.08
$\tau_e^s k_{iS}$, M^{-1}	-	-	-	1.0	2.4	5.1	1.5^a
F_+	-	-	-	0.26	0.19	0.16	-
c_{37}^+, M	-	-	-	0.05	0.01	0.02	-

$a)$ This value was fixed to the experimental value [44].

the energetics of these processes is quite different. It was demonstrated that a lack of free space in the the condensed phase makes the Ore process much less effective than in the gas phase. In the bulk, cores of surrounding molecules (atoms) repel e^+ and e^-, increasing the zero-point energy of the positronium, and surrounding electrons screen e^+-e^- Coulombic attraction, which leads to a decrease of the positronium binding energy. As a result, the Ore gap narrows and may disappear completely. Even for polymers having a rather large free-volume fraction, this argument seems to be valid.

Further developments in this field would probably be forthcoming with more precise studies of the energetics of Ps formation, and measurements of the work functions for e^+ and Ps using low-energy positron beams. Better understanding may come from studies of Ps formation at different temperatures and external electric fields (determination of e^+ mobility, investigation of the positron-blob interaction, e^+ thermalization parameters and its spatial distribution).

Radiation chemistry and Ps chemistry differ in the objects and processes they study. Being a probe of Ps chemistry, the positron delivers information about processes near and inside its terminal blob, while radiation chemistry primarily investigates the processes in isolated spurs.

The theoretical base of the spur process is Onsager's theory of the geminate pair recombination. Contrary to this, the blob model is most appropriate for consideration of early radiation–chemical processes in multiparticle track entities, such as blobs and ionization columns. The main distinction between the spur and blob comes from the large difference in the initial number of ion–electron pairs they contain.

Nevertheless, it seems promising to investigate the problem of Ps formation jointly with the similar problem of intratrack radiolytic hydrogen formation. Despite some distinctions in conditions inside the terminal positron blob and spurs, blobs and ionization columns where reactions of formation of Ps and H_2 occur, these processes have much in common. Their joint description diminishes uncertainty of the parameters involved in the model and leads to more reliable conclusions.

Acknowledgments

This work was supported by the Russian Foundation of Basic Research (Grants 00-03-32918, 01-03-32786).

5.8 Problems

1. Assuming that the distributions of the ion and electron in the spur are Gaussian with dispersion a_{sp}^2 relative to the origin of the coordinates, calculate the mean square ion–electron separation $\overline{(\mathbf{r_i} - \mathbf{r_e})^2}$. Make a numerical estimation in case of water, if $a_{sp} \approx 30$ Å.

2. Starting from the Poisson equation and assuming Gaussian spatial distributions for ions and intrablob electrons with the dispersions a_i and a_e, prove that $a_e - a_i \ll a_i$ and estimate $\Delta a \equiv a_e - a_i$ numerically. Assume the initial number of ion–electron pairs in the blob are $n_0 \approx 30$ and $a_i \approx 40$ Å and $\varepsilon -$ is the dielectric permitivity of the medium. Note that at $r \approx a_i$ out-diffusion flux of electrons is compensated for by their drift in the electric field of the ions (quasi-equilibrium condition).

3. Write equations describing the ambipolar diffusion expansion of the blob species. Construct the solution to these equations assuming that the distribution of electrons is close enough to that of ions. Prove that expansion of the blob species proceeds with the diffusion coefficient equal to $2D_i$, where D_i is the diffusion coefficient of the ions.

4. Calculate electrostatic potential of the blob if the distributions of the blob species are the same as in Problem 1.

5. Prove Eq. (13) assuming that before averaging over initial energy W_0 of the knocked-out electrons, the probability $\varphi_e(W)$ of escaping capture by S during electron slowing-down from energy W to $W - dW$ obeys the following equation

$$\varphi_e(W - dW) = \varphi_e(W)(1 - c_S\sigma_L(W)dx), \qquad dx = dW/|dW/dx|_{vib},$$

where $\sigma_L(W)$ is its e$^-$-capture cross section, $|dW/dx|_{vib}$ is e$^-$ linear energy transfer, and c_S is the concentration of the scavenger. Neglect the effect of the decreasing of c_S inside the blob due to the reaction between S and electrons.

References

[1] H.G. Paretzke *Kinetics of Nonhomogeneous Processes*, G. R. Freeman, Ed., John Wiley & Sons, New York (1987), pp. 89-170.
[2] *Handbook of Radiation Chemistry*, T. Tabata, Y. Ito, S. Tagawa, Eds. CRC Press, Boca Raton, New York (1991).
[3] S.V. Stepanov, V.M. Byakov *J. Chem. Phys.* (March, 2002).
[4] V.M. Byakov, F.G. Nichiporov *Intratrack Chemical Processes*, in Russian, Moscow *Energoatomizdat* (1985).
[5] V.M. Byakov, S.V. Stepanov *J. Radioanal. Nucl. Chem.*, **210(2)** 371 (1996).
[6] S.V. Stepanov *Radiat. Phys. Chem.* **46** 29 (1995).
[7] S.V. Stepanov, V. M. Byakov, Cai-Lin Wang, Y. Kobayashi, K. Hirata *Mater. Sci. Forum*, **363-365** 392 (2001).
[8] P. Debye, E.Hückel *Phys. Ztschr.*, **24** 185 (1923).
[9] A. Ore Univ. of Bergen *Aarb. Naturvit. rekke* **9** (1949).
[10] S.V. Stepanov, V.M. Byakov, O.P. Stepanova *Russ. Phys. Chem.* (English transl.), **74** Suppl. 1, 65 (2000).
[11] V.M. Byakov, S.V. Stepanov *Radiat. Phys. Chem.*, **58(5-6)** 687 (2000).
[12] I.T. Iakubov, A.G. Khrapak *Rep. Prog. Phys.* **45** 697 (1982).
[13] B.E. Springett, J. Jortner, M. Choen *J. Chem.Phys.* **48(6)** 2720 (1968).
[14] E.M. Gullikson, A.P. Mills, Jr., E.G. McRae *Phys. Rev. B*, **37(1)** 588 (1988).
[15] H. Cao, R. Zhang, J.-P. Yuan, et al. *J. Phys.: Condens. Matter* **10** 10429 (1998).
[16] O.E. Mogensen *Positron Annihilation in Chemistry* Springer-Verlag, Berlin, (1995).
[17] Y. Ito in *Positron and Positronium Chemistry*, D. M. Schrader, Y. C. Jean, Eds., Elsevier, Amsterdam (1988) p. 120.
[18] O.E. Mogensen *J. Chem. Phys.* **60** 998 (1974).
[19] V.M. Byakov, V.I. Goldanskii, V.P. Shantarovich *Doklady Akademii Nauk, SSSR* **219** 633 (1974) [*Doklady Phys. Chem.* **219** 1090 (1974)].
[20] V.M. Byakov *Int. J. Radiat. Phys. Chem.* **8** 283 (1976).
[21] S.J. Tao *Appl. Phys.* **10** 67 (1976).
[22] V.M. Byakov, V.I. Grafutin, O.V. Koldaeva *Khimiya Vys. Energii* **17** 506 (1983) [*High Energy Chemistry* **17** 396 (1983)].
[23] Y. Ito *J. Radioanal. Nucl. Chem.*, **210** 327 (1996).
[24] Y. Ito *Mater. Sci. Forum* **255-257** 104 (1997).
[25] I. Billard, T. Goulet, J.-P. Jay-Gerin, A. Bonnenfant *J. Chem. Phys.*, **108** 2408 (1998).
[26] S.V. Stepanov, Cai-Lin Wang, Y. Kobayashi, et al. *Radiat. Phys. Chem.* **58** 403 (2000).
[27] C.L. Wang, Y. Kobayashi, W. Zheng, et al. *Phys. Rev. B* **63** 064204 (2001).
[28] C.L. Wang, Y. Kobayashi, K. Hirata *Radiat. Phys. Chem.* **58** 451 (2000).
[29] V.M. Byakov *Journal de Physique IV*, Colloque C4, Suppl., **3** 85 (1993).

[30] V.M. Byakov, V.I.Grafutin, O.V. Koldaeva, et. al. *Int. J. Radiat. Chem.*, **10** 239 (1977).
[31] T. Sumiyoshi, M. Katayama *Chem. Lett.*, **93** 1021 (1982).
[32] S.V. Stepanov *Journal de Physique IV*, Colloque C4, Suppl., **3** 41 (1993).
[33] W.T. Naff, R.N. Compton, C.D. Cooper *J. Chem. Phys.* **54** 212 (1971).
[34] A.O. Allen, Th.E. Gangwer, R.A. Holroyd *J. Phys. Chem.*, **79** 25 (1975).
[35] A. Henglein *Can. J. Chem.* **55**, 2112 (1977).
[36] O.A. Anisimov, Yu.N. Molin Proc. 4th Int. Conf. Positron Annihilation, Helsinger, Danemark, 1976, G31.
[37] Y. Katsumura, Y. Tabata, S. Tagawa *Radiat. Phys. Chem.*, **19** 267 (1982).
[38] O.A. Anisimov, V.L. Bizyaev, S.V. Vocel, Yu.N. Molin *Chem. Phys. Lett.*, **76** 273 (1980).
[39] R.G. Kokilashvili, V.V. Eletskii, Yu.V. Pleskov *Elektrokhimiya*, **20** 1075 (1984).
[40] K. Hiraoka, M. Nara *Bull. Chem. Soc. Jpn.*, **54** 3317 (1981).
[41] G. Duplatre, et al. *Radiat. Phys. Chem.*, **11** 199 (1978).
[42] C.D. Beling, F.A. Smith *Chem. Phys.*, **81** 243 (1983).
[43] K.Y. Lam, J.W. Hunt *Radiat. Phys. Chem.*, **7** 317 (1975).
[44] V.I. Belevskii, S.I. Belopushkin, L.T. Bugaenko *Khimiya Vys. Energii*, **14**, 315 (1980).
[45] J.-Ch. Abbe, G. Duplatre, A.G. Maggock, et al. *J. Inorg. Nucl. Chem.*, **43** 2603 (1981).
[46] G. Duplatre, Ch.D. Jonah *Radiat. Phys. Chem.*, **24** 557 (1985).
[47] V.M. Byakov, S.V. Stepanov, Y. Katsumura, Y. Ito *Mater. Sci. Forum*, **175-178** 659 (1995).
[48] V.M. Byakov. V.I. Grafutin *Radiat.Phys.Chem.*, **28** 1 (1986).

Answers to the problems

1. $\overline{(\mathbf{r_i} - \mathbf{r_e})^2} = 3a_{sp}^2$.
2. $\Delta a \approx \frac{\sqrt{\pi} e}{2} \frac{a_{bl}^2}{n_0 r_c} \lesssim 1$ Å, where $a_{bl} \approx a_i \approx a_e$, $e = 2.718...$, $r_c = \frac{e^2}{\varepsilon T}$.
4. $\varphi(r) = \frac{e}{\varepsilon r_c} \exp(1 - r^2/a_{bl}^2)$.

Chapter 6
Positrons and Positronium in the Gas Phase

D. M. SCHRADER
*Chemistry Department, Marquette University, P.O. Box 1881,
Milwaukee, WI 53201-1881, USA
david.schrader@marquette.edu*

6.1 Introduction

In this chapter we consider the dynamics of positrons and positronium atoms interacting with atoms and molecules in dilute gases. Our interest is in impact energies less than 100 eV, where the chemistry is interesting. Our emphasis here is on phenomenology rather than theoretical methods or instrumental techniques, both of which are discussed in other chapters.

We do not discuss elastic scattering here because it is not particularly interesting to most chemists, and it has been reviewed recently [1]. Comparisons of electron and positron scattering is treated only briefly here because it is the subject of a recent comprehensive review [2]. We limit the present discussion to topics of most interest to chemists. These inevitably involve molecular (not atomic) targets, and are concerned in particular with electronic (i.e., orbital) structure, vibrational effects, bond breaking, and the formation of compounds that contain positrons.

The topics discussed here reflect their significance, current interest, limitations of space in this book, and the bias of the author. Other authors would have made different selections. Omissions do not reflect adversely.

6.2 Overview of positron scattering

The positron has the charge of a proton and the mass of an electron. One can fruitfully compare and contrast positron scattering with both proton and electron scattering.

It is far easier to do experiments with electrons and protons than with positrons because the former are much more plentiful. Electrons are copiously produced simply by heating a suitable metallic filament, and protons are easily produced by charge-stripping H_2^+. However, the production of positrons requires much more expensive and elaborate methods: either pair production or the decay of a proton-rich nuclide. Electron and proton scattering experiments usually involve the measurements of intensities of currents of scattered incident particles or of ionic products. Positron scattering, however, is done with far fewer incident particles, so much so that the data is gathered by counting individual scattering events.

From a theoretical point of view, many, starting with Massey [3], have observed that the static and polarization parts of the potential generated by a neutral target have the same sign for an incident electron but not for a positron or a proton, and that this tends to make accurate calculations on the positive particles more exacting. It is also true that the polarization potentials for all three particles are the same at long range ($\sim -\alpha/2r^4$) where α is the electric dipole polarizability) but different at short range. This is true because virtual positronium (hydrogen) contributes to the polarization potential for the positron (proton) but not for the electron. All three particles are kept from the nucleus: the positron and proton, by Coulomb repulsion and the electron, by exclusion. Thus all three incident particles have their greatest perturbing effect on a target in its valence region. The positron has a greater effect and a larger polarization potential (in absolute terms) at short range compared to the electron because of the effect of virtual positronium, and the proton has the greatest effect. An accurate theoretical description of positron scattering is impossible without directly addressing the correlation of the positron motion with that of any nearby electron. This creates a formidable computational problem that, after many years of work, is still the object of current research. Calculations of proton scattering, however, are much simpler owing to the existence of computer programs that can calculate, for arbitrary positions of fixed nuclei, electronic wave functions of very high accuracy. For nonrelativisitic energies of interest here, the velocities of protons are about 40 – 50 times smaller than those of target valence electrons, so the adiabatic approximation makes sense, and the use of existing program packages is appropriate.

Incident positrons and protons can each capture a target electron to create a molecular ion and a neutral leaving species, a positronium atom or a hydrogen atom. This is possible for incident energies greater than the ionization potential of the target less the binding energy of Ps (6.80 eV) or H (13.60 eV). If the target ionization energy is less than the binding energy of Ps or H, then the rearrangement is possible at all energies. The effect on the scattering of the leaving neutral atoms depends primarily on the polarizability of the neutral. Since the polarizability of Ps is eight times that of H (see Chapter 1), the influence in the case of leaving Ps is profound. Incident electrons offer no such interesting phenomenology.

6.3 Annihilation

In this section we consider annihilation at two distinct levels. First we examine the essence of the annihilation process itself, starting from one electron and one positron and ending with two photons. Next we consider the process that precedes it, in which the annihilating pair meet each other under the influence of a many-electron environment and the manifold of energy levels provided by an atom or molecule. Finally the relaxation of the post-annihilation system is discussed.

6.3.1 *The primitive annihilation event*

The Feynman diagram for the simplest annihilation event shows that annihilation is possible when the two particles are $\Delta x \sim \hbar/mc \sim 10^{-12.5}$ m apart, and that the duration of the event is $\Delta t \sim \hbar/mc^2 \sim 10^{-21}$ s. The distance is the geometric mean of nuclear and atomic dimensions, which is probably not significant. The distance is so much smaller than electronic wave functions that it may be assumed to be zero in computations of annihilation rates. The time is so short that, during it, a valence electron in a typical atom or molecule moves a distance of only $\sim a_0/10^4$, so that a spectator electron can be assumed to be stationary and the annihilating electron can be assumed to disappear in zero time. Thus the calculation of annihilation rates requires the evaluation of expectation values of the Dirac delta function, and the relaxation of the daughter system (post-annihilation remnant) can be understood with the aid of the sudden approximation [4]. These are both relatively simple computations, providing an accurate wave function is available.

The duration of the annihilation event should not be confused with the positron lifetime, which is the time required for the electron and positron to get to within $\sim 10^{-12.5}$ m from each other. In a condensed phase of ordinary density, the lifetime of a positron is ~ 1 ns. In other words, it takes a long time ($\sim 10^{-9}$ s) for the annihilating particles to find each other, and when they do the annihilation event is very short ($\sim 10^{-21}$ s).

6.3.2 The response of the target

The annihilation rate of a positron in an electronic medium is usually expressed as

$$\lambda = \sigma_a v n = \pi r_0^2 c n Z_{eff} \qquad (6.1)$$

where σ_a is the annihilation cross section, v is the velocity of the incident positron, n is the number density of target species (atoms or molecules) in the annihilation chamber, r_0 is the classical radius of the electron, and Z_{eff} is a parameter that is defined by eqn. (6.1). It is dimensionless and is sometimes confused with the electron-positron contact density ρ_{ep}, which has units l^{-3}:

$$\rho_{ep} = \langle \Psi | \sum_\mu \delta(r_\mu - r_p) | \Psi \rangle \qquad (6.2)$$

where Ψ is the wave function for the positron-atom or -molecule system and the sum is over the electrons. In most molecular environments, a positron experiences an electron density similar to that of positronium, $1/8\pi a_0^3 \sim 0.04 a_0^{-3}$. The intrinsic annihilation rate of a positron-molecule system in a stationary state is:

$$\frac{\Gamma_a}{\hbar} = \pi r_0^2 c \rho_{ep} \qquad (6.3)$$

where Γ_a, the annihilation width, has a value of ~ 1 μeV for most molecular systems.

Z_{eff} was originally introduced to represent the effective number of electrons available to the positron for annihilation, but it is often very much larger than the actual number of electrons Z because it is strongly influenced by the polarizability of the target and by the existence of bound states of the positron-target system. After decades of confusion and conjecture on the nature of annihilation on atoms and molecules, Gribakin has recently provided a physically sensible theoretical framework for the process

[5]. Gribakin's treatment, presented as an evaluation of Z_{eff}, is quite approximate and entails the introduction of several incalculable parameters, so it is not predictive. However, it does succeed in rationalizing the diverse experimental data, which has defied a satisfactory interpretation until now. It is an important and useful advance, and we discuss it at length here.

Gribakin identifies two distinct contributions to the mechanism of annihilation – direct, which does not involve resonant capture of the incident positron, and resonant, which does. The observed annihilation rate and the parameter Z_{eff} are in general the sum of contributions from both mechanisms:

$$Z_{eff} = Z_{eff}^{(\text{dir})} + Z_{eff}^{(\text{res})} \qquad (6.4)$$

6.3.2.1 Direct annihilation

Direct annihilation takes place at low energies as the positron is in the midst of an otherwise elastic scattering event. Gribakin gives an approximate expression for $Z_{eff}^{(\text{dir})}$ [5]:

$$Z_{eff}^{(\text{dir})} \sim 4\pi \rho_{ep}\, \delta R_a \left(R_a^2 + \frac{\sigma_{el}}{4\pi} + 2R_a \Re(f_0) \right) \qquad (6.5)$$

δR_a is the thickness of a thin shell around the target that represents the region of significant overlap of electron and positron density at annihilation time, R_a is the distance of the overlap region from the center of the target, σ_{el} is the elastic scattering cross section, and f_0 is the spherically symmetric part of the scattering amplitude. The three terms on the right above are due, respectively, to the incoming plane wave, the scattered wave, and the interference between them. If the cross section σ_{el} is small or zero, $Z_{eff}^{(\text{dir})}$ is in the range 1 – 10, similar to the actual number of electrons available for annihilation, which is the normal "fly-by" value. If resonant capture is not involved, the measured Z_{eff} is just $Z_{eff}^{(\text{dir})}$. Examples of systems that have measured values of Z_{eff} within a factor of ~ 2 of Z, the actual number of electrons in the systems are the noble gases (except Xenon), some common diatomics (N_2, O_2, CO, and NO), CO_2, and SF_6. For the lighter noble gases, Z_{eff} ranges from $0.6Z$ for Ne to $2.5Z$ for Kr [6].

The low energy elastic scattering cross section may become quite large if there is a weakly bound state or a nearby virtual state. For a nonpolarizable target the cross section becomes [4]

$$\sigma_{el} \approx \frac{2\pi}{\varepsilon + \frac{1}{2}k^2} \tag{6.6}$$

at short range, where ε is the depth of the bound state or the height of the virtual state and k is the momentum of the positron. If the target is polarizable, then ε above is multiplied by the factor [7]

$$\left[1 - \frac{1}{3}\alpha\left(\frac{\pi k}{a} + 4k^2 \ln(\frac{1}{4}Ck\sqrt{\alpha})\right)\right]^2 \tag{6.7}$$

where a is the scattering length $(= \pm 1/\sqrt{2\varepsilon})$ and C is a dimensionless positive constant related to the s-wave phase shift. Now $Z_{eff}^{(dir)}$ can be as large as $\sim 10^3$ for room temperature positrons [5]. This is the largest value of $Z_{eff}^{(dir)}$ for direct annihilation. Systems that annihilate in this way have Z_{eff} values that satisfy:

$$2Z < Z_{eff} < 10^3. \tag{6.8}$$

Systems in this group include Xe (Z_{eff} = 400), H_2 (15), H_2O (320), methane (140), and ethane (700). Many more examples are listed in the compendium of Iwata, et al. [6].

6.3.2.2 Resonant annihilation

Resonant annihilation relies on the capture of an incident positron by the target. The positron-target system must have an excitable state that matches the incident energy of the positron. Capture into such an excited state implies the existence of a true positron-target bound state beneath it; i.e., with an energy below the ground state of the target. Without such a true bound state, the positron would just be re-emitted and the encounter would be indistinguishable from elastic scattering from a target with a virtual state, which we considered immediately above. This necessary condition is illustrated in Fig. 6.1.

A number of neutral atoms have low-lying virtual states without an accompanying true bound state underneath, and some atoms have true bound states but no low-lying capturing resonant state. Thus, resonant annihilation appears to be limited to molecules, and only larger molecules have enough degrees of freedom to trap positrons for a significant length of

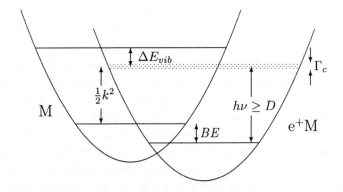

Fig. 6.1 A necessary condition for resonant annihilation. The capture width Γ_c is indicated. It is not to be confused with radiative linewidths, Γ_{rad}, which are generally much smaller. See the text.

time. The resonances are vibrational Feshbach resonances, which were first discussed by Smith and Paul [8]. Rotational levels have not been considered as candidates for capture.

When a capturing state is depopulated by de-excitation to the ground state, an internal energy transfer to another degree of freedom may take place on the time scale of molecular vibrations, $10^{-14} - 10^{-15}$ s, very short compared to annihilation lifetimes. Some longer time τ_c is consumed as the vibrational quantum circulates around the other vibrational degrees of freedom of the system. During this interval, the positron is immersed in the electronic environment of the target system. One of two things ultimately happens: Either the positron annihilates with one of the target electrons, or the vibrational quantum reappears in the capturing state and the positron re-emits. When the time required for this circulation process approaches the annihilation lifetime of the positron, very large values of Z_{eff} can result.

The radiative lifetime, τ_{rad}, of the capturing state of the e^+M system may be in the ms region, giving a radiation width $\sim \hbar/\tau_{rad}$ on the order of feV's. The capture width Γ_c is much larger, and can greatly exceed the annihilation width defined in Eq. (6.3).

From Fig. 6.1 it is clear that the trapping condition can be expressed as:

$$\frac{1}{2}k^2 = h\nu - \Delta E_{vib} \pm \frac{1}{2}\Gamma_c \qquad (6.9)$$

where the terms are, respectively, the incident energy of the positron, the vibrational spacing of the target molecule, the shift of the vibrational levels, and the capture width. The vibrational shift is taken to be a positive quantity. Figure 6.1 shows that, under the assumption that the vibrational spacings are the same for the two molecular systems, the binding energy of the positron to the target molecule ("BE" of chapter 2) is ΔE_{vib}.

As a positron approaches a molecule that has a resonant level, it sees a window of entrance into the resonant manifold. The probability of hitting that window is defined as Γ_c/D, where D is the effective spacing of vibrational levels. The probability of sticking gives a factor of the inverse of the incident energy. D is $h\nu$ in Fig. 6.1, the spacing between resonant states, if only one mode participates in positron capture. Otherwise, D may be smaller than $h\nu$. Gribakin [5] gives the capture cross section as:

$$\sigma_c = \frac{2\pi^2}{k^2} \frac{\Gamma_c}{D}. \tag{6.10}$$

The probability of annihilation is proportional to the probability of capture σ_c and to the ratio of the annihilation width Γ_a (defined in Eq. (6.3)) to the total width:

$$\sigma_a = \frac{2\pi^2}{k^2} \frac{1}{D} \frac{\Gamma_a \Gamma_c}{\Gamma_a + \Gamma_c}. \tag{6.11}$$

Substituting for σ_a from Eq. (6.1), and for Γ_a in the numerator on the right hand side above from Eq. (6.3) gives, in atomic units:

$$Z_{eff}^{(res)} = \frac{2\pi^2}{k} \frac{\rho_{ep}}{D} \frac{\Gamma_c}{\Gamma_c + \Gamma_a}. \tag{6.12}$$

For a capture width Γ_c much greater than the tiny annihilation width Γ_a, the right hand side of Eq. (6.12) becomes independent of both widths:

$$Z_{eff}^{(res)} \to \frac{2\pi^2}{k} \rho_{ep} \rho_r(E_v), \tag{6.13}$$

where we replace D^{-1} by the density of resonant states at the vibrational excitation energy, $\rho_r(E_v)$. Gribakin points out that this quantity increases rapidly with the size of the molecule. It is proportional to $(N_v)^{BE/h\nu}$ where N_v is the number of vibrational modes that participate.

According to Gribakin [5], Z_{eff} does not rise indefinitely with Z, but rather is limited by the relationship between the spacing of different resonant states and their widths: their widths cannot be larger than their spacings. Also, if the incident positron energy rises much above $h\nu$, additional re-emission channels appear that are accompanied by vibrational excitation of the target. Thus at high positron energies, $Z_{eff}^{(res)}$ does not grow proportionately to the density of states $\rho_r(E_v)$. Gribakin estimates that the largest possible value of Z_{eff} at room temperature, which we call the size saturation value, is $\sim 5 \times 10^7$. The largest value of Z_{eff} yet seen is 7.5×10^6 for sebacic acid demethyl esther, $C_{12}H_{22}O_4$.

The theoretical treatment has just begun. The coupling of vibrational and electronic degrees of freedom presents a daunting task to the quantum theoretician. Ab initio calculations are few; most attention has been focussed thus far on empirical investigations [9] and to calculations on simple systems, such as diatomics [10]. Correlations between measured Z_{eff} and elastic scattering cross sections on the one hand, and various chemical features on the other, are being investigated.

In an ambitious calculation on ethylene, da Silva, et al. [11], using 23,000 configurations in a CI calculation, found Z_{eff} to be about 1000 for thermal incident positrons, which compares very well with the measured value, 1200 [6]. The calculation did not contain any vibrational coupling, only correlation between light particles. A recent revision of the calculation resulted in a new value being given: $Z_{eff} \sim 60$ [12]. This constitutes a serious disagreement, even with a lower experimental value of 700 [13], and appears to indicate the importance of vibrational coupling for this molecule.

6.3.3 Annihilation on specific molecular sites

Measurements of the Doppler broadening of the annihilation radiation produced by various molecules has been related to annihilation at specific sites within molecules by Iwata, et al. [15]. From the observed γ-ray spectra, the line width of the dominate peak, which comes from valence electrons, was extracted. Thus, for each molecule there is a single measured quantity, the line width. For a series of hydrocarbons, the observed line widths were found to be linear in the fraction of electrons in C–C (or C–H) bonds. Each type of bond was assumed to contain two electrons. From a linear fit of this data, line widths for the C–C and C–H bonds were extracted and found to be 2.06 and 2.42 keV, respectively. These agree reasonably with an old theoretical estimate in which the positron density was assumed to be constant over the molecule [16].

Next these authors measured line widths for a series of fluorohydrocarbons, $C_nH_lF_m$. The γ-ray spectrum for each was analyzed into components from the hydrogen and fluorine analogs, C_nH_{l+m} and C_nF_{l+m}. The relative weight of the C_nF_{l+m} component was interpreted to be the fraction of annihilation on fluorine in C_nF_{l+m}. This fraction was then normalized by the fraction of valence electrons on F (assumed to be 8) in the molecule to account for annihilation in the C–C bond. The normalized fractions of annihilation on F for a large number of fluorohydrocarbons was found to be equal to the fraction of valence electrons on F in each molecule, to a good approximation. The conclusion of these authors is that annihilation in these molecules takes place with equal probability on any valence electron regardless of what atom it resides on. This is counterintuitive.

The data analysis chosen by these authors departs from that used by Mogensen and others [17, 18], who fit each 1D angular correlation curve to a set of Gaussian functions. The minimum number of Gaussians is used to achieve a good fit, and the width of each is optimized. The momentum components of each γ-ray spectrum are then interpreted in terms of annihilation of core vs. valence electrons without appeal to a preconceived chemical model. The experiment-theory connection can be made if one has an adequate wave function in hand, for then the Doppler profiles or angular correlation curves can be calculated and compared to those measured.

Iwata, et al., use a model in which the electrons are assigned to categories, some of which are bond-related (C–C and C–H) and others, atomic (F). Assignment of electrons to various parts of a molecule is a very old practice in chemistry [19], but one that contains some unavoidable arbitrariness. One wonders whether the authors' conclusion would stand if the interpretation of the data were more rigorous quantum mechanically.

6.3.4 *Relaxation after annihilation*

An interesting old puzzle that has not attracted much attention lately has to do with the relaxation of the post-annihilation system, also called the daughter system. For neutral target molecules, the daughter is necessarily a cation. Extensive fragmentation of several daughter cations has been observed for incident positron energies far below the thermodynamic threshold for fragmentation and ionization [20, 21].

Xu, et al., [21] studied dodecane ($C_{12}H_{26}$) and some alkyl-substituted benzene molecules most extensively. For dodecane, ionization potential is

9.4 eV, so the threshold for Ps formation is 2.6 eV, and the threshold for breaking a C–C bond is about 6.5 eV. Below 2.6 eV, the positron can only excite vibronic modes, and it can annihilate. For a positron incident energy of 1.0 eV, Xu, et al., observed a mass spectrum that looks very much like that for an electron at 70 eV. Fragmentation is widespread except there is no peak above noise for the C_1 ion nor for any ion heavier than C_8. The dominant peaks are for the C_3 and C_4 ions. Clearly annihilation is taking place. At the positronium threshold and just above, the parent ion peak dominates the positron-impact mass spectrum, and by 9.5 eV and above, the PI mass spectrum resembles the EI mass spectrum at 70 eV. Observations on the other molecules studied by these authors showed fragmentation that is specific to a particular bond in each molecule.

These data call for interpretation. We are guided by photoelectron spectroscopy, which shows that ionization takes place from a particular molecular orbital (MO). We believe that annihilation, a very much faster process than ionization by electron impact, vacates a single molecular orbital in a time period of $\sim 10^{-21}$ s; i.e., suddenly. Even the spectator electrons are fixed during the annihilation process. Annihilation creates a daughter system in a nonstationary state. Roughly speaking, it is as if one MO in the parent system is erased in zero time. A quantum mechanical description of the daughter must be a superposition of many of its stationary states, and its fate is determined by the relative weights in the superposition of states. Crawford [22] has shown that annihilation is roughly equally probable from all the valence orbitals of a molecule. Except for the highest occupied MO, annihilation from any others can leave enough energy in the daughter system to fragment it.

An electron in each MO of a molecule exerts forces on each nuclei. The forces from all the electrons add to zero for each nucleus in a stable molecule, but the force from any one MO may be considerable. When one MO is erased, the forces no longer balance, and the observed fragmentation might result from this source. Good subthreshold fragmentation data should be obtained from a molecule that has a Z_{eff} above 1000 but is sufficiently simple to permit a good quantum mechanical treatment. The most attractive candidate molecule in the compendium of Iwata, et al., is methyl chloride, in which there are two kinds of chemical bonds. The C–H and C–Cl bond strengths are about 4.4 and 3.6 eV, respectively, and their breaking can be easily distinguished in a low-energy PI mass spectrum. A complete mechanism for resonant capture and annihilation would be provided by a calculation of the relative probability of annihilation in each MO, and of the forces created on each nucleus by annihilation from that MO. Then the fragmentation and ionization could be worked out theoretically and compared to fragmentation data. This treatment would also give the kinetic energies of the molecular fragments.

A simple mass spectrometric experiment with a well-defined positron beam would give us much useful information. Much more information would be obtained by the application of recoil ion momentum spectroscopy (RIMS) [23, 24] to annihilation from positron-molecule resonances. This would provide the energies and masses of all the ionic fragments. One possible configuration of a RIMS spectrometer involves crossed beams of a supersonic molecular beam of target molecules and a pulsed beam of positrons. This experiment is possible with existing technology [25].

6.4 Bound states

Binding energies of the positron and Ps to many chemical species are listed and discussed in Chapter 2. Here we briefly review some experimental considerations. All the knowledge known to this author that comes from experiment is listed in Table 2.8. The data listed there for much of the work before the 1980's are compromised by the presence of solvents or other condensed hosts. In some cases the results in Table 2.8 rely on an approximate model. Only the data for 1983, 1992, and 2002 in Table 2.8 may be described as the result of direct experimental evidence. We discuss the two cases in which binding energies have been measured.

6.4.1 *Observation of bound states by resonant capture*

Recently Gilbert, et al., [14] measured Z_{eff} as a function of positron incident energy for butane in a scattering experiment and found a strong peak of 23,000 at an energy 0.33 eV. Studies with deuterated butane show a sprectrum with the same structure but shifted by the expected isotope shift for the symmetric stretch mode. This identifies the mode responsible for resonant capture. The spacing of this mode for butane is 0.36 eV, and from Eq. (6.9) it is clear that the binding energy of a positron to this molecule is 0.03 eV. The uncertainty is given as ±0.015 eV. By the same technique, propane binds a positron by 0.020 ± 0.015 eV, and ethane, by

< 0.015 eV. Z_{eff} for acetylene shows some broad structure but no shifts, and 2,2-difluoropropane shows no resonances. Presumably neither binds a positron.

This is an important experiment. Not only is it the first experimental evidence of a bound state of a positron to a neutral molecule, but it also unambiguously establishes resonant capture as the mechanism responsible for the very large observed values of Z_{eff}, thus closing a long debate.

6.4.2 Observation of bound states by dissociative attachment

Compound formation by dissociative attachment,

$$AB + e^+ \to \{e^+AB\} \to A^+ + PsB, \quad (6.14)$$

has been demonstrated for the case A = CH_3 and B = H [26]. The topic has been reviewed recently [25], so we give only a brief discussion here. The thermodynamic threshold for the process above is

$$BE(AB) + IP(A) - 6.80 \text{ eV} - BE(PsB). \quad (6.15)$$

If there are no potential barriers to the dissociation of the intermediate $\{e^+AB\}$ to the products on the right of Eq. (6.14) and if none of the products are excited, then the lowest positron energy that produces the ion fragment A^+ is equal to the expression above, and the binding energy of the compound PsB can be determined. Otherwise, only an upper bound is observed. In the one case in which a compound was made in this way [26], the observed binding energy of PsH was in agreement with the best calculations, but had a large uncertainty (see Chapter 2).

The cross section for the process in Eq. (6.14) is estimated to be 10^{-18} cm^2 [25]. A cross section of about two orders of magnitude larger is expected for dissociative attachment to an anion:

$$AB^- + e^+ \to A + PsB. \quad (6.16)$$

This process can be either exo- or endoenergetic depending on magnitudes of relevant bond energies and electron affinities. The threshold is

$$BE(AB) + EA(AB) - 6.80 \text{ eV} - BE(PsB), \quad (6.17)$$

where EA(AB) is the electron affinity of AB. The kinetic energy of the colliding species in their center-of-mass frame less that for the receding species is equal to the expression above. Both kinetic energies must be known. For uncharged products, translational spectroscopy is appropriate. The experiment is feasible with existing technology, but requires that a positron beam and a keV anion beam be combined in a crossed beam experiment [25].

Mitroy and coworkers [27] have discussed the use of negative ion beams to measure positron affinities of atoms. The idea is to pass a beam of atomic anions through a swarm of thermal positrons in order to make a positron compound by the charge exchange reaction:

$$e^+ + A^- \to e^+A + e^-. \tag{6.18}$$

The threshold is the difference of the electron and positron affinities of the species A. Knowledge of the center-of-mass kinetic energy of the colliding species, and the measurement of that of the receding species yields the positron affinity. A competing reaction is Ps formation, but this produces a neutral product A that will not interfere.

Acknowledgments

The author is indebted to G. F. Gribakin for a critical reading of this chapter in manuscript. Dr. Gribakin suggested many improvements. Any defects that remain are the sole responsibility of the author.

References

[1] M. Charlton and J. W. Humberston, *Positron Physics*(Cambridge University Press, Cambridge, 2001).
[2] M. Kimura, O. Sueoka, A. Hamada, and Y. Itikawa, *Adv. Chem. Phys.* **111**, 537 (2000).
[3] H. S. W. Massey, J. Lawson, and D. G. Thompson, in *Quantum Theory of Atoms, Molecules, and the Solid State*, edited by P.-O.Löwdin (Academic Press, New York, 1966), p. 203.
[4] L. D. Landau and E. M. Lifschitz, *Quantum Mechanics*, 3rd ed. (Pergamon Press, London, 1977).
[5] G. F. Gribakin, *Phys. Rev. A* **61**, 022720 (2000) G. Gribakin, in *New Directions in Antimatter Chemistry and Physics*, edited by C. M. Surko and F. A. Gianturco (Kluwer Academic Publishers, The Netherlands, 2001), p. 413.
[6] K. Iwata, R. G. Greaves, T. J. Murphy, M. D. Tinkle, and C. M. Surko, *Phys. Rev. A* **51**, 473 (1995).
[7] T. F. O'Malley, L. Rosenberg, and L. Spruch, *J. Math. Phys.* **2**, 491 (1961).
[8] P. M. Smith and D. A. L. Paul, *Can. J. Phys.* **48**, 2984 (1970.)
[9] F. A. Gianturco, T. Mukherjee, and A.Occhigrossi, *Phys. Rev. A* **64**, 032715 (2001).
[10] F. A. Gianturco, and T. Mukherjee, *Nucl. Instrum. Methods Phys. Res. B* **171**, 17 (2000).
[11] E. P. da Silva , J. S. E. Germano, and M. A. P. Lima, *Phys. Rev. Lett.* **77**, 1028 (1996).
[12] M. T. do N. Varela, C. R. C. de Carvalho, and M. A. P. Lima, *Nucl. Instrum. Methods Phys. Res. B*, to be published (2002).
[13] M. Charlton, D. P. van der Werf, and I. Al-Qaradawi, *Phys. Rev. A* **65**, 042716 (2002).
[14] S. J. Gilbert, L. D. Barnes, J. P. Sullivan, and C. M. Surko, *Phys. Rev. Lett.* **88**, 043201, 079901(E) (2002).
[15] K. Iwata, R. G. Greaves, and C. M. Surko, *Phys. Rev. A* **55**, 3586 (1997).
[16] S. Y. Chuang and B. G. Hogg, *Can. J. Phys.* **45**, 3895 (1967).
[17] P. Kirkegaard, M. Eldrup, O. E. Mogensen, and N. J. Pedersen, *Comput. Phys. Commun.* **23**, 307 (1981).
[18] P. Kirkegaard, N. J. Pedersen, and M. Eldrup, *PATFIT-88*(RISO-M-2740, Roskilde, Denmark, 1989), unpublished report.
[19] R. S. Mulliken, . *J. Chem. Phys.* **23**, 1833 (1955).
[20] D. L. Donohue, L. D. Hulett, Jr., B. A. Eckenrode, S. A. McLuckey, and G. L. Glish, *Chem. Phys. Lett.* **168**, 37 (1990).
[21] J. Xu, L. D. Hulett, Jr., T. A. Lewis, D. L. Donohue, S. A. McLuckey, and O. H. Crawford, *Phys. Rev. A* **49**, R3151 (1994).
[22] O. H. Crawford, *Phys. Rev. A* **49**, R3147 (1994).
[23] C. L. Cocke and R. E. Olson, *Phys. Rept.* **205**, 153 (1991).

[24] J. Ullrich, R. Moshammer, R. Dörner, O. Jagutzki, V. Mergel, H. Schmidt-Böcking, and L. Spielberger, *J. Phys. B* **30**, 2917 (1997).
[25] D. M. Schrader and J. Moxom, in *New Directions in Antimatter Chemistry and Physics*, edited by C. M. Surko and F. A. Gianturco (Kluwer Academic Publishers, The Netherlands, 2001), p. 263.
[26] D. M. Schrader, F. M. Jacobsen, N.-P. Frandsen, and U. Mikkelsen, *Phys. Rev. Lett.* **69**, 57 (1992).
[27] J. Mitroy and G. G. Ryzhikh, *J. Phys. B* **32**, L411 (1999).

Chapter 7

Positron Porosimetry

Marc H. Weber and Kelvin G. Lynn

Dept. of Physics and the Center for Materials Research, Washington State University, Pullman WA 99164-2711, USA

7.1 Introduction

During the course of the past 6 years, positron annihilation techniques were used and adapted to examine porosity. The driving force was and continues to be the need in the semiconductor industry to lower signal propagation delays caused by the conductor resistance and the insulating layer dielectric value k [1]. The traditional silicon dioxide insulator is being replaced by inorganic and organic materials with lower dielectric value and porosity is being introduced to come even closer to the ultimate value of $k = 1$ for vacuum.

Positrons and positronium (Ps), the bound state of a positron with an electron, are highly sensitive to open volume in a material. Their sensitivity ranges from monovacancies (positrons) to voids to pores of several 10 nm in cross section (positronium). With a variable energy monoenergetic beam, positrons can be implanted in a sample at a chosen depth (or range of depths for a profile). Dependent on the porosity properties of the sample, positronium may be formed there. When positrons and positronium annihilate the emerging gamma rays carry encoded information about the pore structure and their morphology.

In this chapter it will be shown with numerous examples how information on open versus closed porosity, the total porosity, the pore sizes and a measure of the average length of connected pores can be measured with varying degrees of ease. Some of these parameters can be obtained from alternate methods. However, none can provide depth dependent profiles of the parameter without microtoming the sample. No special sample preparation is required. A sample investigated by positrons could easily be reinserted in a device production line to correlate results from positron measurements with device performance.

This chapter is organized by the specific property of porosity rather than by positron technique. This should assist a reader interested in answering a particular question in the evaluation of which method best solves the problem. The sequence is organized in order of increasing complexity of the method needed for the problem. The methods are presented briefly. For more detailed discussions of a technique the reader may refer to a listed reference.

The terminology used here has been defined in preceding chapters. Treaties on positronium can also be found in Berko and Pendleton [2] and Rich [3]. Recent lifetimes are published by Gidley [4]. Jean [5] and Yang and Jean [6] discuss positronium in open volume within polymers. Doppler broadening can be found in Saarinen [7] and Krause-Rehberg [8]. Some aspects of positron porosimetry are presented rather briefly for lack of space. An upcoming review article will go into more details [9].

The separation of open and closed porosity samples is presented. The methods range from observing count rates to measuring the longest lifetime of positronium. In either open or closed porosity, it is useful to know the level of how interconnected pores are. Porosity not only lowers the dielectric value, it also opens the door for impurity intrusion. This is addressed next. Finally the holy grail of porosimetry is addressed: the determination of pore size distributions with the positronium lifetime technique.

Brief discussions of methods beyond the ones presented here will be given. A glimpse is given of what the near future might bring. Finally, the possibility is raised of using positrons and positronium measurements as an inline diagnostics tool. Estimates are made based on the experience gained from the work presented here.

The vast majority of the work on positron porosimetry shown here was carried out as part of a NIST-ATP grant in cooperation with groups at IBM

Yorktown and Almaden. Kenneth P. Rodbell, IBM Yorktown, was instrumental in getting this project off the ground and keeping lofty scientists on solid ground of reality. He coordinated most of the interaction of the WSU positron group and IBM and other partners of the NIST-ATP venture.

Willi Volksen and Robert Miller, both IBM Almaden, diligently manufactured and provided the samples for this study. Their insight into the chemistry of polymers, copolymers and pore generators was invaluable for the understanding of the data. Samples from a wide range of other sources were investigated. Data from some are used here.

The positron group at WSU, Pullman, other than the authors, carried out the brunt of the data acquisition and analysis. Mihail P. Petkov was a fantastic asset during his time at WSU. His effort jump-started the progress. Cai-Lin Wang introduced us to the complexities of lifetime analysis in general and the maximum entropy method (MELT) in particular.

The lifetime results would simply not exist, if it were not for the persistent efforts of Stanislaw Szpala during his thesis project. He single handedly moved the positron beam to its present location and added the lifetime apparatus.

The patient supervision of Jerry Smith, DOE basic energy sciences, and the DOE grant for basic positron science was a great help for this project. The insights won here will flow back into our basic science effort and lead to new results.

Almost all the data originate from a single set of samples, which can be considered representative for the porosity issues discussed. They are ~0.7 μm-thick methyl-silsesquioxane (MSSQ) films, fabricated using a sacrificial porogen technique to incorporate porosity [10]. Si wafers were spin-coated with a solution containing the MSSQ resin and 0—90 wt. % pore generator (porogen) additions (porogen load). This blend was cured in a 300 Torr N_2 ambient at 200°C to achieve vitrification of the MSSQ network. Subsequently at 450°C the porogen decomposed and was driven out of the film. The total porosity is proportional to the porogen load.

7.2 Open versus closed porosity

Several versions of positronium annihilation measurements are sensitive to the differences between open and closed porosity. The detection geometry, the annihilation types and the lifetime change when closed pores with no path to the sample surface change to open pores, which are linked to the sample surface. The experimental options are discussed in order of increasing complexity of the measurement and analysis.

7.2.1 Detector count rates

In a sample with closed porosity, where no pores are linked to the surface of the sample all annihilation events occur within the sample. In the case of open porosity some or all have an unobstructed path of connected pores to the surface. Just like in gas adsorption and desorption measurements [11] positronium, trapped in pores can leave the sample only in the latter case. The detector geometry can be arranged such that positronium annihilations can only be observed, if they occur within the sample (with aperture), or no such restrictions are made (no aperture). Data for either configuration are shown in Figure 7.1 as a function of concentration of pore forming agent (porogen load) and a fixed implantation energy.

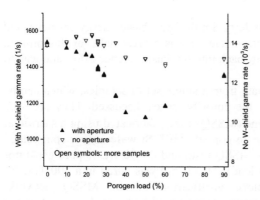

Figure 7.1: Detected annihilation rates with and without an aperture to restrict the "field of view" to events from within the sample.

The strength of the signal drops dramatically in the apertured case at about 23% porogen load. In the unshielded case, only a gradual small change is detectable. When the porogen load increases above 50% positronium also encounters the interface to the silicon substrate and annihilates more efficiently within the sample.

7.2.2 3-to-2 photon ratio

The 3-to-2 photon ratio technique observes the ratio of 3 versus 2 photon annihilations of positronium (and positrons). In vacuum positronium will annihilate via 3-photon decays only. Trapped inside closed pores, both annihilation paths are possible. This change can be used to detect the onset of open porosity as shown in Figure 7.2. A change in slope (the slope is shown as a solid line) occurs at about 23% porogen load, indicative of an increased likelihood for positronium to escape from the sample.

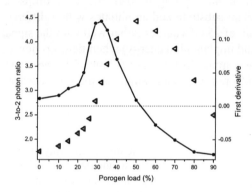

Figure 7.2: Ratio of 3 to 2 photon annihilations (◁, left scale) when positrons are implanted at a mean depth of 100 nm as a function of porogen load. At a porogen load of ~23% the slope increases, indicating positronium escapes from the samples. Beyond 50% positronium can also encounter the Si interface. The first derivative is shown as a solid line (right scale).

At porogen loads in excess of 50% the network of connected pores extends across the sample depth. Positronium encounters the interface to the silicon substrate, where it annihilates fast and almost exclusively via pick-off into two photons. The rollover in the ratio between 40 and 60% porogen load is an indication of percolation. Percolation, when the first channel extends across the sample occurs sooner [12, 13]. A number of such channels must exist before the measured ratio changes visibly. Hence, percolation occurs at a somewhat lower porosity. This will come up again later, when pore sizes are discussed.

7.2.3 Positronium lifetime

The lifetime of positronium also changes as a function of closed vs. open porosity. The lifetime of positronium in vacuum is much longer than that of positronium in pores.

$$\tau_{open} = 1/(\lambda_{escape} + \lambda_{pore}) \gg 1/\lambda_{pore} = \tau_{pore}$$

The intensity of this component indicates open porosity as shown in Figure 7.3 as a function of porogen load (which is proportional to the total porosity) of MSSQ films. The intensity is small for porogen loads up to 15% and then increases sharply. Fitting a constant to the values for low porogen loads and a line for the data for porogen loads from 23 to 40% yields an open-porosity-threshold at 22.6±1% porogen load. The shown intensities are the fractions of all observed annihilations that occur with this lifetime. The intensity reaches a maximum at 40% porogen load and then drops as positronium encounters the silicon substrate and annihilates with a different lifetime. The intensity maximum of 25±1% agrees rather well with the signal rate loss of the same sample (and implantation energy) when field-of-view of the detector is restricted to the sample. In that case (Figure 7.1) the signal rate drops by 23% from the 20% porogen load sample to the 40% porogen load sample.

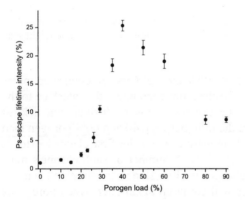

Figure 7.3: Intensity of the lifetime associated with positronium escape from the sample via open pore channels as a function of porogen load.

7.3 Pore connectivity

The ratio of 3-to-2 photon annihilations is not only a measure for open porosity; it also changes with the concentration of pores and their size. A typical ratio measurement can be performed in several seconds on standard positron beams of 10^5 to 10^6 positrons per second. This method is ideal for depth profiles. The incident energy of the positrons is ramped up to implant the positrons at increasing mean depths, where positronium forms and samples pores.

7.3.1 The concept of the 3-to-2 photon ratio technique

Various gas and liquid adsorption techniques are used to determine the porosity of a specimen. They are mostly based on the Brunauer-Emmett-Teller method (BET) [14]. Atoms or molecules penetrate into a sample through interconnected pores with links to the sample surface. The adsorbed volume and temperature and pressure dependent data are used to quantify the porosity and surface to volume ratios, which contain information about the pore size distribution [15]. A recent review is published by Schneider [11]. Care must be taken that the used probe (gas or liquid) does not react with the sample. When pores become too small, the probe may not penetrate into them. Pores or interconnected pores are not detected, when no connection to the sample surface exists. For example, thin capping layers would close all pores and render the technique useless, even though the pores may be totally interconnected below the cap.

Roughly speaking, 3-to-2 photon annihilation ratio measurements can be considered as a BET technique, which is sensitive to both open and closed pores. Positronium can be considered as the smallest atom possible. No pore will be too small. Positrons are implanted rather than adsorbed and forms positronium. Positronium annihilates into 2 or 3 photons from within pores, or into 3 photons only after escape (desorption) out of the sample through open porosity. In addition, depth dependent information can be provided.

Positronium can pick-off an electron during a collision with a pore wall and annihilate into two photons. Between collisions, only three photon annihilations occur, just as in vacuum. Quantum mechanically, the overlap with the wall-electron wave functions decreases with the distance from the wall and pick-off (two photons) becomes less likely. With increasing pore size collisions become less frequent. The ratio of 3 photon annihilations to 2 photons probes the combination of pore size and total pore volume as well as their link to the sample surface, and can be measured by examining the energy distribution of annihilation photons. This 3-to-2 photon ratio can be calibrated to absolute fractions of positronium in the annihilation spectrum [16, 17].

7.3.2 Positronium range—depth profiles

When a monoenergetic beam is used, positrons are implanted into a sample with an implantation profile resembling the derivative of a Gaussian function [18, 19], which depends on the incident positron energy and the density of the sample. For example, $E = 2$ keV energy positrons stop at a mean depth of $T_{mean} = 120$ nm in a sample with a density of $\rho = 1$ g/cm^3 following the empirical relation [20, 21]. In the case of the MSSQ based samples, used here as an example, the positron diffusion length is about 10 nm [22] and positrons annihilate with a lifetime of about 0.5 ns [23]. The diffusion length of positronium is much smaller with an estimated value of 1 to 2 nm for amorphous SiO$_2$ and polymers [24, 25]. The binding energy of positronium inside of pores (vacuum) is bigger than in the matrix material. After an inelastic interaction with the pore wall, positronium is trapped in it and can now move only along connected pores. The measured effective range of positronium is a combination of its diffusion to pores and in pores, i.e. the extent of interconnected pores.

The difference in the annihilation ratio for positronium at the surface and in the sample is used to measure the effective range of positronium. Implantation profiles for a range of incident energies (density 1 g/cm^3) were calculated. In this simulation the fraction that stops within a diffusion length of the surface can reach it and annihilates into 3 photons; the remainder annihilate 10% into three photons and 90% into two photons, as shown in Figure 7.4. The fractions are chosen as an example. A short effective range appears as a sharp transition from surface measurement to bulk measurement value.

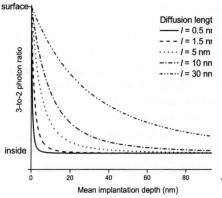

Figure 7.4: Simulation of 3-to-2 photon ratio for increasing diffusion lengths in the sample as a function of mean implantation depth into the sample (density $\rho = 1$ g/cm^3).

For a fixed temperature the diffusion length is constant, while the introduction of pores permits the motion of positronium to the surface from deeper in the sample. For isolated and closed pores this change in range is determined by the size of the pores. When the pores connect, the length of the connected chain becomes the dominant value. As percolation is reached, this length rises sharply.

For atoms with a finite lifetime the diffusion length and the diffusion coefficient are linked via the lifetime of the particle. Here, we refer to the mean depth from which positronium can leave the sample as the escape depth.

$$L = \sqrt{D\tau}; \qquad L_{eff}^2 = D_{matrix}\tau_{matrix} + D_{pore}\tau_{pore} = L_{matrix}^2 + L_{pore}^2$$

In addition to the effective diffusion length, which determine the curvature in the depth dependent data, the amount of positronium that is formed plays a role. Two effects contribute, one of which is energy dependent [26].

The energetic positron slows down on its track to it's implantation depth, it ionizes the sample and leaves a spur of free electrons behind [27, 28]. The number of electrons at the terminal of the spur and their mobility determine the formation likelihood for positronium. The cross section for positronium formation becomes constant independent of incident energy. The second path to positronium formation is the Øre process [29]. When the potential energy needed to ionize an electron from a molecule is less than the binding energy of positronium liberated when an electron and the positron combine, positronium formation is possible and efficient.

Both positronium formation contributions can be approximated by an exponential association function provided the incident energy is large compared to the binding energy of positronium and is added to the simulation in Figure 7.5. Also included in this simulation is the fact that positronium annihilates more likely into three photons as the size of pores or their concentration increases. In the case of a short effective diffusion length ($l \leq 10$ nm), this effective diffusion length controls the shape of the curve. In the case of larger values, the energy dependent formation becomes obvious. The ratio increases above the surface (low implantation energy) value.

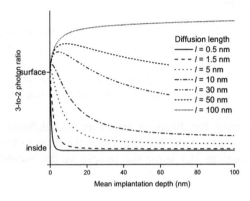

Figure 7.5: 3-to-2 photon ratio when considering the diffusion length, the formation probability and the increased chance for three-photon annihilation in larger pores.

7.3.3 Data and results

Data were accumulated for a set of samples made of MSSQ with porogen loads from 0 to 90% (Figure 7.6). The samples have different thickness (350 to 780 nm) and the density drops as the porogen load increases. At the silicon interface the density changes and positronium annihilates fast into two photons, causing the near vertical drop in the ratio to the equivalent of no 3-photon events.

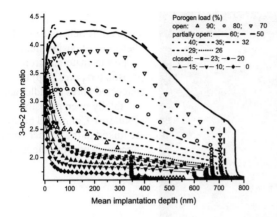

Figure 7.6: 3-to-2 photon ratio versus mean implantation depth for samples with porogen load from 0 to 90%. Samples with isolated and closed pores are shown with lines and symbols (0 to 23%); partially open porosity sample data are shown as lines (26 to 60%); totally open porosity sample data are shown with open symbols (70 to 90%).

The data for samples with 0 to 60% porogen load resemble the simulation rather well. The influence of the interface to the substrate becomes important at larger loads. Almost all pores are interconnected and the channels extend across the sample. The probability for positronium to encounter the interface to silicon is high. Positronium moving towards the surface escapes rapidly without loosing all of its kinetic energy in wall collisions. It can leave the region just outside of the sample surface and move beyond the volume in which it is detected efficiently. Thus, the measured value of the 3-to-2 photon ratio drops dramatically to levels comparable with partially open porosity. These changes are illustrated schematically in Figure 7.7.

7.3.4 Analysis

A simple fit of the data with the product of an exponential association and an exponential decay to estimate the escape depth, overestimates the escape depth by folding the positron implantation profile and diffusion into the fitting parameters [30]. A more appropriate numerical fitting method based on the diffusion equation was used to take both the implantation profile and diffusion into account [31]. When it is applied to the 3-to-2 photon ratio data suitable absorbing boundary conditions need to be included. The results for the escape depth are shown in Figure 7.8 [30]. In addition to the diffusion-like motion of positronium in connected pores, positrons and positronium diffuse to the pores.

178 *Principles and Applications of Positron and Positronium Chemistry*

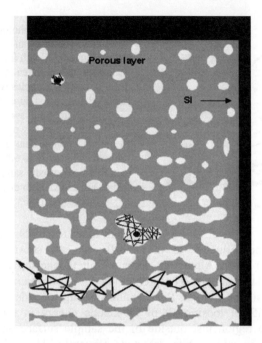

Figure 7.7: Schematic of the behavior of positronium in porous materials from isolated pores (top) to connected pores (middle) and totally open porosity (bottom). Connected pores are linked to the surface from a mean depth (middle left).

The contribution of positron diffusion length (L_+ = 10 nm [22]) was removed from the escape depth values. The diffusion constant in a material is a function of diffusion length and annihilation rate $D = L^2 \lambda$. Here, the rates for positrons and positronium are similar ($\lambda \approx 2$ ns^{-1}). Thus the measured combined effective diffusion length of positrons L_+ and positronium escape L_{esc} is $L_{eff}^2 = L_+^2 + L_{esc}^2$ [30].

At porogen loads above 23% the escape depth rises more rapidly. Pores begin to connect at this porogen load. Percolation, when the first channel of connected pores spans the entire sample thickness, occurs at a higher load. This will be discussed later. In samples with porogen loads in excess of 50% the escape length becomes far greater than the sample thickness.

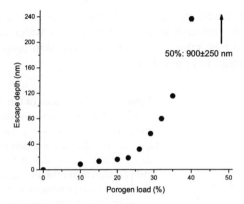

Figure 7.8: Escape depth from which pores are connected to the surface versus porogen load. The contribution of positron diffusion was removed from the fitted diffusion length to obtain the positronium escape depth. At 50% porogen load the escape depth exceeds the thickness of the samples. The present model does not include this and becomes

The fit results from 3-to-2 photon annihilation data can also be used to determine the fraction of the total porosity, which is open to the surface. The fraction is determined by comparing the asymptotic values of the fit for the total signal and the constant part. Here, too the implantation profile and the motion of positrons and positronium have to be taken into account. The results are shown in Figure 7.9.

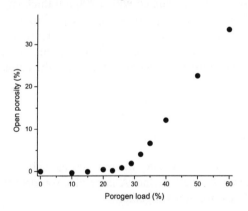

Figure 7.9: Fraction of the porosity, which is open to the sample surface versus porogen load.

7.3.5 Skin layers and buried open porosity

In some processes the porous sample may show depth dependent variations in porosity. An extreme case is when the porosity is large in the sample, but no pores exist in a thin skin layer near the surface. The skin layer prevents deterioration of the sample from intrusion of impurities such as water or metal atoms. However, care must be taken not to puncture the skin layer. This is illustrated on special samples in Figure 7.10.

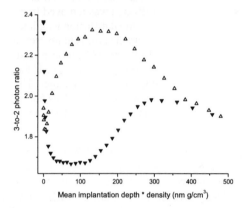

Figure 7.10: Two cases of the 3-to-2 photon ratio versus mean implantation depth ($\rho \sim 1$ g/cm^3). Open symbol data show a thin skin layer, which "seals" the surface and prevents positronium escape. In the layer no positronium can exist. The closed symbols represent a sample with a 200 nm thick capping layer deposited on top of the film. It, too, does not allow the existence of

In the first case, open symbols, a thin skin layer was created by an etch treatment. The second case shows a sample capped with a 200 nm thick cap. Either treatment seals the surface and prevents the escape of positronium. Below the surface layer, significant porosity can be observed. However, because the implantation profile is spread across larger depth regions in the second case, a careful data analysis must be undertaken to extract the actual porosity.

7.4 Impurity intrusion and chemical effects

The chance for pick-off annihilation can be altered dramatically by impurities and paramagnetic elements like fluorine. A large electron density either prevents the formation of positronium altogether.

7.4.1 All materials are not equal

The models that relate lifetimes to pore sizes contain no material dependence. However, it is easy to imagine that the interaction of positronium with the pore wall does indeed depend on the wall material.

Further the range of positrons and positronium in the matrix is certainly material dependent. Silk™ [32, 33] (an organic thermosetting polymer, trademarked by Dow Chemical) was introduced a while ago as a highly suitable candidate to replace silicon dioxide as an intralayer dielectric. 3-to-2 photon ratios were measured on a variety of Silk™ materials with and without pores as shown in Figure 7.11. FE-SEM cross-section images indicate that the pore sizes and porosity are similar to a MSSQ film with about 35% porogen loading.

It turns out that the positronium signal from the pores is very small and quite comparable to MSSQ with less than 10% porogen. Either positronium rarely forms and traps the pores, or the positronium-wall interaction is much stronger and causing pick-off annihilation within a few bounces, rather than thousands of bounces as in the case of MSSQ.

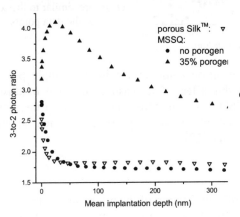

Figure 7.11: 3-to-2 photon ratio of samples containing Silk™. As a reference a porous MSSQ sample dataset is also shown. The density if Silk™ is assumed to be 1 g/cm^3.

7.4.2 Chemistry

Oxygen plasma treatments are common processes during electronic device manufacturing. Figure 7.12 shows the effect of oxygen plasma treatment on a film of HSSQ [22]. Similar to porogen free MSSQ this material contains some open volume, which is measurable with the 3-to-2 photon technique. When the material is exposed to oxygen plasma for short times (see case of 10 minutes) voids become larger below the surface. However the process is more than offset by a decrease near the surface. In the case of the fully oxidized sample (after 2 hours of O-plasma exposure) all beneficial effects have been reversed, consistent with the dielectric values for the samples, which increase from 2.8 to 6.83 with oxygen treatment.

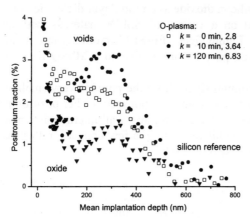

Figure 7.12: Depth-profile of positronium fractions, extracted from the 3-to-2 photon ratio, for O-plasma treated samples. The scale emphasizes the relatively small Ps fractions. Legend: black boxes – reference material: red circles – 10 min O-plasma; and blue downward triangles – 2 hours O-plasma. Statistical errors are similar to the scatter of the data points.

The Doppler shifts of positron annihilations in the sample provide a more sensitive measure of this effect. The plasma treatment oxidizes the rather open structure of HSSQ. Oxygen atoms are included in the network. These have high momentum electrons in p-orbitals, which lower the sharpness of the annihilation line [22].

7.4.3 Uptake of water

In porous layers with a significant fraction of open porosity water will enter hydrophilic materials like MSSQ and cause an increase in the dielectric properties or oxidize metal interconnects in a device. The water intrudes the sample during a 4-month shelf time and lowers the 3-to-2 photon ratio (Figure 7.13). An in situ bake at about 300C drives all water out of the film; the as received state is recovered.

7.4.4 Metallic impurities

Metal ion or atom in-diffusion potentially poses a threat to the integrity of porous intra-layer dielectrics. Linked pores might provide a path for metal motion and ultimately cause short circuits. Diffusion barriers are being developed to prevent this from happening. However, their high dielectric values can potentially offset any gain from the introduction of pores.

Pinholes pose another threat. The effect of the presence of metals can be seen in Figure 7.14. Here silica based Xerogels were prepared with various amounts of metals. The near surface change in the 3-to-2 photon ratio was used to estimate the range of positronium. The same model as in the case for porous MSSQ was used. The range dropped from 1.55±0.05 mm in the plain silica sample to 0.15±0.07 mm in the case of tantalum and to 0.8±0.09 µm in the iron containing sample.

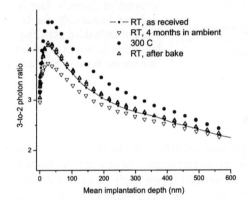

Figure 7.13: Change in the 3-to-2 photon ratio due to water uptake. Black boxes are the "received sample" data; green circles are after 4 months in air. During an in-situ bake at 300 C (red upward triangles) a temperature induced increase is observed. After cooling back to room temperature the returns to the as received state (open blue downward triangles). Symbol sizes exceed the errors.

It is expected that the sensitivity of positrons and positronium to changes in the matrix material will be used extensively in the near future. For example, tantalum-silicon-nitride was found to be an effective diffusion barrier between copper and silicon dioxide [34]. The applicability to porous materials needs to be checked. Work has been carried out, for example, by Gidley et al. on TiN diffusion barriers [35].

184 *Principles and Applications of Positron and Positronium Chemistry*

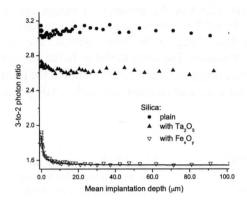

Figure 7.14: Depth profiles of the 3-to-2 photon ratio for plain silica Xerogel and versions that include tantalum (blue upward triangles) or iron (green downward triangles). In the plain and tantalum case, positronium can move far through the open sponge like network of channels. This is reduced dramatically for the case of embedded iron. The line stems from a fit to extract the escape depth of 880 nm.

7.5 Pore size distributions

The most crucial parameter for porosity is, arguably, the distribution of pore sizes in a given sample. A number of techniques have been developed to answer this question. Gas adsorption-desorption techniques reviewed for example by Schneider [11] rely on the Brunauer-Emmett-Teller method (BET) [14] to extract pore size distributions [15]. They probe interconnected pore structures. Difficulties arise when pores with sizes smaller than 2 nm are present. Isolated pores are invisible. Small angle X-ray scattering (SAXS) and Small angle neutron scattering (SANS) are used extensively to evaluate small particles [36]. Both techniques were extended to the case where the particles are pores in thin films [37]. Neither method can distinguish between open and closed pores and the scattered intensity from the pores is small. To date, very small pores (<1 nm) have not been resolved. Sample preparation is a major difficulty for microscopy techniques like transmission electron (TEM) and scanning electron (SEM) microscopy when extreme care is taken the results are encouraging. The preparation may alter the pore distribution and thin slices might not be representative of the sample. Other techniques used include ellipsometric porosimetry [38] and surface acoustic wave spectroscopy (SAWS) [39, 40].

7.5.1 Pore sizes determine lifetimes

The lifetime of positrons and positronium is a function of the electron density at the site of the positron (positronium atom). Obviously, the electron density is low inside pores and much higher in the material. As was shown in previous parts, positronium, in particular, can trap in pores with dimensions from several angstroms to many tens of nanometers. In a classical picture, the positronium atom will bounce from the walls of a pore with a thermal velocity (v). During the periods of each bounce the positron in positronium may annihilate with an electron from wall atoms and molecules with the probability $P_{collision}$. In the pore, only annihilations with its bound electron are possible (self-annihilation) with the rate $\lambda_{self} = 1/142$ ns^{-1}. This classical picture works when the de Broglie wavelength of positronium (6 nm) is small compared to the pore dimensions [41]. Quantum mechanically speaking the positronium wave function overlaps with wall electron wave functions. With increasing size of the pore, the bouncing rate or the wave function overlap decrease. The lifetime increases. Eventually the pore is so large that the annihilation rate due to wall bounces becomes comparable with the self-annihilation rate and size dependent changes in the combined annihilation rate are not measurable any more.

In a measurement, lifetimes are observed. They correspond to the inverse of the sum of the self-annihilation the wall bounces dependent annihilation rate (λ_{wall}).

$$\frac{1}{\tau} = \lambda_{self} + \lambda_{wall}(size) = \frac{1}{142ns} + \frac{v \cdot P_{collision}}{\ell}$$

The product $v \cdot P_{collision}$ was estimated for highly porous silica powder to be 0.021±0.002 nm/ns [42]. Semi-empirical models were developed to translate measured lifetimes into pore sizes for sub nm pores [43-45]. The sizes of pores considered here are, however, much larger than what commonly occurs in polymers. While lifetimes in polymers typically do not exceed several nanoseconds, the pores encountered here are responsible for lifetimes up to 100 ns. The model developed for small pores had to be extended to cover the larger range [43-45]. Gidley et al. accomplished this by taking into account that positronium can also exist in excited states [46, 47]. A more empirical approach was taken by Ito et al.[48] It is assumed that the positronium thermalizes long before annihilation, that material dependencies can be neglected and that the lifetime is a measure of a mean free path of the positronium atom in a pore. The shape of pores cannot be determined by lifetime measurements. Typically spherical or cubic shapes are assumed. If pores are connected to form channels, it is assumed that they have circular or square cross sections.

The extension of lifetime measurements to detect lifetimes that are 100 to 1000 times larger than what is common in standard lifetime systems is a very challenging task. Considerable demands on precision and stability of an apparatus have to be met, in particular when positron beams are to be used.

7.5.2 The apparatus

Monoenergetic beams of positrons are required. A suitable time start pulse has to be extracted either by pulsing the beam or by extracting and detecting secondary electrons, which are created when the positron impinges on the sample. This latter method is used for the measurements discussed here. The apparatus is described by Szpala et al.[49]. It has been improved since to greatly reduce systematic effects [50]. General discussions of beams can be found in a book edited by Coleman [51]. Only a handful of facilities exist across the world that can perform beam based lifetime measurements. Suzuki et al.[52] and Bauer-Kugelmann et al.[53] operate pulsed beams to generate a lifetime start signal. Gidley et al.[54] based on an earlier apparatus by Lynn et al.[55] use secondary electrons to start the clock in an electrostatically guided beam.

It should be noted that in a magnetic field the m=0 spin triplet component and the spin singlet component of positronium mix. The annihilation rate of the triplet increases with increasing field. In the present apparatus with a magnetic field of about 0.02 Tesla this lowers the vacuum lifetime of the triplet state from 142 ns to 140.8 ns. This is well within experimental uncertainties and will be ignored [2, 17].

7.5.3 Raw data

Typical spectra are shown for test samples of graphite and non-porous MSSQ and MSSQ with 70% porogen added in Figure 7.15. Part (A) shows the full dynamic range for the spectrometer and Part (B) a narrow time window near time zero.

On the long time scale the featureless background is important when determining long lifetimes of up to 142 ns. The slope for background data

>400 ns in the case of graphite and porogen free MSSQ is zero within statistical uncertainties. Both graphite and porogen free MSSQ show a very small long-lived component, which is independent of pores in the material. Rather, positrons can diffuse to the surface of the sample and then escape after forming positronium at the surface.

A small fraction of positrons will not enter the sample but be backscattered from the surface. Some are confined by the electron acceleration grid and returned to the sample. For a positron with 2 keV energy and a potential of the grid just below 2 kV the turn around time to the sample is ≤2 ns. The magnitude of this backscatter bump is material dependent and <6% in low Z, low density samples.

A sharp peak at about 6 ns, occurs when backscattered positrons pass the grid and reach the CEMA and trigger timing pulses without the secondary electron time of flight. This 6 ns peak vanishes in statistical noise in the case of samples that cause longer lifetimes. In the lifetime analysis, the data in the 6 ns peak region are ignored in the present discussion.

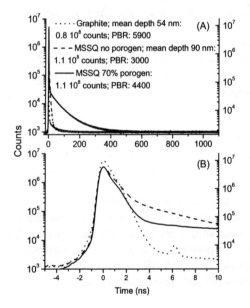

Figure 7.15: Part (A): Typical lifetime spectrum for graphite and MSSQ without pores at 54 and 90 nm mean implantation depth respectively. Shown are raw data. Graphite counts per bin are shown on the left scale and MSSQ counts on the right side scale (the 70% porogen set (green) is scaled up by 1.3 to match the non porous MSSQ background). Indicated are net counts (without background) and the peak to background ratio. The time bins are 78 ps wide. A total of ~15 000 channels are used. Part (B): Same data as in figure 7.15(A), but with a narrow time window around 0 ns.

188 *Principles and Applications of Positron and Positronium Chemistry*

7.5.4 Data analysis and fitting methods

As can be seen in Figure 7.15 very different lifetimes occur and each dataset shows different lifetimes. In principle the sample material and the size distribution of pores and the open porosity component each generate an annihilation rate (λ_i) with some relative intensity (I_i). The spectrum is then convoluted with the experimental response function (R). Random statistical noise and a certain background (B) level are added. The measured time spectrum M(t)

The reversal of this process (i.e. the extraction of annihilation rates and intensities) is ill conditioned. The discrete lifetime analysis program PATFIT [56] is used here to obtain the system resolution and the true start time of the exponential decay (t_{zero}). MELT [57, 58] lifetime distributions are obtained. The data sets contain between 10 and 20 million counts per spectrum, acquired in less than an hour, which greatly surpasses the statistical precision of typical lifetime spectra. In each case the incident positron energy was 2 keV. With increasing porosity (porogen load) the density drops and the mean implantation depth increases from about 90 nm to several hundred nm. The data for some of the samples are shown in Figure 7.16. The time range to 80 ns is shown in parts (A) and (B).

The resolution function and short lifetimes remain rather constant throughout the porosity range. The longer lifetimes increase steadily with porosity up to about 50% porogen load. At larger loads the lifetimes remain similar while their intensities drop.

7.5.5 Average lifetimes—mean pore dimensions

A fast and reliable measure of the mean pore sizes is to extract a mean lifetime from the data. The values are shown in Figure 7.17 from data collected with and without the aperture. For samples with porogen loads in excess of 20% the values are smaller when the aperture is in place, indicating that positronium escapes from the samples from the mean depth of positron implantation. Approximate values are shown on the top axis. The mean lifetimes drop dramatically for porogen loads above 50%, independent of aperture configuration, because fast pick-off at the silicon interface, consistent 3-to-2 photon ratio measurements presented earlier.

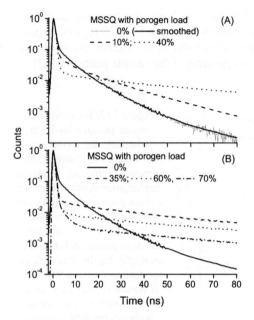

Figure 7.16: Part (A): Lifetime data for MSSQ containing 0% to 40% porogen load in the time window from −10 to 80 ns. Part (B): 0%, 35%, 60% to 80%. The background for each set was subtracted and the spectra are normalized to 1 in the peak channel. For times greater than 50 ns a 50pt running average was applied for clarity in the figure. The original statistical fluctuations are shown in the 0% case. The data are scaled to 1 in the peak channel and the constant background was subtracted.

The right side scale shows the pore diameter associated with a given lifetime. These values should be considered with caution as significant contributions to the mean lifetime stem from the matrix MSSQ material, which contains inherent open volume. The values apply only to data with the aperture in place and for porogen loads up to 50%.

7.5.6 Average lifetimes and 3-to-2 photon ratio

The average lifetime is compared to the positronium fraction (after calibration of the 3-to-2 photon ratio [16, 17]) in Figure 7.18. The data accumulated with and without the aperture are shown.

For porogen loads up to 20% the relation is linear independent of the aperture placement. When the aperture is not in place (downward triangles) the same linear relation continues to a porogen load of about 35%. In the case of the apertured lifetime setup the slope changes to a lower value at about 20% and then remains constant to a porogen load of 50%. At higher loads the mean lifetime values drop sharply while the positronium fraction drops less sharply (open symbols). The trends are consistent with the onset of open porosity at 20 to 23% porogen load (the aperture configuration becomes important) and percolation in the range from 35 to 40%, where the linear relationship breaks down. Variations in the sample thickness do not matter because percolation is independent of the sample geometry [12].

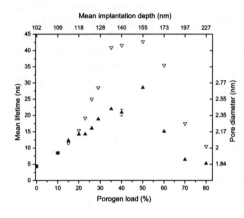

Figure 7.17: Mean lifetime versus porogen load. Up triangles: the aperture is inserted; down triangles: the aperture is removed. The mean implantation depth chosen for a given porogen load is shown on the top axis. The pore diameters (right axis) are meant as guides and should be used only for the data with the aperture and up to a 50% load. A typical error (including an estimate of systematic errors) is shown for 40% porogen load.

The positronium fraction (F) for a given mean lifetime is $F = \tau_{mean} \cdot (71.5 \pm 1.2)^{-1}/\text{ns}$. After this calibration, the 3-to-2 photon ratio can be related to a mean pore dimension. This may be premature because no data exist to date on the dependence of this relation on the concentrations of specific pore sizes, which will probably alter the slope. Given fixed pore generating and matrix materials (as in this case) the relation appears to be valid.

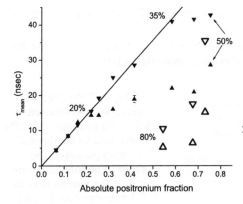

Figure 7.18: Mean lifetime vs. the positronium fraction, obtained from the 3-to-2 photon ratio. Down triangles: lifetime data without the aperture; up triangles: with the aperture. Open symbols: samples with porogen load > 50% (positronium reaches the Si interface). Line: a liner fit to "no-aperture" for <37% load. Numbers: porogen loads in %. A typical error (incl. systematic uncertainties) is shown for 40% porogen load.

7.5.1 Full analysis—Pore size distributions

More information can be extracted when the data are deconvoluted from the experimental resolution and the backscattering component and separate lifetimes are extracted with PATFIT to obtain the experimental resolution function and MELT to obtain lifetime distributions as shown in Fig 7.19 for the case of 23% and 80% porogen load. The probabilities were scaled to the peak value for the component at 0.5 ns and are enhanced by factors of 10 and 200 for lifetimes larger than 2 ns and 7 ns respectively.

A number of comments should be made.

(a). The full data spectra with a range from —5 ns to 900 ns were used for the analysis. When cutting the data at some arbitrary time (for example at 60 ns) the vast majority of collected annihilation events are discarded [46, 59]. Residual influences of shorter lifetime components may influence the fitting of the remainder of the data. The response of the system cannot be determined reliably.

(b). With a time resolution of ~500 ps FWHM the 125 ps para-positronium component cannot be resolved. However, when positronium is formed, ¼ of its intensity populates the para state. A 125 ps fixed lifetime was included. The intensity was a fitting parameter and was consistent (within uncertainties) with the ¼ to ¾ ratio of para and ortho state populations.

(c). Four more or less distinct lifetime components appear in the result (Figure 7.19). In cases of open porosity component, an additional

component appears on the long lifetime side with about 100 ns mean lifetime.
(d). The question of a bimodal distribution or a single mode in the range from 10 to 60 ns will be discussed in more detail later.

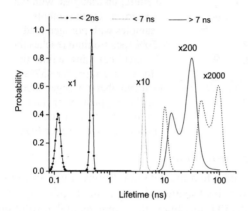

Figure 7.19: Lifetime distribution as obtained by MELT for a sample with 23% porogen load (at the threshold to open porosity). The probabilities of lifetime are scale to the maximum of the 0.5 ns component. Lifetime probabilities for 2 to 7 ns are increased by a factor of 10 and for even longer lifetime by a factor of 200. The longest lifetime components for a 80% porogen load sample are also shown (dashed).

The 2 dominant components are due to the annihilation of positrons in the sample MSSQ material independent of pores (~0.5 ns) para-positronium (~0.1 ns). Ortho-positronium annihilations in the MSSQ cage structure occur with a ~4 ns lifetime. Lifetimes of 10 ns and greater are due to positronium in pores and tend to increase with increasing porogen load. Open porosity is associated with a lifetime of ~100 ns (80% case, dashed line).

In the following the double peak shape will considered as two distinct lifetimes. However, they may simply represent lower and upper bounds of a broad single distribution. The lifetime analysis was performed on the full range of porogen loads. The results are shown in Figures. 20 and 21.

The positron lifetime hovers around 0.5 ns (●). All other lifetimes are due to positronium. The para-positronium (□) with nominally 0.125 ns and the smallest ortho positronium lifetime (Λ) of about 3.6 ns originate the cages inherent to MSSQ. This result agrees well with measurements on thick samples published earlier by Li et al.[23] The remaining three larger lifetimes originate from positronium in small (B) and large (γ) closed pores and pores, connect to channels which link to the surface (Ω open porosity).

Except for the case of no porogen load when a small amount of positrons diffuse to the sample surface and escape after forming positronium there (≈1%), the longest lifetime (≈95 to 108 ns) appears only when the porogen load exceeds the threshold beyond which pores begin to connect. Their intensity rises sharply up to 25% for 40% porogen load, agreeing well to the results obtained when data were collected with and without the aperture to restrict the field of view. The subsequent decline in the intensity of the open porosity component (Ω) can be explained by the 3-to-2 photon ratio data. The pores are interconnected sufficiently well (mean escape depth) such that significant amounts of positronium reach the interface to the substrate and annihilate fast. This effect is also responsible for the increase in the intensity of the *positron* lifetime. Differences cannot be resolved here.

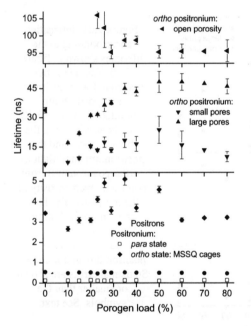

Figure 7.20: Lifetime results versus porogen load shown on three separate time scales. The shortest lifetimes on the bottom frame are due to annihilations of positrons and positronium in the MSSQ material. The middle frame shows the positronium annihilations from closed pores and from open pores in the top frame. Statistical errors are shown or smaller than the symbols. See text.

The lifetime of the *small* and *large* pores components increase linear with porogen load to 18 ns and 45 ns respectively at 40% porogen load and then level off or decline somewhat. Their intensities evolve in opposite directions; the *small* pores (B) signal decreases, while *large* pores (7) becomes more intense until 29% porogen load, followed by a sharp collapse in intensity approximately exponential in shape just as open porosity takes off. Predominantly large pores connect and turn to open porosity.

At >29% porogen load the *large* pores feed the network of connected pores and at >40% porogen load the intensity of open porosity levels off because of the fast pick-off at the interface to the substrate, indicating that percolation occurs within the porogen load range of 29 to 40%.

As a check that all positronium that was originally formed is accounted for, the ratio of the para positronium and the sum of all ortho positronium intensities is compared. Within uncertainties of the experiment, the relative intensities add up. In the intensity plot 1/3 of the sum of all ortho positronium is included as a line.

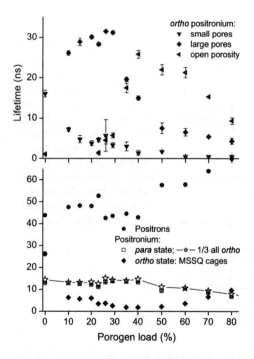

Figure 7.21: Intensities corresponding to the lifetimes shown in figure 7.20. The intensities associated with positrons and positronium annihilation in the MSSQ matrix are shown in the bottom panel and the positronium annihilations (ortho positronium) from pores and open porosity in the top panel. The line-and-star in the bottom panel indicates 1/3 of the sum of all ortho positronium annihilations. Statistical errors are shown or smaller than the symbols. See text.

7.5.8 From lifetimes to pore sizes

In order to convert the measured lifetimes into dimensions (and shapes) of the pores, additional information is required. A basic shape of pores has to be assumed. The single parameter (lifetime) is not sufficient to determine a three dimensional object. Here simple spherical shapes for isolated pores and channels (tubes) with circular cross sections are assumed. The diameter of the pores can be related to the classical mean free path (ℓ) of a particle in such objects of volume V and surface area A [46].

$$\ell = 4\frac{V}{A}$$

This relation applies generally and specifically to spheres and cylinders. Empirical relations have been developed to convert the measured lifetime (τ) to this mean free path (ℓ). In case of small pores (<2 nm) an earlier version is adequate [43, 44]. The model is based on the assumptions of spherical potential wells with infinite depth and radius r that traps the positronium.

$$\frac{1}{\tau(r)} = \lambda(r) = \frac{1}{4}(\lambda_p + 3\lambda_o)\left[1 - \frac{r - \Delta R}{r} + \frac{1}{2\pi}\sin\left(\frac{2\pi \cdot (r - \Delta R)}{r}\right)\right];$$

$\tau \leq 10 n \sec.$

The parameter, ΔR, is determined empirically from calibration measurements. The spin-averaged vacuum annihilation rates λ_p (para) and λ_o (ortho) enter as the pre-factor $\frac{1}{4}(\lambda_p + 3\lambda_o) = (0.5$ ns$)^{-1}$. Dull, Gidley and coworkers extended the range of the model to include all positronium lifetimes up to $1/\lambda_o = 142$ ns—the lifetime of ortho positronium in vacuum—by including thermal populations of excited states of positronium in the pores [47]. The resulting calibration curve is shown in Figure 7.22. This extended model includes a temperature dependence, which was confirmed in a limited set of measurements. Other models on the pore size-lifetime correlation have been published [48]. Their predictions are similar within about a factor of 2. Estimates on the influence of a specific pore shape (spheres or cubes for example) resulted in shifts on the same scale as the model dependent change.

The shape of the pores plays a minor role for the lifetime. One cannot distinguish between a channel and a number of spheres of the same diameter. However, by combining lifetime data and depth dependent measurements (which could include lifetime depth profiles), the two cases can be separated. This said, finally, the pore sizes (diameters) obtained from these samples are shown in Figure 7.23. The lifetime linked to positronium in the open cages of MSSQ remains roughly constant and translates into a mean volume diameter of 1.2±0.053 nm. When the porosity decreases towards the inherent value of MSSQ, when no porogen is added, both pore sizes appear to merge to a single size of 1.47 nm. If the volume of two cages is merged to one larger sphere the diameter increases by $2^{-1/3}$=1.26, which is the ratio (within uncertainties) of the single cage size to that of the pores when no porogen is added.

7.5.9 Porosity

Given a pore size distribution (in this case two values) and an assumed shape of pores, their relative contributions to porosity can be predicted. Obviously, large pores with high concentrations will dominate. If a positron is implanted within a diffusion length (10 nm [22]) from a pore or positronium is formed within its diffusion length (1-2 nm [24, 25]) of the pore, they contribute to the positronium signal from the pore. Both combine to an effective range (d). The intensity (I) observed for a particular pore size is proportional to the range (d), the surface area (A) of the pore and the concentration of pores of this size (n). The proportionality factor is the trapping probability (σ). The volume of the pores is (V) as determined by the lifetime for a given shape. The channels are assumed to be cylindrical in shape and their diameter to length ratio is small D/L<<1.

$$I_{sphere} = n\sigma \cdot d(\pi \cdot D^2) \qquad V = \frac{\pi}{6} \cdot D^3$$

$$I_{channel} = n\sigma \cdot d\left(\pi \cdot DL \cdot (1+\frac{D}{2L})\right) \qquad V = \frac{\pi}{4}D^2 L$$

The porosity (P) is $P = nV$ for each pore size. With pores of the size i=1,2,3 (1 and 2 are spheres and 3 are channels) that yields

$$P = \frac{1}{\sigma \cdot d} \sum_{i=1,2,3} f_i I_i D_i .$$

A shape dependent factor ($f=1/6$ for spheres, $f=1/4$ for channels when D<<L) is introduced [9]. This model is rather simple and excludes effects like detrapping or quantum mechanical tunneling among neighboring pores [60].

The result is shown in Figure 7.24 for the case of $\sigma d = 1$ nm. Plotted are the products $p = f \cdot I \cdot D$ for each pore size. For porogen loads above 40% the interface to the substrate influences the measured values. The plot is restricted to loads up to 40%. For the case of channels the size based on the channel lifetime includes a contribution from positronium in vacuum. Hence, the lifetime of the larger pores is used to estimate the diameter. The lifetime is approximately independent of the length of the channel. Also shown in the figure is the porosity based on 69±0.05% (SAWS) [39, 40]. Except for the case of no porogen load, the agreement is very good. The agreement is even better when a constant offset of 3.5% is used, a reasonable assessment of the porosity inherent to porogen free MSSQ. The excellent agreement with SAWS data justifies the above choice of $\sigma d = 1$ nm.

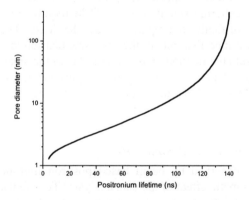

Figure 7.22: Calibration curve from measured lifetimes to pore diameters, based on the mean free path of positronium in spheres [47].

198 *Principles and Applications of Positron and Positronium Chemistry*

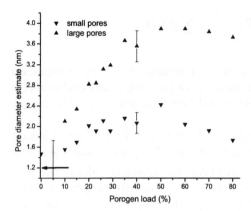

Figure 7.23: Pore diameter estimates versus the porogen load. Only the values for the "*small*" and "*large*" pores are shown. The MSSQ cage related lifetime translates into about 1.2±0.5 nm pores. The large error bar represents the scatter in the lifetime values across the porogen load range and should be considered representative for all results. Typical errors are shown at 40% porogen load.

From the individual porosity contributions the open and closed porosity fractions can be calculated and is shown in Figure 7.25 together with the complement $P_{closed}=(1-P_{open})$, the closed porosity.

Other techniques were used to investigate the pore size distribution. Small angle neutron scattering (SANS) [61], transmission electron microscopy (TEM) and small angle X-ray scattering (SAXS) are among them. Neutron scattering is performed on special samples with and without deuterided raw materials. The scattering signal is weak. TEM requires thin slices of samples, which are difficult to prepare to say the least. The preparation technique may cause modifications in the pore size distribution. SAXS appears to be the most reliable method among these. Again the small signal requires stacks of samples. An average among the samples and across each sample is measured. Measurements on similar films to the ones discussed here yielded a single pore size distribution.

7.5.10 Bimodal or not bimodal—critical comments

Is the bimodal distribution as observed by beam based positron lifetime analysis (BPALS) real or a systematic effect of the data analysis? To date the answer cannot be given with certainty. Arguments could be made why such a distribution is not observed by SAXS and is observed by BPALS in data shown here [62], and in work by Gidley *et al.*[46] Positrons are implanted at specific depths and only after measuring at different mean depth can one compile a sample average. The cages in MSSQ are observed directly. The smaller pore size contributes only a small fraction to the total porosity.

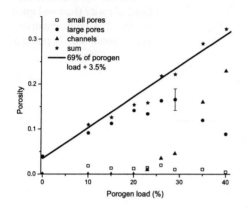

Figure 7.24: Contributions of each pore size to the porosity. For channels the diameter of the larger pores is used. See text. Also shown estimates of pore formation based densities surface acoustic wave measurements 69% plus an offset due to porosity in plain MSSQ (line). A representative estimate of the errors is shown at 29%.

Simulations have revealed a systematic tendency of the lifetime analysis technique to split broad distributions into two or possibly more narrow ones. The separation of the pore size distributions shown in Figure 7.19 is not as clear as for the smaller lifetime components. As an example the bimodal result shown in Figure 7.19 was used to create a noisy dataset. A second set was simulated from a lifetime distribution where the two pore size distributions were smeared into one broad component. Care was taken not to shift the mean lifetime and the intensity share of the distribution compared to the total was kept constant. The simulation input is shown in Figure 7.26.

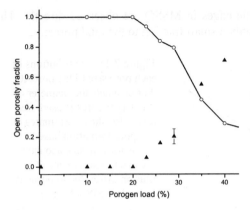

Figure 7.25: Open porosity fraction of the total porosity (7) and the complement, the closed porosity (line and open symbols). A representative error is shown at 29% porogen load.

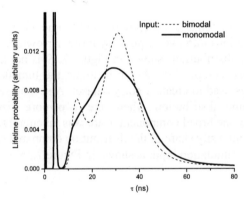

Figure 7.26: Simulated lifetime distributions based on a 23% porogen load sample. The bimodal distribution was smeared out into a monomodal one. The magnitude of the shorter lifetimes exceeds the displayed range.

Even though the lifetime distributions appear to be quite different, the recreated data are almost identical except for a small deviation near 200 ns and less obvious ones at shorter times. If the relative differences are plotted, systematic differences beyond the statistical noise are noticeable up to 200 ns, particularly when several channels are binned together. Given sufficient statistics, in principle, one can tell the difference between a bimodal and a monomodal distribution. The shown simulated spectra are based on 10^8 counts, five to 10 times the amount collected for the data discussed here.

The need to extract lifetimes from 125 ps to 142 ns from a single spectrum pushes positron lifetime technology into new realms. To date the vast majority of encountered lifetimes fall below 10 ns. The data analysis was optimized to deal with this situation. Great care was and has to be taken to obtain the correct value of t_{zero}. This is the true time when positrons enter the sample and commence annihilating with respective sample dependent annihilation rates. In practice the value of t_{zero} and the amplitudes of the exponentially decaying functions are highly correlated.

The relatively small numbers of counts per channel at large times contribute little to the overall quality of the fit (or lack thereof). For the case of porous materials where the longest lifetimes are of interest the current methods may not be appropriate.

It should be emphasized that for lifetimes up to about 10 ns, the established methods have worked remarkably well. Comparisons of the results on the same samples studied at different laboratories under a range of experimental and analysis conditions yield similar results [63]. All provide the same trends if not exactly the same absolute values.

That said, the established methods including those not mentioned above have their limits. PATFIT deals with discrete lifetimes. CONTIN [64, 65] and MELT provide continuous distribution of lifetimes. The merits and demerits of MELT and CONTIN were debated extensively [66-68]. Kansy developed an algorithm that can fit both, discrete lifetimes and log-normal distributions of lifetimes [69]. These challenges become increasingly severe as the range of lifetimes included in the data increases. To date, no method has addressed this issue satisfactorily. A number of approaches have been taken to overcome or circumvent the problems. However, their detailed discussion exceeds the scope of this chapter and will be presented elsewhere [9].

For now, the bimodal distribution *may* be an artifact. The two lifetimes can be considered as lower and upper bounds of the pore size distribution. This technique is the only one available that can provide non-destructive depth profiles without sample preparations other than mounting them in the vacuum system. Depth profiled lifetime data are currently being collected. This is practical due to the high data acquisition rate of $3 \cdot 10^3$ to 10^4 lifetime events per second, depending on the implantation depth.

7.6 Beyond count rates, ratios, and lifetimes

A couple of measurement techniques can be used on porous materials to gain further insights into the physics and chemistry of porous materials. The first is a Doppler broadening technique with a greatly improved signal to noise ratio to explore chemistry. The second measures the small angular deviations from antiparallel photon emission with momentum resolution far superior to Doppler broadening techniques. In principle, the shapes of pores and the energy loss process of positronium can be determined with this technique. Finally, beam lifetime measurements can be expanded into depth profiled lifetime measurements.

7.6.1 High sensitivity Doppler broadening

In a typical Doppler measurement only one of the two annihilation photons with on average half of the Doppler shift is observed. With second detector opposite to the first one and operated in coincidence with the first one the full Doppler shift is observed. The signal to noise ratio improves by a factor of ~1000. The shell structure of tightly bound core electrons of atoms does not change much when the atoms form a solid. Doppler shifts from these electrons can be detected and permit the identification of specific elements next to the annihilation site [70].

Since the positron traps in vacancy like open volume and positronium in pores, it will be possible to selectively detect impurities next to vacancies [71]. For metal indiffusion experiments one can design test structures of silicon, metal, low-k layer samples. Two detector coincident measurements would be performed as a function of temperature and time to observe the chemical signature of the metal in the low-k layer. The effectiveness of diffusion barriers can be tested by depositing the barrier prior to the low-k layer.

This method can also detect 3 photon decays. If the solid angle for detection of photons is large, the probability is high to observe the third photon (E_3) with the lowest energy in one detector together with one of the others (E_a or E_b). Due to symmetry and the broad distributions of the three photons, such events appear as if Doppler events with very large broadening occur (Figure 7.27). In addition to coincident events, the sum energy of all detected photons $E_{photons}$ has to equal the sum energy of the electron and positron E_{rest} rest masses. The Doppler shift momentum (p) is proportional to the difference in the photon energies.

$$E_{photons} = E_a + E_b + E_3 = 2mc^2 = 1.022 MeV$$
$$p \propto \Delta E = E_a - (E_b + E_3)$$

Para positronium two photon annihilations cause the narrow sharp feature in the center at zero momentum. The events at large momentum (> 4a.u.) are due to three photon annihilations. By adding lead filters between the sample and the detectors, the third photon is absorbed and only two photon events meet the sum energy restriction.

The very large signal to noise ratio of 1 in 10^6 and better makes this method highly sensitive to small fractions of three photons, where the standard 3-to-2 photon method fails. It has been proposed to use this technique to observe rare 3 photon decays of positrons in metals [72].

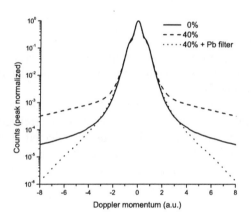

Figure 7.27: Doppler broadening spectrum from two detectors in coincidence as a function of the Doppler shift momentum in atomic units. MSSQ samples with 0% and 40% porogen are shown. The lead filter is used to stop the low energy third photon from reaching a detector. In that case, only two photon events are observed. Statistical errors are of the order of the line width and smaller.

7.6.2 Angular correlation measurements

With position sensitive detectors the deviations from antiparallel emission of two annihilation photons can be observed. The small angles on the order of millirad are translated into two dimensional electron (or positronium) momentum distributions. The resolution is sufficient to distinguish positron from para-positronium annihilations and to determine the velocity distribution of positronium.

Early angular correlation (AC) experiments on materials with closed and open porosity demonstrated that positronium indeed emerges from the surface by observing a net momentum away from the sample. The random pore and channel orientation in the sample resulted in zero net momentum in the case of closed porosity [73].

7.6.3 Lifetime depth profiles

Lifetime depth profiles will be useful for the detection of inhomogeneous pore size distributions and in tracking impurities. For microelectronic device fabrication it is crucial that subsequent processing steps do not alter the deposited porous layers. The case presented above of oxygen plasma treatment is just one example. A lifetime depth profile could provide direct evidence for the changes in pore sizes discussed in the work on HSSQ samples and oxygen plasma treatment [22]. Gidley *et al* have carried out similar depth profiles [74].

7.6.4 Combination techniques

Temperature dependent work with positrons on porous materials is still in their infancy. Doppler broadening measurements have been published but suffer from the uptake of water (ice) [75].

It is possible to combine lifetime measurements with Doppler broadening techniques. In so-called Age-Momentum-Correlated measurements (AMOC), Doppler broadening information is collected as a function of time since positron implantation into the sample [76]. The energy loss mechanism for positronium as it traps in pores can be investigated. This might reveal material dependent effects that have not been included in any porosity studies by positrons.

7.7 Inline diagnostics with positronium

The data and methods discussed in the previous sections show the power of positron and positronium annihilation methods for the characterization of porous materials and low-k dielectrics in particular. The obvious question is, whether this power can be harnessed for an online diagnostic tool in a semiconductor production line. Such a tool should be reliable, compatible with existing processes, rapid, and not more complex than any other system.

The 3-to-2 photon technique, simple counting setups and, possibly mean lifetime measurements could fulfill these criteria. A simple setup, shown schematically in Figure 7.28, is suitable for the first two applications. Positrons are implanted into the sample. Focusing into micron-sized areas is possible. Positronium forms, traps in pores and annihilates in closed pores or escapes through open porosity. Two detectors, one behind the sample and a second with an aperture on the side, observe all positronium (and positron) annihilations and only those from within the sample, respectively. The former detector is also set up to provide 3-to-2 photon ratios.

In the simplest measurement, data collected for two energies can provide a measure of the escape depth, the porosity and the open porosity fraction. The values are extracted by comparing the measurements to a set of calibration curves. With a positron beam flux of 0.2 pA, 100 locations on a wafer could be checked in 30 minutes. Mean lifetime measurements can be carried out with a pulsed positron beam. An intermediate time resolution of about 1 ns will be sufficient. A single measurement can be accomplished in about 1 minute.

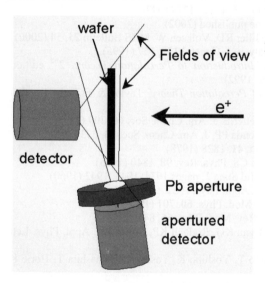

Figure 7.28: Schematic view of a possible positronium inline diagnostic tool to check low-k porous layers after deposition.

Laboratory beams take advantage of radioisotopes like ^{22}Na to generate positrons. Beams with 0.2 pA flux and mm-sized cross sections are standard. Given standard accelerator technology, a 3 MeV accelerator can replace the long half-life radioactive (2.7 years for ^{22}Na) source with a short-lived isotope (minutes) and deliver 2 pay positrons. Radiation issues would be comparable to RBS machines or MeV ion implanters. No radiation damage is caused in the wafer.

References

[1] Semiconductor Research Corporation, International Technology Roadmap for Semiconductors (2001); http://public.itrs.net/Files/2001ITRS/Home.htm.
[2] Berko S, Pendleton HN, Ann. Rev. Nucl. Part. Sci. **20**, 543 (1980).
[3] Rich A, Rev. Mod. Phys. **53**, 127 (1981).
[4] Gidley DW, Rich A, Sweetman E, West D, Phys. Rev. Lett. **49**, 525 (1982).
[5] Jean YC, Mater. Sci. Forum **175-178**, 59 (1995).
[6] Yang H, Jean YC, Mater. Sci. Forum **255-257**, 41 (1997).
[7] Saarinen K, Hautojärvi P, Corbel C, Semicond. & Semimetals **51**, 209 A (1998).
[8] Krause-Rehberg R, Leipner HS, *Positron Annihilation in Semiconductors*, Springer Series in Solid-State Sciences **127** (1999).
[9] Weber MH, Lynn KG, to be published (2002).
[10] Hawker CJ, Hedrick JL, Miller RD, Volksen W, MRS Bulletin **25**, 54 (2000)
[11] Schneider P, Applied Catalysis A-General **129**, 157 (1995).
[12] Stauffer D, Aharony A, *Introduction to Percolation Theory*, 2nd edition, Taylor & Francis, London (1992).
[13] Sahimi M, *Applications of Percolation Theory*, Taylor & Francis, London, (1994).
[14] Brunauer S, Emmett PH, Teller E, J. Am. Chem. Soc. **60**, 309 (1938).
[15] Barrett EP, Joyner LG, Halenda PP, J. Am. Chem. Soc. **73**, 373 (1951).
[16] Mills AP Jr, Phys. Rev. Lett. **41**, 1828 (1978).
[17] Hughes VW, Marder S, Wu CS, Phys. Rev. **98**, 1840 (1955).
[18] Makhov AF, Sov. Phys. Solid State **2**, pages 1934, 1945, 1942 (1960).
[19] Valkealahti S, Nieminen R, Appl. Phys. A **32**, 95 (1983).
[20] Schultz PJ, Lynn KG, Rev. Mod. Phys. **60**, 701 (1988).
[21] Puska MJ, Nieminen RM, Rev. Mod. Phys. **66**, 841 (1994).
[22] Petkov MP, Weber MH, Lynn KG, Rodbell KP, Cohen SA, Appl. Phys. Lett. **74**, 2146 (1999).
[23] Li H-L, Ujihira Y, Yoshino T, Yoshino K, Yoshii K, Yamashita T, Horie K, Polymer **39**, 4075 (1998).
[24] Brandt W, Paulin R, Phys. Rev. B **5**, 2430 (1972).
[25] Hirata K, Kobayashi Y, Ujihira Y, J.Chem. Soc. Faraday Trans. **92**, 985 (1996).
[26] Eldrup M, Vehanen A, Schultz PJ, Lynn KG, Phys. Rev. B **32**, 7048 (1985).
[27] Mogensen OE, J. Chem. Phys. **60**, 998 (1974).

[28] Mogensen OE, Appl. Phys. **6**, 315 (1975).
[29] Dupasquier A, *Enrico Fermi School summer course on Positrons in Solids*, Varenna, Italy, 1981, Brandt W, Dupasquier A, eds. (North Holland, Amsterdam, 1983), p. 510.
[30] Petkov MP, Weber MH, Lynn KG, Rodbell KP, Appl. Phys. Lett. **79**, 3884 (2001).
[31] van Veen A, Schut H, de Vries J, Hakvoort RA, IJpma MR, *Positron Beams for Solids and Surfaces*, London, Ont. Canada, Schultz PJ, Massoumi GR, Simpson PJ, eds. (AIP New York, 1990), p.171.
[32] Shaffer II EO, K. E. Howard KE, M. E. Mills ME, Townsend PH, Advanced Electronic Materials MRS Spring 2000.
[33] Postava K, Yamaguchi T, Horie M, Appl. Phys. Lett. **79**, 2231 (2001).
[34] Angyal MS, Shachamdiamand Y, Reid JS, Nicolet MA, Appl. Phys. Lett. **67**, 2152 (1995).
[35] Sun JN, Gidley DW, Dull TL, Frieze WE, Yee AF, Ryan ET, Lin S, Wetzel J, J. Appl. Phys. **89**, 5138 (2001).
[36] Ballauff M, Current Opinion in Colloid and Interface Science **6**, 132 (2001).
[37] Bauer BJ, Lin EK, Lee HJ, Wang H, Wu WL, J. Electron. Mat. **30**, 304 (2001).
[38] Baklanov MR, Mogilnikov KP, Polovinkin VG, Dultsev FN, J Vac. Sci. Technol. B **18**, 1385 (2000).
[39] Hietala SL, Smith DM, Hietala VM, Frye GC, Martin SJ, Langmuir **9**, 249 (1993).
[40] Flannery CM, Murray C, Streiter I, Schulz SE, Thin Solid Films **388**, 1 (2001).
[41] Gidley DW, Frieze WE, Dull TL, Yee AF, Ryan ET, Ho H-M, Physical Review B **60**, R5157 (1999).
[42] Gidley DW, Marko KA, Rich A, Phys. Rev. Lett. **36**, 395 (1976).
[43] Tao SJ, J. Chem. Phys. **56**, 5499 (1972).
[44] Eldrup M, Lightbody D, Sherwood JN, Chem. Phys. **63**, 51 (1982).
[45] Wang YY, Nakanishi Y, Jean YC, Sandreczki TC, J. Polymer Sci. B **28**, 1431 (1990).
[46] Gidley DW, Frieze WE, Dull TL, Sun J, Yee AF, Nguyen CV, Yoon DY, Appl. Phys. Lett. **76**, 1282 (2000).
[47] Dull TL, Frieze WE, Gidley DW, Sun JN, Yee AF, J. Phys. Chem. B **105**, 4657 (2001).
[48] Ito K, Nakanishi H, Ujihira Y, J. Phys. Chem. B **103**, 4555 (1999).
[49] Szpala S, Petkov MP, Lynn KG, Rev Sci. Instrum. **73**, 147 (2002).
[50] Weber MH, Lynn KG, unpublished work (2001).
[51] Coleman P, editor, *Positron Beams and their applications*, World Scientific, Singapore (2000).
[52] Suzuki R, Ohdaira T, Shioya Y, Ishimaru T, Jpn. J. Appl. Phys. Part 2 **40**, L414 (2001), and references therein.

[53] Bauer-Kugelmann W, Sperr P, Kögel G, Triftshäuser G, Mater. Sci. Forum **363-365**, 529 (2001) *Positron Annihilation ICPA-12*, Munich Germany. Triftshäuser W, Kögel G, Sperr P eds. (Trans Tech. Publ. Switzerland 2001).

[54] Xie L, Demaggio GB, Frieze WE, Devries J, Gidley DW, Hristov HA, Yee AF, Phys. Rev. Lett. **74**, 4947 (1995).

[55] Lynn KG, Frieze WE, Schultz PJ, Phys. Rev. Lett. **52**, 1137 (1984).

[56] Kirkgaard P, Pedersen NJ, Eldrup M, PATFIT-88, Risø-M-2740 (Roskilde, Denmark,1989).

[57] Shukla A, Peter M, Hoffmann L, Nucl. Instrum. Meth. Phys. Res. A **335**, 310 (1993).

[58] Shukla A, Hoffmann A, Manuel AA, Peter M, Mater. Sci. Forum, **255-257**, 233 (1997), and references therein.

[59] Petkov MP, Weber MH, Lynn KG, Rodbell KP, Appl. Phys. Lett. **77**, 2470 (2000).

[60] McGervey JD, Yu Z, Jamieson AM, Simha R, Mater. Sci. Forum 175-178, 727 (1995).

[61] Wu WL, Wallace WE, Lin EK, Lynn GW, Glinka CJ, Ryan ET, Ho H-M, J.Appl. Phys. **87**, 1193 (2000).

[62] Wang CL, Weber MH, Lynn KG, Rodbell KP, submitted to Appl. Phys. Lett. (2002).

[63] Wästlund C, Eldrup M, Maurer FHJ, Nucl. Instrum. Meth. Phys. Res. B **143**, 575 (1998).

[64] Provencher SW, EMBL Technical Report DA05, European Molecular Biology Laboratory, Germany (1982); and Comp. Phys. Comm. **27**, 229 (1982).

[65] Gregory RB, Zhu Y, Nucl. Instrum. Meth. Phys. Res. A **290**, 172 (1990).

[66] Gregory RB,. J. Appl. Phys. **70**, 4665 (1991).

[67] Dannefaer S, Kerr D, Craigen D, Bretagnon T, Taliercio T, Foucaran AJ, Appl. Phys **79**, 9110 (1996).

[68] Dlubek G, Hübner Ch, Eichler S, Nucl. Instrum. Meth. Phys. Res. B **142**, 191 (1998).

[69] Kansy J, Mater. Sci. Forum **363-365**, 652 (2001) *Positron Annihilation ICPA-12*, Munich Germany. Triftshäuser W, Kögel G, Sperr P eds. (Trans Tech. Publ. Switzerland 2001).

[70] Asoka-Kumar P, Alatalo M, Ghosh VJ, Kruseman AC, Nielsen B, Lynn KG, Phys. Rev. Lett. **77**, 2097 (1996).

[71] Petkov MP, Weber MH, Lynn KG, Crandall RS, Ghosh VJ, Phys. Rev. Lett. **82**, 3819 (1999).

[72] Weber MH, Lynn KG, Radiation Phys. Chem. **58**, 749 (2000).

[73] Gessmann T, Petkov MP, Weber MH, Lynn KG, Rodbell KP, Asoka-Kumar P, Stoeffl W, Howell RH, Mater. Sci. Forum **363-365**, 585 (2001) *Positron Annihilation ICPA-12*, Munich Germany. Triftshäuser W, Kögel G, Sperr P eds. (Trans Tech. Publ. Switzerland 2001).

[74] Gidley DW, Frieze WE, Dull TL, Yee AF, Ryan ET, Ho HM, Phys. Rev. B **60**, R5157 (1999).
[75] Uedono A, Chen ZQ, Suzuki R, Ohdaira T, Mikado T, Fukui S, Shiota A, Kimura S, J. Appl. Phys. **90**, 2498 (2001).
[76] Stoll H, Bandzuch P, Siegle A, Positron Annihilation—ICPA-12, Mater. Sci. Forum **363-365**, 547 (2001) *Positron Annihilation ICPA-12*, Munich Germany. Triftshäuser W, Kögel G, Sperr P eds. (Trans Tech. Publ. Switzerland 2001).

Chapter 8

Positron Annihilation Studies on Superconducting Materials

C. S. Sundar

Materials Science Division Indira Gandhi Centre for Atomic ResearchKalpakkam 603 102, Tamil Nadu, India

8.1 Introduction

The discovery of superconductivity in Ba doped La_2CuO_4 by Bednorz and Muller [1] in 1986 has heralded an unprecedented research activity [2, 3] to understand the mechanism of superconductivity in high temperature superconductors (HTSC). The fascination continues as there is still no definitive theory for high temperature superconductivity, and newer and interesting facets of strongly correlated electron systems are being unearthed [4]. The last 15 years has also seen the discovery of several new families of superconductors that include, the alkali doped C_{60} [5], borocarbides [6], MgB_2 [7] and more recently the hole injected C_{60} in a field effect geometry [8]. The announcement of each of these new superconducting materials is predictably followed by a series of investigations that include the determination of structure, and theoretical calculations of the electronic structure and vibrational properties. Experiments on the isotope effect, and pressure dependence of transition temperature are carried out to clarify the role of phonons, if any, in mediating superconductivity. A variety of doping experiments are carried out in an attempt to increase the transition temperature or the critical current density. Experiments on the determination of critical fields, the superconducting gap, coherence length and penetration depth help to characterize the superconducting state, and studies leading to information on the carrier density, density of states at the Fermi level, plasma frequency, scattering rate etc., characterize the normal state, and help to build a comprehensive picture on the mechanism of superconductivity. In the study of these new superconducting materials, which are complex systems, wherein the structure, composition and defects have significant influence on superconductivity, several new experimental techniques have been used, apart from the traditional measurements involving resistivity, magnetisation, specific heat, tunneling etc. However, it is to be realized that not all the resulting information is germane to the understanding of superconductivity per se, or even put significant constraints in building the model for superconductivity. It is against this background, we explore the role of positron annihilation spectroscopy [9, 10] to the study of superconducting materials.

The annihilation characteristics of a positron in a medium is dependent on the overlap of the positron wavefunction with the electron wavefunction [9]. From a measurement of the two photon momentum distribution, information on the electron momentum distribution can be obtained and this forms the basis of extensive studies on electron momentum distribution and Fermi surface of solids [9]. In the presence of defects, in particular, vacancy type defects, positrons are trapped at defects and the resultant annihilation characteristics can be used to characterize the defects [9, 10]. Given these inherent strengths of the technique, in the years following the discovery HTSC, a large number of positron annihilation experiments have been carried out [11, 12]. These studies can be broadly classified into three categories: (1) Studies on the temperature dependence of annihilation characteristics across T_c, (2) Studies on structure and defect properties and (3) Investigation of the Fermi surface. In this chapter we present an account of these investigations, with focus mainly on the Y 1: 2: 3 system (for an exhaustive review, see Ref. 11).

8.2 Positron annihilation in superconductors

The earliest investigations [13, 14] on the possibility of application of positron annihilation spectroscopy to the study of superconductivity predates BCS theory, and even before a detailed understanding of positron annihilation in metallic systems was known. Within the framework of the BCS theory [15], it is well known that the superconducting transition is associated with the pairing interaction, mediated by phonons, of electrons near the Fermi surface, leading to a gap $2\Delta = 3.5k_BT_c$ in the excitation spectrum. This results in the smearing of the electron momentum distribution [16] $\delta k/k_F = 2/\pi\xi_0 k_F$, where k_F is the Fermi momentum and ξ_0 is the coherence length. One of the earliest experiments to look for this smearing in the electron momentum distribution was carried out by Briscoe et al [17] on superconducting Pb (see Figure 8.1). However, no discernable change in the annihilation characteristics across T_c was detected—an observation that is rationalized in terms of the large coherence length in Pb, leading to a smearing that is below the resolution of angular correlation experiments. Lifetime measurements [12, 14] in Pb also did not indicate any changes. The absence of changes across T_c arises because the superconducting pairing affects only a small portion of electrons near the Fermi surface, whereas the contribution to the annihilation rate arises from all the valence and core electrons in the system.

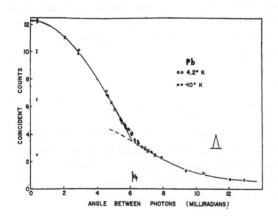

Figure 8.1 Angular correlation of positrons annihilating in normal and superconducting Pb. The arrow at approximately 6 mrad indicates the position of the Fermi cutoff for a free electron Fermi sphere, and no detectable smearing is seen in the superconducting state. The triangle indicates the angular resolution. From Briscoe et al. [17].

Formal theory of positron annihilation in a BCS superconductor has been worked out more recently [16, 18, 19], following the discovery of HTSC, and these substantiate the insensitivity of positron annihilation parameters to the superconducting transition in conventional superconductors. The influence of the opening of the gap in the superconducting state on the thermalisation of positrons has been analysed by Perkins and Woll [20], and this is also seen to be very small. Thus, it appears that positron annihilation is of limited value in the investigation of the changes in electronic structure associated with the superconducting transition in the conventional BCS superconductors. It is against this background that the large changes in annihilation characteristics seen in the cuprate superconductors have attracted considerable attention.

8.3 Temperature variation of positron annihilation characteristics in HTSC

The first measurement of the temperature dependence of annihilation characteristics across T_c in $YBa_2Cu_3O_{7-x}$ (Y 1: 2: 3) was carried out by Jean et al [21]. In the ceramic superconductor, the positron lifetime and the Doppler lineshape parameter, I, were observed to decrease below T_c, whereas no such change was seen in the oxygen deficient non-superconducting compound. Figure 8.2 shows the results [12] on single crystal of Y 1: 2: 3. It is seen that the lifetime which is constant in the normal state, decreases below the superconducting transition temperature. This significantly large change in lifetime is in contrast to the conventional superconductors, wherein there is no change in lifetime across T_c. We show in Figure 8.3, the results of positron lifetime measurements in $La_{1.85}Sr_{0.15}CuO_4$ [22]. In the superconducting sample, an increase in lifetime is observed below T_c (33K), a feature not seen in the non-superconducting sample. These initial studies, were followed by several investigations, mostly on the Y 1: 2: 3 system [23-31]. While most of these measurements in Y 1: 2: 3 indicated a decrease in positron lifetime below T_c, in some experiments [25, 28] different kinds of temperature dependencies have also been observed. For example, Harshman et al[25] observed an increase in lifetime below T_c.

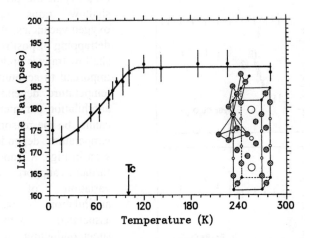

Figure 8.2 Variation of positron lifetime as a function of temperature in single crystal of Y 1: 2: 3. The arrow indicates the T_c. The inset shows the unit cell of Y 1: 2: 3. From Sundar et al. [12].

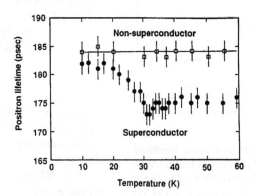

Figure 8.3 Positron lifetime vs temperature for superconducting $La_{1.85}Sr_{0.15}CuO_4$ with $T_c = 33$ K. The lifetime variation in the non-superconducting sample, obtained by heating at 950 K in vacuum is also shown. From Jean et al. [22].

In some of the experiments [30, 31], variation of lifetime uncorrelated with T_c, has been observed. This has prompted explanations for the observed temperature dependence of lifetime in terms of positron trapping behaviour [32, 33] in the presence of shallow traps such as oxygen vacancies. While the detrapping of positrons from shallow traps is undoubtedly important to account for the temperature dependence of annihilation parameters, this cannot be the reason for the temperature dependence as seen in Figs. 8.2 and 8.3. In further support of the variation of positron annihilation parameters associated with the superconducting transition, we present in Figure 8.4, the recent results of Doppler broadening lineshape, S, measurements in $DyBa_2Cu_3O_7$ (T_c=91.6K), $Y_{0.5}Pr_{0.5}Ba_2Cu_3O_7$ (Tc=27.2K) and the non-superconducting $PrBa_2Cu_3O_7$ [34, 35]. Clearly, a correlated variation of the lineshape parameter with the superconducting transition is observed.

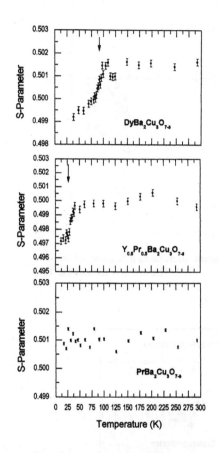

Figure 8.4 Doppler broadening lineshape parameter vs temperature in $DyBa_2Cu_3O_7$ $Y_{0.5}Pr_{0.5}Ba_2Cu_3O_7$ and $PrBa_2Cu_3O_7$. The arrow indicates T_c. From Jung et al [34,35].

The different temperature dependencies of positron annihilation parameters observed in Y1: 2: 3 and $La_{1.85}Sr_{0.15}CuO_4$ (cf. Figs 8.2 and 8.3), and even within the same system of Y

1: 2: 3 [21-31] has caused considerable confusion in identifying the intrinsic temperature dependence of lifetime across T_c. Several plausible explanations have also been put forward to account for the observed experimental results: These include (1) smearing of the momentum distribution in the superconducting state [29] to account for the decrease in lineshape parameter in Y 1: 2: 3, (2) change in the positron-hole correlation due to the real-space pairing of holes in the Cu-O planes [36], and (3) the stiffening of the charge response to the positron within the resonating valence bond model [37]. The later models predicted an increase in lifetime below T_c.

One of the key experiments [38] that has helped to elucidate the different temperature dependencies in HTSC's is the controlled study on undoped, as well as on Zn and Ga doped Y 1: 2: 3. Figure 8.5 shows the results of the experiments on Y 1: 2: 3 doped with various levels of Zn. It is seen that the decrease in bulk lifetime seen in undoped Y 1: 2: 3 reverses to an increase in lifetime in Zn doped Y 1: 2: 3. The variation of lifetime in all the cases is seen to be correlated with T_c, which decreases with increase in Zn content. In the following, we show [39] that the different temperature dependencies in undoped and Zn-doped Y 1: 2: 3 are related to the difference in the positron density distribution with respect to the superconducting Cu-O planes.

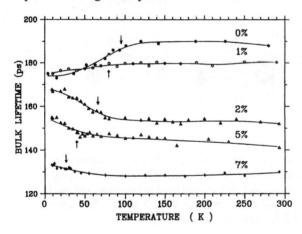

Figure 8.5 Variation of bulk lifetime as a function of temperature in undoped and Zn doped Y 1:2:3. Arrows are marked at T_c. From Jean et al. [38].

8.3.1 Positron distribution and the electron-positron overlap

The crystal structure of Y 1: 2: 3 (see inset of Figure 8.2) is characterized by two dominant structural features: The superconducting Cu-O planes and the Cu-O chains. The latter are believed [2] to play an important role in controlling the doping of the Cu-O planes. Using the known structural parameters, the positron density distribution (PDD) in Y 1: 2: 3 has been calculated [39] by solving the Schrodinger equation, with the positron potential obtained as a sum of the Hartree potential obtained from OLCAO calculations, and the correlation potential in the local density approximation. The resulting [39] PDD in the (010) plane of Y 1: 2: 3 are shown in Figure 8.6(a). From this figure, it is seen that the maximum of the positron density is between the Cu(1) atoms of the Cu-O chain in the basal plane, and there is very little positron density in the Cu(2)-O plane. These features of the PDD are in agreement with those obtained using the

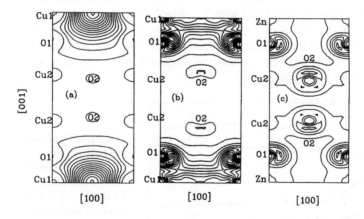

Figure 8.6 (a) Contour plots of the positron density distribution in $YBa_2Cu_3O_7$. The maximum of the positron density is in between the Cu(1) atoms. (b) Contour plot of the overlap of the electron and positron densities in the (010) plane of undoped Y 1:2:3, and (c) Zn- doped Y 1:2:3. From Bharathi et al. [39].

the linearized augmented plane wave (LAPW) method [27] and superposed atomic potentials [40, 41]. Further insight is obtained from a calculation of the electron-positron overlap function, that provides information on the relative contribution of various atoms in the unit cell to the observed annihilation rate. The overlap function in Y 1: 2: 3 [see Figure 8.6(b)] is seen to be centered around the apical oxygen atom, O(1), associated with the Cu-O chains and there is very little contribution from the Cu (2)-O plane. However, in the case of Zn-doped Y 1: 2: 3 (see Figure 8.6c), it is seen that in addition to the contribution from the O(1), there is a significant contribution from the O(2) atom of the Cu(2)-O plane. Based on this difference in contribution to the annihilation rate from the planar and apical oxygen atoms, the difference in temperature dependence of annihilation parameters undoped and Zn-doped Y 1: 2: 3, (cf. Figure 8.5) has been explained by invoking charge (electron) transfer from the plane to the chains [38, 39]. This is schematically shown in Figure 8.7. An electron transfer from the planar to apical O atom can account for the decrease in lifetime in undoped Y 1: 2: 3, since the annihilation is dominated by the apical O atom. In the case of Zn-doped Y 1: 2: 3, such a depletion of electrons from the Cu-O layers contributes to an increase in lifetime, since there is significant annihilation from the planar O atoms of the Cu-O layer.

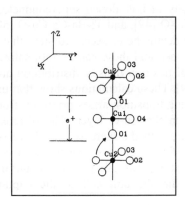

Figure 8.7 Schematic diagram of the local electron transfer in Y 1:2:3. From Bharathi et al. [39].

While the above-mentioned physical picture may appear too simplistic, at least at the present stage of development in the theory of superconductivity in HTSC [4], we must mention that it provides a broad qualitative framework to account for the observed temperature dependence of annihilation parameters, even in other high-temperature superconductors [39], provided the correct positron distribution within the unit cell is taken into account. The variation of positron lifetime with pressure has also been interpreted in terms of the model based on charge transfer [42].

It may be remarked that the notion of charge transfer between the Cu-O planes and the Cu-O chains was first introduced [43] to account for the variation of lattice parameter and Cu-O bond lengths with oxygen stoichiometry in Y 1: 2: 3. We have extended it to account for the positron results, and the interpretation presented above suggests that there is a charge transfer and associated changes in Cu-O bond lengths when Y 1: 2: 3 enters the superconducting phase. Evidence for changes in the Cu-O bond lengths across the superconducting transition has been obtained in EXAFS measurements [44]. The charge transfer excitation between the apical oxygen and the plane has also been used to interpret the infrared [45] and Raman data [46] in Y 1: 2: 3. The mechanism of charge transfer forms the basis of several theoretical models of superconductivity [47].

8.3.2 Studies on other high-temperature superconductors

Positron annihilation measurements across T_c, coupled with the calculations of PDD have been carried out in a variety of hole-doped superconductors that include $YBa_2Cu_4O_8$ [48], Bi-Sr-Ca-Cu-O [49], and Tl-Ba-Ca-Cu-O [50, 51] systems. We will not labor with the details here, except to state that a variety of temperature dependencies are seen and these can be rationalized when the results are analysed in terms of positron density distribution and the electron-positron overlap function [39]. These calculations show that the positron's sensitivity to the superconducting transition arises primarily from the ability to probe the Cu-O network in the Cu-O layer. The different temperature dependencies of lifetime, i.e., both the increase and decrease, can be understood in terms of a model of local electron transfer from the planar oxygen atom to the apical oxygen atom, after taking into account the correct positron density distribution within the unit cell of the cuprate superconductor.

As another example of positron lifetime measurements as a function of temperature, we present the results [52] in electron-doped superconductor $Nd_{2-x}Ce_xCuO_{4-y}$. As seen in Figure 8.8, the lifetime in the superconducting (T_c = 22K) oxygen-deficient $Nd_{1.85}Ce_{0.15}CuO_{3.98}$ sample exhibits a pronounced decrease with the lowering of temperature, whereas no such change is seen in the stochiometric nonsuperconducting sample. With support from theoretical calculations [52], the decrease in lifetime at low temperatures in the superconducting sample has been understood in terms of an increase in the positron trapping at oxygen vacancies, which act as shallow traps. This example illustrates the importance of shallow traps in determining the temperature dependence of positron lifetime in HTSC systems. It is also seen from Figure 8.8, that no changes in lifetime occurs across T_c in the Nd-Ce-Cu-O system whose structure (T' phase) is characterized [52] by a planar CuO_2 layer, in contrast to the octahedral and pyramidal Cu-O environments in other cuprate superconductors. This may be taken as evidence for the importance of the apical oxygen atom in determining the charge transfer and the consequent temperature variation of annihilation parameters across T_c (cf. Figure 8.7). Positron lifetime is also not seen to change across the superconducting transition in Ba-K-Bi-O [53], which is a phonon-mediated superconductor.

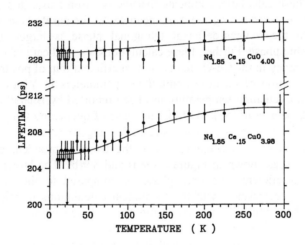

Figure 8.8 Variation of positron lifetime with temperature in the superconducting $Nd_{1.85}Ce_{0.15}CuO_{3.98}$ and non-superconducting $Nd_{1.85}Ce_{0.15}CuO_4$ samples. The arrow is drawn at $T_c = 22K$. From Sundar et al. [52].

8.4 Defects in HTSC

It is well known that in the HTSC materials, the structural defects and, in particular, the oxygen vacancies play an important role in controlling the superconducting properties [2]. On quenching the superconducting orthorhombic Y 1: 2: 3 from elevated temperatures, the oxygen atoms are known to be depleted from the basal plane containing the Cu-O chains, resulting in formation of the nonsuperconducting tetragonal phase. Since in the Y 1: 2: 3 system the positron density is in the region of the Cu-O chains [cf. Figure 8.6(a)], it can be anticipated that the annihilation characteristics will be sensitive to the oxygen content in the basal plane. Theoretical calculations of the postiron distribution and lifetimes have been carried out [40, 41] in the ordered structures having oxygen stoichiometry of O_6, $O_{6.5}$, and O_7. These calculations indicate that the positron exists in a delocalized state in the ordered arrangement of vacancies in the basal plane and that the lifetime in the oxygen-deficient tetragonal phase is larger than in the orthorhombic phase. Further, these calculations also indicate that an isolated oxygen vacancy in the Cu-O chain is not an efficient trap of positrons.

The variation of positron annihilation parameters in $YBa_2Cu_3O_{7-x}$ as a function of quenching temperature has been reported by Bharathi et al. [54] and Smedskjaer et al [55]. With the increase of quenching temperature, the oxygen deficiency, x, is seen to increase (cf. Figure 8.9c). Associated with this is the increase in lifetime, τ, and the Doppler broadening lineshape parameter , I, as shown in Figures. 8.9(a) and 8.9(b). The larger lifetime in the oxygen-deficient tetragonal phase as compared to the orthorhombic phase, is in accordance with the theoretical calculations mentioned above [40, 41]. The continuous increase in lifetime with quenching temperature, as seen in Figure 8.9, can be attributed to the increase in oxygen vacancy concentration. However, given that the oxygen vacancies are weak traps [40], the variation of lifetime with quenching temperature may reflect changes in the electronic/structural properties of the Cu-O chain region associated with the change in oxygen deficiency, rather than the vacancy concentration itself, directly. As indicated earlier [43], the changes in structural parameters with increasing oxygen deficiency has been explained in terms of electron transfer from the Cu-O chain to the Cu-O plane. Given that the positron probes the chain region [cf. Figure 8.6(a)], such an electron transfer can also account for the observed increase in lifetime with the oxygen deficiency.

In view of the fact that the positron annihilation characteristics in the various ordered phases are different [40, 41], it can be expected that PAS can contribute to understanding phase transformations in Y 1: 2: 3. One interesting aspect of the Y 1: 2: 3 system relates to the nature of the low-temperature part of the phase diagram. Theoretical calculations by Katchaturyan et al. [56] indicate decomposition of off-stoichiometric Y 1: 2: 3 into orthorhombic O_7 and tetragonal O_6 phases, whereas calculations by de Fontaine et al. [57] suggest decomposition into orthorhombic $O_{6.5}$ and O_7 phases. To resolve this issue, positron lifetime experiments have been carried out [58] as a function of aging time at various aging temperatures, and we present illustrative results in Figure 8.10. It is seen that in a sample having a nominal composition of $YBa_2Cu_3O_{6.75}$ with aging at 200°C, while the average oxygen stoichiometry as determined by weight measurements remains constant, the positron lifetime is seen to increase. Such an increase in lifetime can be understood if the system phase-separates into oxygen-rich and oxygen-deficient regions, with the positron probing the open volumes of the latter. From a comparison of the increased lifetime with the theoretically calculated

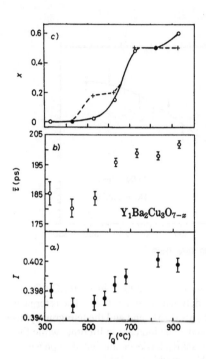

Figure 8.9 Variation with quenching temperature of, (a) positron lifetime, (b) Doppler broadened lineshape parameter, I, and (c) oxygen deficiency, as obtained from weight loss (+) and T_c (o) measurements. From Bharathi et al. [54].

[40] values for $O_{6.5}$ and O_6 phases, it is inferred [58] that the oxygen deficient region corresponds to the tetragonal O_6 phase.

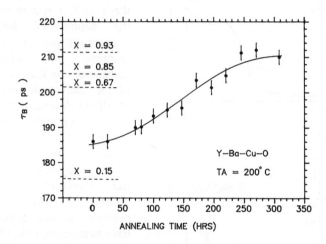

Figure 8.10 Lifetime versus annealing time at 200° C in a $YBa_2Cu_3O_{7-x}$ sample having the average oxygen deficiency of x = 0.25. The reference lifetime values in orthorhombic phase (x= 0.15), and and the tetragonal phase (x = 0.67, 0.85 and 0.93) are indicated. The increase in lifetime, while the average stoichiometry remains the same, has been attributed to phase separation. From Vasumathi et al. [58].

Further examples of positron study of defects in HTSC are studies carried out to understand the nature of flux-pinning defects that lead to an increase in critical-current density on neutron-irradiated Y 1: 2: 3. Experiments [59] on positron lifetime and critical-current density measurements on various neutron-irradiated samples of Y 1: 2: 3 indicate that the critical current density is correlated with the micro-void density, as obtained from the analysis of positron lifetime measurements. Investigation of defects in other HTSC superconductors, such as La-Sr-Ca-Cu-O [60], Bi-Sr-Ca-Cu-O [49], and Nd-Ce-Cu-O [52], have also been carried out.

8.5 Fermi surface studies

The angular correlation of positron annihilation radiation, in particular 2D-ACAR, has been successfully used to investigate Fermi surfaces in metals and alloys and in several A15 superconductors [9]. In the early days of HTSC, one of the key issues that had a direct bearing on understanding the normal state of cuprates and in constructing the theory of high-temperature superconductivity, relates to the question of the existence of the Fermi surface. Band structure calculations [61, 62] in Y 1: 2: 3 predict a Fermi surface (see Figure 8.11) made of four principal sheets: A ridge due to the electronic states of the Cu-O chains, two barrels due to Cu-O planes, and a pill box due to the chains and apical oxygens. Given the fundamental importance of this problem, several groups have carried out [63-66] 2D-ACAR studies in Y 1: 2: 3. An unequivocal answer to the question of the existence of the Fermi surface was not obtained for long time due to the difficulties associated with the preparation of good quality single crystals free from defects. In fact, in some of the experiments, Fermi surface-like features were seen even in the insulating phase. The search for the Fermi surface in Y 1: 2: 3 by 2D-ACAR was successful as soon as untwined samples became available. The ridge features, which are associated with the electronic states of the Cu-O chain, the region probed by the positron (cf. Figure8.3), was first observed by Haghighi

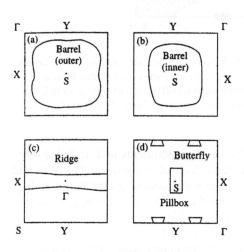

Figure 8.11 The calculated Fermi surface of $YBa_2Cu_3O_7$ in the Γ-X-S-Y plane. From Bansil et al. [62].

et al. [67] and was soon confirmed by two other independent measurements [68, 69]. The results from 2D-ACAR experiments, summarized in Figure 8.12, are seen to be in conformity with band structure calculations. These positron experiments, coupled with angle-resolved photoemission experiments, have helped to clarify the much-debated controversy with respect to the description of the electronic structure of cuprates.

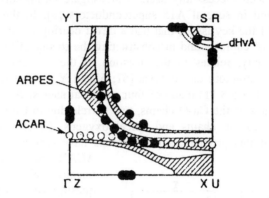

Figure 8.12 Cross section in the (k_x, k_y) plane of the calculated Fermi surface in Y 1:2:3. The experimental results from angle-resolved photoemission (ARPES), de Haas van Alphen measurements (dHvA), and angular correlation of annihilation radiations (ACAR) are indicated. From Pickett et al. [61].

Following the success of the 2D-ACAR experiments in Y 1: 2: 3, further detailed experiments to investigate the effects of oxygen stoichiometry and the influence of rare earth and transition metal doping on the Fermi surface of the Y 1: 2: 3 system have been carried out [70]. In addition, experiments in other superconducting cuprates, such as Nd-Ce-Cu-O [71], La-Sr-Cu-O [72], Bi-Sr-Ca-Cu-O [73], and Tl-Ba-Ca-Cu-O [74], have also been carried out. While Fermi surface signatures, in agreement with LDA calculations, are seen in all the cuprate superconductors, some inconsistencies remain, but these are generally attributed to the complications arising out of positron trapping at defects. Efforts at improving the resolution, statistics, and single-crystal quality continue for more detailed studies on the electronic structure of cuprates. In principle, 2D-ACAR could provide direct measurements of the superconducting gap, but there are no convincing results as yet. In contrast, the ARPES experiments have made great strides in providing detailed information on the superconducting gap and its symmetry [75].

8.6 Studies on other novel superconductors

8.6.1 Fullerenes

Fullerenes, the close-caged carbon clusters C_{60} and C_{70}, have attracted considerable attention, both due to their elegant molecular symmetry and to the observation of superconductivity in alkali-doped C_{60} [5]. Another dominant interest [76] in these molecular solids has been in their structural transformations with respect to temperature and pressure which are related to the orientational correlations amongst the molecules. From the point-of-view of the use of positron annihilation spectroscopy in the study of C_{60}, it is first of interest to know if the positron probes the interior of the hollow cage structures, or if the annihilation is from the interstitial regions of the solid. From the measurements [77] of positron lifetime in C_{60} as a function of pressure, coupled with theoretical calculations of positron density distribution (see Figure 8.13), it has been established that positron probes the interstitial regions. The annihilation characteristics are hence sensitive to the effects of intercalation of K [78]. For example, the lifetime is seen to decrease [78] from 374 ps in pristine C_{60} to 343 ps in K_3C_{60}. While experiments on variation of positron lifetime across T_c in alkali-doped C_{60} have not been carried out so far, studies of structural transition in pristine C_{60} and the metal-insulator transition in RbC_{60} have been investigated using positron annihilation spectroscopy [79].

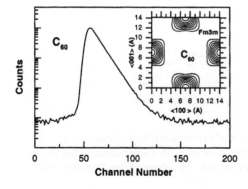

Figure 8.13 Positron lifetime spectrum in solid C_{60}, characterized by a single component with $\tau = 374$ ps. The inset shows the positron distribution in the basal plane of C_{60}, with maxima in the interstitial regions of the fcc lattice. From Sundar et al. [79].

8.6.2 Borocarbides

With the discovery of superconductivity (Tc = 15.5 K) in the Y-Ni-B-C system [6, 80], a new class of quaternary borocarbide superconductors has emerged. Superconductivity has been observed in several rare earth (Lu, Tm, Er and Ho) nickel borocarbides[80], and with transition metals such as Pd and Pt. The superconducting phase having the composition of YNi_2B_2C, crystallizes [81] in a tetragonal structure with alternating Y-C and Ni_2B_2 layers. Band structure calculations [82] indicate that these materials, unlike cuprate superconductors, are three-dimensional metals.

Figure 8.14(a) shows the results [83] of positron lifetime measurements in a series of rare earth borocarbides. The lifetime is seen to increase linearly with the lattice volume, as obtained from x-ray diffraction measurements. Calculations [83, 84] of positron density distribution in these rare earth borocarbides indicate that the positron samples the unit cell uniformly (unlike many of the cuprate superconductors) and the calculated lifetime is seen to increase slightly with the unit cell volume as shown in the top panel. It is also noted from Figure 8.14(a) that the lifetime in YNi_2B_2C is significantly larger than this linear trend.

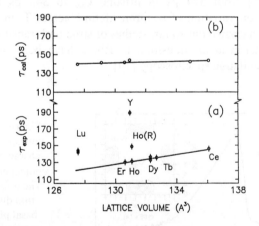

Figure 8.14 (a) Variation of experimental bulk lifetime as a function of unit cell volume in various rare earth borocarbides. (b) Computed bulk lifetime vs unit cell volume. From Bharathi et al. [83].

In order to account for this, further calculations [84] have been carried out envisaging various possible defect structures, such as monovacancies at carbon site, Y site, and clusters of carbon vacancies, etc. Based on these calculations, it is inferred [84] that the measured lifetime at room temperature in YNi_2B_2C is best accounted for in terms of positron trapping at clusters of carbon vacancies

The results of the experiments [84] on the variation of positron lifetime as a function of temperature across T_c in YNi_2B_2C are shown in Figure 8.15. This large decrease in lifetime cannot be accounted for in terms of thermal contraction of the lattice. Based on a positron trapping model, incorporating detrapping from carbon vacancies, it is possible to account [84] for the observed variation of lifetime as shown in Figure 8.15. It is also noted from Figure 8.15 that in the YNi_2B_2C system, no change in lifetime is seen across T_c. This behavior, which is similar to the behavior seen in conventional BCS superconductors, is different from that seen in the cuprate superconductors. This may be taken as evidence for a phononic mechanism. However, it is also possible that the absence of lifetime change across T_c is related to positron trapping at carbon vacancies, and further experiments are called for.

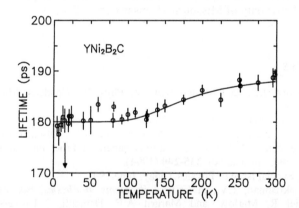

Figure 8.15 Positron lifetime vs temperature in YNi_2B_2C. The arrow indicates T_c. The continuous line is the fit based on a model incorporating the detrapping of positrons from shallow traps at carbon vacancies. From Sundar et al. [84].

8.7 Summary

Starting from the days when positron annihilation spectroscopy was not considered a useful technique to explore superconductivity in BCS superconductors, we have come a long way. The extensive studies carried out in cuprate superconductors, both in the temperature variation of annihilation parameters across T_c and the investigation of defects in these systems, bear testimony to this. The 2D-ACAR studies leading to the elucidation of the Fermi surface, at a time when most needed, certainly forms the most notable and enduring contribution of positron annihilation spectroscopy to the HTSC saga. However, experiments below the superconducting transition, providing detailed information on the superconducting gap and its symmetry at the level of detail as done with ARPES [75], still needs to be done.

Acknowledgments

The contents of this chapter are based on a long-standing collaborative work done with Dr. A. Bharathi, Materials Science Division, Kalpakkam and Prof. Y.C. Jean, University of Missouri—Kansas City, USA..

References

[1] Bednorz, J.G., and Muller, K.A., Z. Phys. B.- Condens. Mater. **64**, 189 (1986).
[2] High Temperature Superconductors and Materials and Mechanisms of Superconductivity, Eds. J. Muller and J.L. Olsen, North-Holland, Amsterdam, 1988, See also various papers in Proc. Int. Conf. M^2HTSC, Grenoble, Physica, vol. **235-240** (1994).
[3] See, Special issue of Physics Today, June 1991.
[4] Ramakrishnan, T.V., in *Critical Problems in Physics*, Ed. Val. L. Fitch, Daniel R. Marlow, and Margrit A.E. Dementi, (Universities Press, Hyderabad, India, 2000), p 75.
[5] Hebard, A.F., Rossiensky, M.J., Haddon, R.C., Murphy, D.W., Glarum, S.H.,. Palstra, T.T.M , Ramirez, A.P. and Kortan, A.R., Nature **350**, 600 (1991).

[6] Nagarajan, R., Mazumdar, C., Hossain, Z., Dhar, S.K., Gopalakrishnan, K.V., Gupta, L.C., Godart, C., Padalia, B.D., and Vijayaraghavan, R., Phys. Rev. Lett. **72**, 274 (1994).
[7] Nagamutsu, J., Nagakawa, N., Muranaka, T., Zenitani Y., and Akimutsu, J., Nature **410**, 63 (2001).
[8] Schon, J.H., Ch. Kloc, and Batlogg, B., Science **293** 2432 (2001)
[9] Brandt W. and Dupasquier A., Eds., *Positron Solid State Physics*, 1983, North Holland, Amsterdam.
[10] Puska, M.J., and Nieminen, R.M., Rev. Mod. Physics **66**, 41 (1994).
[11] Manuel, A.A., Chapter 192 in the *Handbook on the Physics and Chemistry of Rare Earths*, Ed. K.A. Gschneidner, Jr., L. Eyring and M.B. Maple, (Elsevier, North-Holland, 2000), p. 417.
[12] Sundar, C.S., Bharathi, A., Hao, L., Jean, Y.C., Hor, P.H., Meng, R.L., Huang, Z.J., and Chu, C.W., in *Superconductivity and Its Applications*, Ed. H.S. Kwok, Plenum, New York, 1990, p.335.
[13] Dresden, M., Phys. Rev. **93**, 1413 (1954).
[14] Shafroth, S.M., and Marcus, J.A., Phys. Rev. **103**, 585 (1956).
[15] Schreiffer, J.R., *Theory of Superconductivity* (Addisson-Wesley, USA, 1988).
[16] Gyorffy, B.L., Majssnerowski, J., Suvasini, M.B., Szoteck, Z.,and Temmerman, W., in *Positron Spectroscopy of Solids*, Ed. A. Dupsaquier and A.P. Mills, Jr., (North-Holland, Amsterdam, 1995), p. 145.
[17] Briscoe, C.V., Beardsley, G.M., and Stewart, A.T., Phys. Rev. **141**, 379 (1966).
[18] Barnes, S.E., and Peter, M., Phys. Rev. B **40**, 10958 (1989).
[19] Benedek, R., and Schlutter, H.B., Phys. Rev. B **41**, 1789 (1990).
[20] Perkins, A., and Woll, Jr., A.J., Phys. Rev. **178**, 530 (1969).
[21] Jean, Y.C., Wang, S.J., Nakanishi, H., Hardy, W.N., Hayden, M.Y., Kiefl, R.F., Meng, R.L., Hor, P.H., Huang, Z.J., and Chu, C.W., Phys. Rev. B **36**, 3994 (1987).
[22] Jean, Y.C., Kyle, J., Nakanishi, H., Turchi, P.E.A., Howell, R.H., Wachs, A.L., Fluss, M.J., Meng, R.L., Hor, P.H., Huang Z.J., and Chu, C.W., Phys. Rev. Lett., **60**, 1069 (1988).
[23] Usmar, S.G., Sferlazzo, P., Lynn, K.G., and Moodenbaugh, A.R., Phys. Rev. B **36**, 8854 (1987).
[24] Sundar, C.S., A.K. Sood, Bharathi, A., and Hariharan, Y., Physica C **153-155**, 155 (1988).
[25] Harshman, D.R., Schneemeyer, L.F., Waszzak, Y.V., Jean, Y.C., Fluss, M.J., Howell, R.H., and Wachs, A.L., Phys. Rev. B **38**, 848 (1988).
[26] Usmar, S.G., Lynn, K.G., Moodenbaugh, A.R., Suenga M., and Sabatini, R.L., Phys. Rev. B **38**, 5126 (1988).
[27] von Stetten, E.C., Berko, S., Li, S.S., Lee, R.R., Brynestad, J., Singh, D., Krakauer, H., Pickett, W.E., and Cohen, R.E., Phys. Rev.Lett. **60**, 2198 (1988).

[28] Corbel, C., Bernede, P., Pascard, H., Rullier-Alberque, F.,Korman, R., and Mariccp. J.F., Appl. Phys. A **48**, 335 (1989).
[29] Smedskjaer, L.C., Veal, B.W., Leginni, D.G., Paulikas, A.P.,and Nowicki, L.J., Phys. Rev. B **37**, 2330 (1988).
[30] Usmar, S.G., Biasini, M., Moodenbaugh, A.R., Xu, Y., and Fretwell, H.M., J. Phys. Condens. Mater. **6**, 10487 (1994).
[31] Ishibahi, S., Suenega, K., Yamamoto, R., Doyama, M., and Matsumoto, T., J. Phys. Condens. Mater. **2**, 3691 (1990).
[32] Nieminen, R.M., J. Phys. Chem. Solids **52**, 1577 (1991).
[33] McMullen, T., Jena, P., Khanna, S.N., Li, Y., and Jensen, K.O., Phys. Rev. B. **43**, 10422 (1991).
[34] Jung, K., Byrne, J.G., de Anrade, M.C., and Maple, M.B., J. Appl. Phys. **78**, 5534 (1995).
[35] Jung, K., Byrne, J.G., de Anrade, M.C., and Maple, M.B., Physica B **217**, 23 (1996).
[36] Chakraborty, B., Phys. Rev. B **39**, 215 (1989).
[37] McMullen, T., Phys. Rev. B **41**, 877 (1990).
[38] Jean, Y.C., Sundar, C.S.,, Bharathi, A., Kyle, J., Nakanishi, H., Tseng, P.K., Hor, P.H., Meng, R.L., Huang, Z.J., Chu, C.W., Wang, Z.Z., Turchi, P.E.A., Howell, R.H., Wachs, A.L., and Fluss, M.J., Phys. Rev. Lett. **64**, 1593 (1990).
[39] Bharathi, A., Sundar, C.S., Ching, W.Y., Jean, Y.C., Hor, P.H., Xue, Y.Y., and Chu, C.W., Phys. Rev. B **42**, 10199 (1990).
[40] Bharathi, A., Sundar, C.S., and Hariharan, Y., J. Phys.Condens. Matter. **1**, 1467 (1988).
[41] Jensen, K.O., Nieminen, R.M., and Puska, M.J., J.Phys.Cond.Matter. **1**, 3727 (1989).
[42] Chu, C.W., Hor, P.H., Lin, J.G., Xiong, Q., Meng, R.L., Xue, Y.Y., and Jean, Y.C., in *Frontiers of High Pressure Research*, Ed. H.D. Hochheimer and R.d. Etters (plenum, New York, 1991), p. 383.
[43] Cava, R.J., Batlogg, B., Rabe, K.M., Rietman, E.A., Gallagher, P.K., and Rupp Jr., K.W., Physica C **156**, 523 (1988).
[44] Conradson, S.D., and Raidstrick, I.D., Science **243**, 1340 (1989).
[45] Batistic, I., Bishop, A.R., Martin, R.L., and Tesanovic, Z.,Phys. Rev. B **40**, 6896 (1989).
[46] Monien, H., and Zawadowski, A., Phys. Rev. Lett. **63**, 911 (1989).
[47] Bishop, A.R., Martin, R.L., Muller, K.A., and Tesanovic, Z., Z. Phys. B **76**, 17 (1989).
[48] Sundar, C.S., Bharathi, A., Jean, Y.C., Hor, P.H., Meng, R.L., Xue, Y.Y., Huang, Z.J., and Chu, C.W., Phys. Rev. B **41**, 11685 (1990).
[49] Sundar, C.S., Bharathi, A., Ching, W.Y., Jean, Y.C., Hor, P.H., Meng, R.L., Huang, Z.J., and Chu, C.W., Phys. Rev. B **43**, 13019 (1990).
[50] Sundar, C.S., Bharathi, A., Ching, W.Y., Jean, Y.C., Hor, P.H., Meng, R.L., Huang, Z.J., and Chu, C.W., Phys. Rev. B **42**, 2193 (1990).

[51] Jean, Y.C., Nakanishi, H., Fluss, M.J., Wachs, A.L., Turchi, P.E.A., Howell, R.H., Wang, Z.Z., Meng, R.L., Hor, P.H., Huang, Z.J., and Chu, C.W., J. Phys.: Condens. Matter. **1**, 2696 (1989).
[52] Sundar, C.S., Bharathi, A., Jean, Y.C., Hor, P.H., Meng, R.L., Huang, Z.J., and Chu, C.W., Phys. Rev. B **42**, 426 (1990).
[53] Sundar, C.S., Bharathi, A., Jean, Y.C., Hinks, D.G., Dabrowski, B., Zheng, Y., Mitchell, A.W., Ho, J.C., Howell, R.H., Wachs, A.L., Turchi, P.E.A., Fluss, M.J., Meng, R.L., Hor, P.H., Huang, Z.J., and Chu, C.W., Physica C, **162-164**, 1379 (1989).
[54] Bharathi, A., Hariharan, Y., Sood, A.K., Sankara Sastry, V., Janawadkar, M.P., and Sundar, C.S.,, Europhys. Lett. **6**, 369 (1988).
[55] Smedskjaer, L.C., Veal, B.W., Legnini, D.G., Paulikas, A.P., and Nowicki, L.J., Physica B+C **150**, 56 (1988).
[56] Khatchaturyan, A.G., Semenovskya, S.V., and Morris Jr., J.W., Phys. Rev. B **37**, 2243 (1988).
[57] Berera, A., and deFontaine, D., Phys. Rev. B **39**, 6727 (1989).
[58] Vasumathi, D., Sundar, C.S., Bharathi, A., Sood, A.K., and Hariharan, Y., Physica C **167**, 149 (1990).
[59] Lu, X., Wang, S.J., Sundar, C.S., Bharathi, A., Lyu, Y., Ching, W.Y., and Jean, Y.C., Materials Science Forum **105-110**, 755 (1992).
[60] Sundar, C.S., Bharathi, A., Vasumathi, D., and Hariharan, Y., Materials Science Forum **105-110**, 1253 (1992).
[61] Pickett, W.E., Kraueker, H.E., Cohen, R.E., and Singh, D.J., Science **255**, 46 (1992).
[62] Bansil, A., Pankaluoto, R., Rao, R.S., Mijnarends, P.E., Dlugosz, W., Prasad, R., and Smedsjkaer, L.C., Phys. Rev. Lett. **61**, 2480 (1988) ; Bansil, A., Mijnearends, P.E., and Smedsjkaer, L.C., Phys. Rev. B **43**, 3667 (1991).
[63] Manuel, A.A., Singh, A.K., Jarlborg, T., Genoud, P., Hoffman, L., and Peter, M., in *Positron Annihilation*, Ed..Dorikens-vanpraet, M.Dorikens and D.Segers, World Scientific, Singapore, 1989, p.109.
[64] Hoffman, L., Manuel, A.A., Peter, M., Walker, E., and Damento, A., Europhys. Lett. **6**, 61 (1988).
[65] Smedskjaer, L.C., Liu, J.Z., Benedek, R., Leginini, D.J., Lam, D.J., Stahulek, M.D., and Claus, H., Physica C **156**, 269 (1988).
[66] Haghighi, H., Kaiser, J.H., Rayner, S., West, R.N., Fluss, M.J., Howell, R.H., Turchi, P.E.A., Wachs, A.L., Jean, Y.C., and Wang, Z.Z., J. Phys.: Condens. Matter, **2**, 1911 (1990).
[67] Haghighi, H., Kaiser, J.H., Rayner, S., Liu, J.Z., Shelton, R., Howell, R.H., Solal, F., Sterne, P.A., and Fluss, M.J., Phys. Rev. Lett. **67**, 38 (1991).
[68] Smedskjaer, L.C., Bansil, A., Welp, U., Wang, Y., and Bailey, K.G., J. Phys. Chem. Solids **52**, 1541 (1991).
[69] Peter, M., Manuel, A.A., Hoffmann, L., and Sadowski, W., Europhys. Lett. **18**, 313 (1991).

[70] Hoffman, L., Manuel, A.A., Peter, M., Walker, E., Gauthier, M., Shukla, A., Barbiellini, B., Massida, S., Adam, Gh., Adam, S., Hardy, W.N., and Liang, R.N., Phys. Rev. Lett. **71**, 4047 (1993).
[71] Manuel, A.A., Shukla, A., Hoffmann, L., Jalborg, T., Barbiellini, B., Massida, S., Sadowski, W., Walker, E., Erb, E., and Peter, M., J. Phys. Chem. Solids **56**, 1951 (1995).
[72] Howell, R.H., Sterne, P.A., Fluss, M.J., Kaiser, J.H., Kitazawa, K., and Kojima, H., Phys. Rev. B **49**, 13127 (1994).
[73] Chan, L.P., Harshman, D., Lynn, K.G., Massida, S., and Mitzi, D.B., Phys. Rev. Lett. **67**, 1350 (1991).
[74] Barbiellini, B., Gauthier, M., Hoffmann, L., Jarlborg, T., Manuel, A.A., Massida, S., Peter, M., and Triscone, G., Physica C **229**, 113 (1994).
[75] Ding, H., Norman, M.R., Campuzano, J.C., Randeria, M., Bellman, A.F., Yokoya, T., Takahashi, T., Mochiku, T., and Kadowaski, K., Phys. Rev. B **54**, R9678 (1995).
[76] Heiney, P.E., J. Phys. Chem. Solids **53**, 1333 (1992).
[77] Jean, Y.C., Lu, X., Lou, Y., Bharathi, A., Sundar, C.S., Lyu, Y., Hor, P.H., and Chu, C.W., Phys. Rev. B **45**, 12126 (1992).
[78] Lou, Y., Lu, X., Dai, G.H., Ching, W.Y.M, Xu, Y.N., Huang, M.Z., Tseng, P.K., Jean, Y.C., Meng, R.L., Hor, P.H., and Chu, C.W., Phys. Rev. B **46**, 2644 (1992).
[79] Sundar, C.S., Premila, M., Gopalan, P., Hariharan, Y., and Bharathi, A., Fullerene Science and Technology **3**, 661 (1995); Sundar C.S., in *Nuclear & Radiation Chemical Approaches to Fullerene Science*, Ed. T. Braun (Kluwer Academic, The Netherlands, 2000), p.3.
[80] Cava, R.J., Takagi, H., Zandbergen, H.W., Krajewski, J.J., Peck, Jr., W.F., Siegrist, T., Batlogg, B., van Dover, R.B., Felder, R.J., Mizuhashi, K., Lee, J.O., Eisaki, H., and Uchida, S., Nature (London) **367**, 252 (1994).
[81] Siegrist, T., Zandbergen, H.W., Cava, R.J., Krajewski, J.J., and Peck, Jr., W.F., Nature (London) **367**, 254 (1994).
[82] Pickett, W.E., and Singh, D.J., Phys. Rev. Lett. **72**, 3702 (1994).
[83] Bharathi, A., Sundar, C.S., Hariharan, Y., Radhakrishnan, T.S., Hossain, Z., Nagarajan, R., Gupta, L.C., and Vijyaraghavan, R., Physica B **223&224**, 123 (1996).
[84] Sundar, C.S., Bharathi, A., Hariharan, Y., Radhakrishnan, T.S., Hossain, Z., Nagarajan, R., Gupta, L.C., and Vijyaraghavan, R., Physical Review B **53**, R2971 (1996).

Chapter 9

Positronium in Si and SiO$_2$ Thin Films

R. Suzuki

*National Institute of Advanced Industrial Science and Technology AIST
Central 2, Tsukuba-shi, Ibaraki 305-8568, Japan*

9.1 Introduction

Since positron annihilation spectroscopy is highly sensitive to atomic defects in solid materials, positron annihilation experiments have been carried out extensively on silicon (Si) and silicon dioxide (SiO$_2$), both of which are extremely important for the microelectronic device industry. While several reviews are available [1], those reviews are mainly focused on positron (not positronium) annihilation behavior because positronium (Ps) formation dose not occur in bulk crystalline Si. Recent positron annihilation experimental studies revealed that Ps formation occurs in some Si-based thin films, such as porous Si and hydrogenated amorphous Si; furthermore, Ps formation is dominant in high-purity amorphous SiO$_2$ thin films. In this chapter, Ps annihilation characteristics in Si and SiO$_2$ thin films will be discussed from the experimental point of view.

9.2 Amorphous Si

Amorphous silicon (a-Si) has promoted considerable interest because of its potential for inexpensive solar cell applications. However, it is not easy to investigate a-Si samples by means of conventional positron annihilation spectroscopy because a-Si samples are only available as thin films. In spite of this difficulty, several positron annihilation experiments were carried out for a-Si and hydrogenated amorphous silicon (a-Si:H) films by means of conventional positron annihilation spectroscopy. Dannefaer et al. [2] investigated a-Si films (7--12 μm in thickness) evaporated or sputtered onto crystalline Si substrates. Jung et al. [3] and He et al. [4] reported the results of a-Si and a-Si:H films (70--100 μm in thickness) deposited by an rf-sputter method. Schaefer et al. [5] investigated a crushed a-Si:H film deposited by a dc glow discharge method. These experimental studies revealed that a-Si:H samples have a long-lifetime component which could be attributed to ortho-Ps (o-Ps) pick-off annihilation in micro-voids.

In order to investigate positron and Ps annihilation behavior in a-Si:H in greater detail, positron annihilation lifetime spectroscopy (PALS) with a monoenergetic pulsed positron beam was applied to a-Si:H thin films (~1 μm in thickness) deposited by a plasma-enhanced chemical-vapor deposition (PECVD) method [6]. Figure 9.1 shows the positron lifetime spectra of the films grown under four different RF-power densities (0.03 W/cm^2, 0.13 W/cm^2, 0.51 W/cm^2, and 0.76 W/cm^2) at the incident positron energy of 6 keV. In the films prepared at 0.13 W/cm^2 and 0.51 W/cm^2, a long-lived o-Ps component of ~9 ns was found. From the known relationship between the o-Ps pick-off lifetime and void size [7], the void diameter was estimated to be ~1.3 nm. On the other hand, the films prepared at 0.03 W/cm^2 and 0.76 W/cm^2 showed significantly lower long-lived component intensity, indicating that the structure of a-Si:H strongly depends on the growing conditions.

Figure 9.2 shows the annealing temperature dependence of the positron lifetime parameters [8]. From 400°C to 500°C, the long-lived component intensity, I_3, dramatically increases but the lifetime τ_3 does not significantly change. Since I_3 is a relative intensity, there are two possibilities for this phenomenon; one is increase in micro-void concentration, the other is reduction of trapping at small vacancies where Ps cannot form. In this case, the former possibility could be ruled out because it requires larger energy than the latter one. In Figure 9.2(b), the intermediate lifetime τ_2 after the annealing at 400°C--500°C is significantly higher than that of the initial sample. This also suggests that the concentration of small vacancies, which contributes to short lifetime components, is reduced. After 600°C annealing, I_3 suddenly decreases in all the films. Since hydrogen desorption from monohydride species of Si surfaces is known to occur in the range between 500°C and 600°C [9], [10], the sudden decrease in the o-Ps component at 600°C could be due to hydrogen desorption from void surfaces.

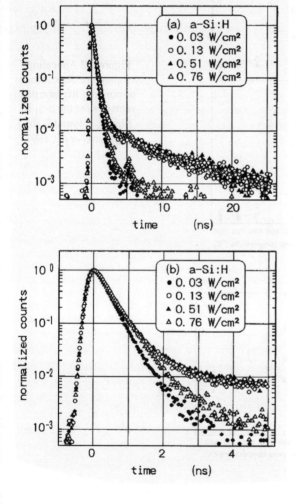

Figure 9.1 Positron lifetime spectra of the PECVD grown a-Si:H films prepared at power densities of 0.03 W/cm^2, 0.13 W/cm^2, 0.51 W/cm^2, and 0.76 W/cm^2. (Suzuki et al., 1991)

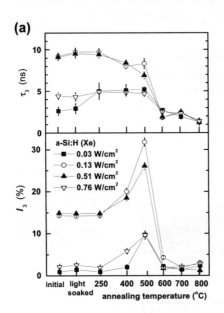
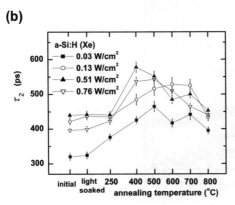

Figure 9.2 Annealing temperature dependence of positron lifetime parameters for the a-Si:H films; (a) τ_3 and I_3 (b) τ_2. (Suzuki *et al.*, 1997)

9.3 Porous Si

Positronium formation was also found in porous Si obtained by anodization of crystalline silicon in HF acid solutions. Itoh, Murakmi and Kinoshita [11] found a long-lived (>10 ns) component in the positron lifetime spectrum measured by conventional PALS. The authors investigated Ps behavior in porous Si at various temperatures by means of PALS with a monoenergetic pulsed positron beam [12],[13].

Figure 9.3(a) shows positron lifetime spectra of a porous Si thin film at the sample temperatures of 25°C (initial), 300°C, and 500°C, and Figure 9.3(b) shows positron lifetime spectra measured at 500°C and at 200°C after 500°C annealing. Strong temperature dependence was observed in the long-lived component.

In this experiment [12], the temperature dependence of the long-lived component was found to be similar to that of the a-Si:H experiment described in the previous section--the intensity of the long-lived component annealed at 300°C--500°C is significantly higher than that of the initial sample; furthermore, a sharp decrease in the intensity of the long-lived component was found between 500°C and 600°C annealing. As discussed in the previous section, the increase in the o-Ps component after 300°C--500°C annealing could be attributed to micro-structural relaxation. This relaxation could reduce small vacancy-type defects, at which positrons annihilate via the free state. The latter phenomenon could be due to the hydrogen desorption from monohydride species at the void surface.

As shown in Figure 9.3(b), the lifetime of the long-lived component measured at 200°C after annealing at 500°C is significantly larger than that measured at 500°C. The temperature dependence was found to be reversible. This suggests that a thermally activated process, such as increase in dangling bond density induced by surface reconstruction, affects the lifetime shortening of the o-Ps component [12].

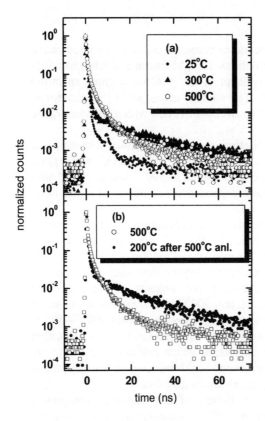

Figure 9.3 Positron lifetime spectra for porus Si (a) measured at the sample temperature of 25°C, 300°C, and 500°C, and (b) mesured at 500°C and at 200°C after 500°C annealing. (Suzuki *et al.*, 1998)

9.4 Amorphous SiO$_2$ thin films

9.4.1 Thermally grown SiO$_2$ on Si

Several positron annihilation studies on amorphous SiO$_2$ (not only bulk but also thin films) revealed that Ps formation occurs with high efficiency in intrinsic micro-voids of amorphous SiO_2 network. Ps formation in bulk amorphous SiO_2 was confirmed by PALS, angular correlation of annihilation radiation (ACAR) [14], and age-momentum correlation (AMOC) [15]. Ps formation in amorphous SiO_2 thin films was confirmed by PALS ([16], [17]), ACAR ([18]) and AMOC ([19]).

Figure 9.4 shows typical positron lifetime spectra for thermally grown SiO_2 (500 nm in thickness) on Si, measured at the incident positron energy of 2 keV and 15 keV. The SiO_2 film was grown by a thermal method with a dry oxygen condition (Si + O_2(gas) -> SiO_2) on Si(100). Almost all positrons annihilate in the SiO_2 film at 2 keV whereas about 90 % of positrons annihilate in Si substrate at 15 keV. In the spectrum at 2 keV, both long-lived component (τ~1.5 ns), which could be attributed to o-Ps pick-off annihilation, and short-lived component, which could be attributed to para-Ps (p-Ps) self-annihilation, are observed. The intensity of the long-lived component at 2 keV is ~65%.

The Ps formation in SiO_2 can be seen in AMOC data more clearly ([20]). From the two-dimensional (2D) AMOC spectrum, we can obtain S parameter (low-momentum parameter) and/or W parameter (high-momentum parameter) as a function of annihilation time. Figure 9.5 shows the S(t) plot and S-W map of the SiO_2 film obtained with an AMOC apparatus with a monoenergetic pulsed positron beam. At the positron energy of 2 keV, the high S value in the young age region

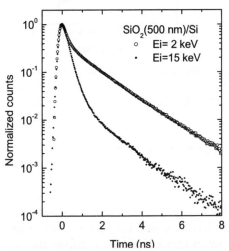

Figure 9.4 Positron lifetime spectra of amorphous SiO_2 (500 nm) on Si(100) at the positron incident energy of 2 keV and 15 keV.

indicates narrow momentum distribution, while the low S value in the old age region indicates broad momentum distribution. This time dependency is similar to AMOC results of bulk amorphous SiO_2 ([15]). Therefore, the annihilation behavior of positrons in thin films is basically the same as that in bulk amorphous SiO_2.

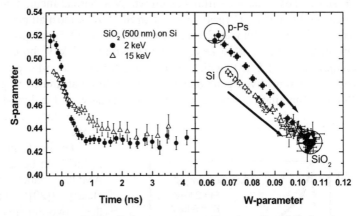

Figure 9.5 (Left) S parameter as a function of annihilation time and (Right) S-W plot for amorphous SiO_2 (500 nm) on Si(100) at the positron energy of 2 keV and 15 keV.

There should be at least three annihilation processes, p-Ps self-annihilation, o-Ps pick-off annihilation, and free positron annihilation in SiO_2. In the S(t) curve of 2 keV, the p-Ps component mainly contributes to the region t<~0.5 ns, free positron contribution should be in the region between 0.5 ns and 1.5 ns, and o-Ps contribution is dominant after ~1.5 ns. As can be seen in Figure 9.5, the S(t) curve of 2 keV is almost flat after ~0.5 ns, indicating that the momentum distribution of o-Ps is almost the same as that of free positron annihilation. This suggests that both o-Ps atoms and free positrons annihilate with the same electrons; i.e., predominantly annihilate with electrons of oxygen in the SiO_2 network, because oxygen is dominant at the first layer of the micro-void wall.

At 15 keV, about 90% of positrons are implanted in the Si substrate. In the young age region, annihilation in the Si substrate is dominant. Thus, from the data of 2 keV and 15 keV, we can obtain three characteristic points-of-annihilation processes in the S-W map; p-Ps self-annihilation in SiO_2, annihilation in the Si substrate, and annihilation with the electrons of the SiO_2 network. Moreover, the contribution of annihilation in SiO_2 can be clearly seen in the trajectory of 15 keV while the fraction of positrons annihilating in SiO_2 is only ~10%.

It should be noted that the S parameters of both o-Ps pick-off and free-positron annihilation are lower than that of the Si substrate, because positrons predominantly annihilate with electrons of oxygen in the SiO_2 network. Only p-Ps self-annihilation has a higher S value than that of Si. The S parameter observed in conventional Doppler- broadening-of-annihilation radiation is the average of p-Ps, o-Ps, and free-positron annihilation. Therefore, if the Ps fraction decreases due to the presence of defects, impurities, etc., the intensity of the narrow momentum component due to p-Ps self-annihilation decreases, and as a result the averaged S parameter decreases.

9.4.2 Detection of defects in amorphous SiO_2

As discussed in the previous section, Ps formation occurs with high probability in amorphous SiO_2. In recent years various amorphous SiO_2 samples have been investigated using conventional positron spectroscopy techniques as well as slow-positron beam techniques. Several positron annihilation experiments have been applied to the study of defect properties created by ion implantation ([21], [22]), gamma (or X) ray ([23], [24],[14]), excimer laser ([25]), beta ray ([26]), neutron ([14]), etc. From these studies it was found that positronium intensity is highly sensitive to the radiation-induced bond-breaking-type defects, and these defects suppress Ps formation.

Figure 9.6 shows the long-lived component intensity, I_L, as a function of positron energy for dielectric-coated cavity mirrors, Ta_2O_5/SiO_2 on amorphous SiO_2 substrate, which is used for ultraviolet (350 nm) free-electron laser experiments ([27]). The mirror has alternating dielectric layers of $\lambda/4$ thickness with a $\lambda/2$-thick SiO_2 layer on top for surface protection. The initial loss of the new mirror was observed to be 250 ppm at the optimized wavelength of 350 nm. The sample was exposed to undulator radiation of 3.5 mAh at room temperature in an electron storage ring. Then, these mirrors were exposed to RF-induced O_2 plasma to remove carbon at the surface. Finally, the samples were annealed for 180 min at 230°C.

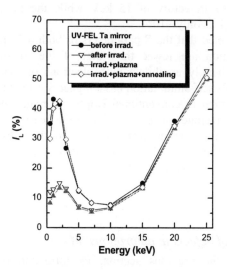

Figure 9.6 Intensity of the long-lived o-Ps component as a function of positron incident energy for SiO_2/Ta_2O_5 multi-layer mirror. (Suzuki et al., 1996)

A long-lived o-Ps component of ~1.5 ns was observed in the low- energy region corresponding to the first amorphous SiO_2 layer, as well as in the high-energy region corresponding to the substrate. After undulator radiation, the loss in the mirror increased from 250 ppm to 3849 ppm and the o-Ps intensity decreased. After O_2 plasma treatment, no significant change was observed in the long-lived component, while the loss in the mirror was restored to 802 ppm. This restoration is due to carbon removal from the surface by O_2 plasma. After 230°C annealing, the intensity of the long-lived component was mostly restored to that of the initial sample, and the loss in the mirror was also restored to 297 ppm. This result suggests that defects of some kind, which affect both the loss in the mirror and the intensity of the long-lived component, were created in the first SiO_2 layer by the exposure to the undulator radiation.

It should be noted that the restoration temperature of the long-lived component is considerably lower than the temperature in ion-implanted amorphous SiO_2 and gamma- (or X-) ray irradiated SiO_2 [21]. Since ions and gamma rays have much higher energies than undulator photons, they will create multiple types of defects that suppress Ps formation. On the contrary, since the intrinsic absorption edge of high-purity SiO_2 is located at around 9 eV, high-purity SiO_2 is transparent to the fundamental of undulator radiation. Fujinami and Chilton ([25]) reported that no significant change was observed in UV excimer laser-irradiated thermally grown SiO_2 itself, while dramatic change was observed in excimer laser-irradiated CVD-grown SiO_2, which has a high concentration of Si-O-H. The results are quite similar to the results of CVD-grown SiO_2. Thus, the defects created by undulator photons could be due to breakage of the weak bonds, such as Si-O-H.

Positronium formation is also sensitive to ion-implanted amorphous SiO_2. Figure 9.7 shows the intensity of the long-lived component, I_L, as a function of the positron incident energy for Xe ion-implanted amorphous SiO_2 ([22]). The sample was obtained by a vapor-phase axial deposition (VAD) method. Xe ions of 400 keV were implanted into the sample to doses of 1×10^{14} and 5×10^{15} ions/cm^2 at room temperature. While there is a small difference between I_L of 1×10^{14} and I_L of 1×10^{15}, both have a minimum at around 4-5 keV, corresponding to the mean positron implantation depth of ~200 nm at which the ions are implanted.

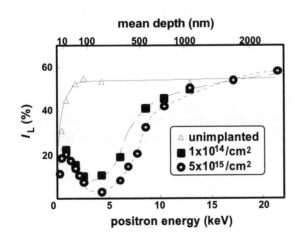

Figure 9.7 Intensity of the long-lived component as a function of positron energy for Xe ion-implanted amorphous SiO_2. (Suzuki et al., 1994b)

9.5 Porous SiO_2

For future high-speed microelectronic devices, copper interconnection with low dielectric constant (low-k) interlayer films is required to decrease RC (R: interconnect resistance, C: interlayer dielectric capacitance) delay. Recently, porous SiO_2 and silica-based films, developed for low-k films, have been extensively studied by positron annihilation spectroscopy [28], [29], [19]. Since Ps formation occurs with high probability, and the o-Ps annihilate via pick-off process in SiO_2-based materials, positron annihilation spectroscopy (especially PALS) gives useful information on the size of the pores.

Figure 9.8 shows the results of PALS on low-k films, grown by a PECVD method with a source gas of hexamethyldisiloxane (HMDSO, $(CH_3)_3SiOSi(CH_3)_3$), at an incident energy of 4 keV (Suzuki, 2001). The PECVD method uses two frequencies, LF (380 kHz) and RF (13.56 MHz). In Figure 8(a), the positron lifetime spectra of the films grown under different LF power are shown. The lifetime of the long-lived o-Ps component strongly depends on the LF power. Figure 9.8(b) shows the result of an inverse Laplace transformation analysis CONTIN [30]. The average cavity volume calculated from the distribution of Figure 9.8(b) ranges from 0.33 nm^3 to 1.05 nm^3. There is apparent correlation between the dielectric constant value (k = 2.66--4.13) and the pore size [19].

By the use of other methods, such as spin coating and sputtering techniques, one can obtain porous SiO_2 films, which have larger pores than CVD grown films. One problem in the analysis of such porous SiO_2 films is Ps emission from the surface if the pores are interconnected. Figure 9.9 shows the positron lifetime spectra of porous SiO_2 films, with and without a cap SiO_2 layer, grown by a sputtering method ([20]). As can be seen in Figure 9.9(a), the film with a cap layer shows significantly higher intensity of the long-lived component than the film without a cap layer. Furthermore, three-gamma annihilation in the film without a cap layer is significantly higher than that in the film with a cap layer (Figure 9.9(b)), indicating Ps emission from the surface of the film without a cap layer. Therefore, capping is important for the precise pore-size measurements on porous films, which contain interconnected pores.

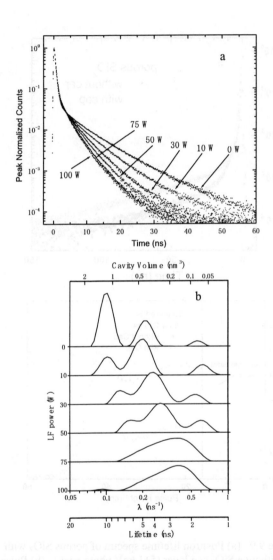

Figure 9.8 (a) Positron lifetime spectra and (b) annihilation rate probability function for low-k films grown by a double-frequency PECVD method with different LF powers. (Suzuki et al., 2001)

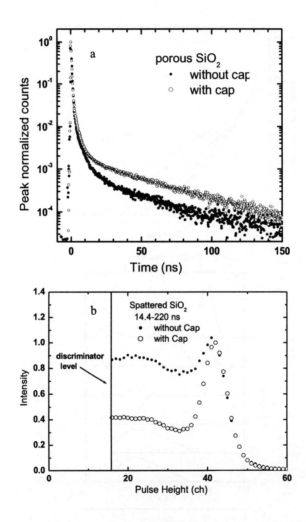

Figure 9.9. (a) Positron lifetime spectra of porous SiO_2 with and without a SiO_2 cap layer (511 keV photo peak). (b) Pulse height spectra of the γ–ray detector for the long-lived component in the annihilation time range between 14.4 ns and 220 ns from the peak.

9.6 Summary

Positronium formation and annihilation behavior in Si and SiO_2 thin films are reviewed. Positronium is highly sensitive to pore (or void) sizes, surface properties of pores, defects near pore surfaces, etc., in various Si and SiO_2 samples. Therefore, not only positron annihilation spectroscopy but also "positronium" annihilation spectroscopy is useful for characterization of Si and SiO_2 materials.

References

[1] Asoka-Kumar, P., Lynn K.G. and Welch, D.O. (1994) "Characterization of defects in Si and SiO_2-Si using positrons", J. Appl. Phys. **76**, 4935.
[2] Dannefaer, S., Kerr, D. and Hogg, B.G. (1983) "A study of defects in amorphous silicon films" J. Appl. Phys. **54**, 155.
[3] Jung, A.L., Wang, Y.H., Liu, G., Xiong, J.J. et al. (1985) "Defects and microvoids in a-Si and a-Si:H", Non-Cryst. Solids **74**, 19.
[4] He, Y.J., Hasegawa, M., Lee, R., Berko, S. et al. (1986) "Positron-annihilation study of voids in a-Si and a-Si:H", Phys. Rev. B **33**, 5924.
[5] Shaefer, H.-E., Wurschum, R. Schwarz, R., Slobodin, D. et al. (1986) "Amorphous hydrogenated silicon studied by positron lifetime spectroscopy", Appl. Phys. A **40**, 145.
[6] Suzuki, R., Kobayashi, Y., Mikado, T., Matsuda, A. et al. (1991) "Characterization of hydrogenated amorphous silicon films by a pulsed positron beam", Jpn. J. Appl. Phys. **30**, 2438.
[7] Nakanishi H., Wang, S.J. and Jean, Y.C. (1988) "Positrons and positronium in liquids", Positron and Positronium Chemistry, edited by Schrader, D.M., and Jean Y.C.; Elsevier, Amsterdam, p.159.
[8] Suzuki, R., Ohdaira, T., Uedono, A., Ishibashi, S. et al. (1997) "Positron Lifetime Study on Semiconductor Thin Films", Mater. Sci. Forum **255-257**, 714.
[9] Gupta, P., Colvin, V.L. and George, S.M. (1988) "Hydrogen desorption kinetics from monohydride and dihydride species on silicon surfaces", Phys. Rev. B **37**, 8234.
[10] Ito, T., Yasumatsu, T., Watanabe H., and Hiraki A. (1990) "Structural change of crystalline porous silicon with chemisorption", Jpn. J. Appl. Phys. **29**, L201.
[11] Itoh, Y., Murakami, H. and Kinoshita, A. (1993) "Positron annihilation in porous silicon", Appl. Phys. Lett. **63**, 2798.

[12] Suzuki, R., Kobayashi, Y., Mikado, T., Ohgaki, H. et al. (1994) "Positron-lifetime study on porous silicon with a monoenergetic pulsed positron beam", Phys. Rev. B **49**, 17484.
[13] Suzuki, R., Ohdaira, T. and Mikado, T. (1998) "Low-energy pulsed positron beam at the ETL linac facility" Proc. Int. Workshop on Advanced Techniques of Positron Beam Generation and Control, RIKEN, Wako, Japan (Committee of Crossover Research Program for Basic Nuclear Science).
[14] Hasegawa, M., Tabata, M., Miyamoto, T., Nagashima, Y. et al. (1995) "Positron and positronium in free volume in oxides: silica glass and neutron-irradiated alumina", Mater. Sci. Forum **175-178**, 269.
[15] Stoll, H., Wesolowski, P., Koch, M., Maier, K., Major, J., et al. (1992) "$\beta^+\gamma$ age-momentum-correlation measurements with an MeV positron beam" Mater. Sci. Forum **105-110**, 1989.
[16] Suzuki, R., Kobayashi, Y., Mikado, T., Ohgaki, H. et al. (1992) "Investigation of near surface defects by variable-energy positron lifetime spectroscopy", Mater. Sci. Forum **105-110**, 1459.
[17] Uedono, A., Wei,L., Tanigawa, S., Suzuki, R. et al. (1993) "Characterization of SiO_2 Films Grown on Si Substrates by Monoenergetic Positron Beams" Journal de Physique II **3**, 177.
[18] Rivera, A., Montilla, I., Alba Garcia, A., Escobar Calindo, R. et al. (2001) "Native and irradiation-induced defects in SiO_2 Structures studied by positron annihilation techniques", Mater. Sci. Forum **363-365**, 64.
[19] Suzuki, R., Ohdaira, T., Shioya, Y. and Ishimaru, T. (2001) "Pore characteristics of low-dielectric-constant films grown by plasma-enhanced chemical vapor deposition studied by positron annihilation lifetime spectroscopy", Jpn. J. Appl. Phys. **40**, L414.
[20] Suzuki, R., Ohdaira, T., Uedono, A. and Kobayashi, Y. (in press) "Positron annihilation in SiO_2-Si studied by a pulsed slow positron beam", Appl. Surf. Sci.
[21] Fujinami, M. and Chilton, N.B. (1993) "Ion implantaion induced defects in SiO_2: The applicability of the positron probe" Appl. Phys. Lett. **62**, 1131.
[22] Suzuki, R., Kobayashi, Y., Awazu, K., Mikado, T. et al. (1994) "Positron lifetime study on ion-implanted amorphous SiO_2 with a variable-energy pulsed positron beam" Nucl. Instrum. Methods Phys. Res., Sect. B **91**, 410.
[23] Khatri, R. Asoka-Kumar, P. Nielsen, B. Roellig, L.O. et al. (1993) "Positron trap centers in X-ray and gamma-ray irradiated SiO_2", Appl. Phys. Lett. **63**, 385.
[24] Uedono, A., Watauchi, S., Ujihira, Y. and Yoda, O. (1994) "Defects in electron irradiated amorphous SiO_2 probed by positron annihilation", Hyperfine Interact. **84**, 225.
[25] Fujinami, M. and Chilton, N.B. (1994) "Study on excimer laser induced defects in SiO_2 films on Si by variable-energy positron annihilation spectroscopy" Appl. Phys. Lett. **64**, 2806.

[26] Saito, H., Nagashima, Y., Hyodo, T., Chang, T.B. (1995) "Detection of paramagnetic centers on surfaces of amorphous-SiO_2 fine grains using positronium", Mater. Sci. Forum **175-178**, 769.

[27] Suzuki, R., Ohdaira, T., Yamada, K., Yamazaki, T. et al. (1996) "Slow Positron Study on Dielectric-coated Mirror for Free-Electron-Laser Experiments", J. Radioanalytical & Nucl. Chem. **211**, 47.

[28] Gidley, D.W., Frieze, W.E., Dull, T.L. Yee, A.F. et al. (1999) "Positronium annihilation in mesoporous thin films", Phys. Rev. B **60**, R5157.

[29] Petkov, M. P., Weber, M. H., Lynn K.G. and Rodbell, K. P. (2000) "Probing capped and uncapped mesoporous low-dielectric constant films using positron annihilation lifetime spectroscopy", Appl. Phys. Lett. **77**, 2470.

[30] Gregory, R.B. and Zhu, Y. (1990) "Analysis of positron annihilation lifetime data by numerical laplace inversion with the program CONTIN", Nucl. Instrum. Methods Phys. Res., Sect. A **290**, 1172.

Chapter 10

Application to Polymers

P.E. Mallon

Division of Polymer Science, Department of Chemistry, University of Stellenbosch, Private Bag X1, Matieland 7602, South Africa

10.1 Introduction

The aim of this chapter is to introduce the reader to the application of positron annihilation techniques to polymers. An extensive review of the large volume of publications related to positron studies in polymers will not be presented. Rather it is intented to introduce the reader to the theory and techniques used in polymer studies and indicate the types of information that can be obtained about different polymer systems. The main focus of this chapter will be on the use of positron annihilation lifetime spectroscopy (PAL) in polymer studies. Chapter 11 discusses the use of monoenergetic slow positron beams used to study polymers surfaces. One of the interesting new developments in the application of positron annihilation techniques in polymers is the positron age-momentum correlation technique (AMOC). This technique promises to shed new light on the mechanisms of positronium formation and annihilation in polymer systems. A more detailed discussion of this technique can be found elswhere in this text.

10.1.1 Free volume in polymers

The concept of free volume is critical in explaining and understanding the physical behavior of polymers. The free volume theory of materials was first developed by Eyring and others [1-3]. It is based on the idea that molecular motion in the bulk state depends on the presence of holes, or places where there are vacancies or voids. When molecules move into the hole, the hole exchanges place with the molecule. In the case of the motion of polymer chains, more than one hole may be required for chain movement. Thus for a polymeric segment to move from its present position to an adjacent site a critical void volume must first exist before the segment can "jump". The important point is that molecular motion (and therefore many of the unique physical properties which polymers exhibit) can not occur without the presence of holes. In polymers the local free volume holes or cavities of atomic and molecular dimensions arise due to the irregular packing of the chains in the amorphous state (static) and molecular relaxation of the polymer chains and terminal ends (dynamic and transient state). The presence of these holes mean that the density of the polymer samples are about 10% less than the densities of their crystalline state. A simple expression of the free volume (V_f) can be written as the total volume (V_t) minus the "occupied volume" (V_0):

$$V_f = V_t - V_0 \qquad (1)$$

The holes from the free volume affect the mechanical, thermal and relaxation properties of polymers. Despite the importance of free volume only limited experimental data about the free volume of polymers has been reported. This is mainly due to the lack of suitable probes for open volumes of molecular dimensions of a few Å and the short time scale of many of the dynamic holes (from as short as 10^{-13}s).

A number of techniques have been employed to examine free volume properties of polymers. These include; small angle x-ray scattering and neutron diffraction that have been used to determine denisty fluctuations to deduce free volume size distributions [4-7]. Photochromic labelling techniques by site specific probes have been developed to monitor the rate of photoisomerizations of the probes and from this deduce free volume distributions [8-11]. Additional probing methods used to probe voids and defects in materials such as scanning tunneling microscopy (STM) and

Application to Polymers   255

atomic force microscopy are sensitive to angstrom size holes, but are limited to static holes on the surface, which limits their use in polymers. Scanning electron microscopy (SEM) and transmission electron microscopy (TEM) are more sensitive to static holes at a size of 10 Å or larger. Figure 10.1 shows a comparison of the techniques used to probe voids in terms of their ability to resolve defect size and defect concentration.

Positron annihilation lifetime spectroscopy (PAL) has been developed to be, probably, the most succesful technique for the direct examination of local free volume holes in polymers. Due to the small size of the positronium probe (1.59 Å) compared to other probes, PAL is particulary sensitive to small holes and free volume of the order of angstrom magnitude. Because of the relatively short lifetime of the o-Ps (typically about 2-4 ns in polymers) PAL can probe holes due to molecular motion from 10^{-10}s or longer. In addition unlike other methods, PAL is capable of determining the local hole size and free volume in a polymer without being significantly interfered with by the bulk. PAL has also been developed to be a quantitative probe of free volume for polymers. Not only does it probe the free volume size and fractions of free volume, but it also gives detailed information on the distribution of free volume hole sizes in the range from 1 to 10 Å.

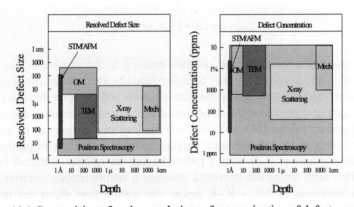

Figure 10.1 Comparision of various techniques for examination of defects and voids in materials (OM-optical microscopy, TEM- transmission electron microscopy STM-Scanning tunneling microscopy, AFM-atomic force microscopy, Mech- mechanical techniques).

10.2 Theoretical aspects

Positrons emitted for a radioactive source (such as ^{22}Na) into a polymeric matrix become thermalized and may annihilate with electrons or form positronium (Ps) (a bound state of an electron and positron). The detailed mechanism and models for the formation of positronium in molecular media can be found in Chapters 4 and 5 of this book. The para-positronium (p-Ps), where the positron and electron have opposite spin, decays quickly via self-annihilation. The long-lived ortho positronium (o-Ps), where the positron and electron have parallel spin, undergo so called "pick-off" annihilation during collisions with molecules. The o-Ps formed in the matrix is localized in the free volume holes within the polymer. Evidence for the localization of o-Ps in the free volume holes has been found from temperature, pressure, and crystallinity-dependent properties [12-14]. In a vacuum o-Ps has a lifetime of 142.1 ns. In the polymer matrix this lifetime is reduced to between 2 - 4 ns by the so-called pick-off annihilation with electrons from the surrounding molecule. The observed lifetime of the o-Ps (τ_3) depends on the reciprocal of the integral of the positron ($\rho_+(r)$) and electron ($\rho_-(r)$) densities at the region where the annihilation takes place:

$$\tau_3 = const \bullet \left(\int \rho_+(r) \rho_-(r) dr \right)^{-1} \quad (2)$$

where *const* is a normalization constant relating to the number of electrons involved in annihilation. In order to understand the annihilation lifetimes and molecular volumes one must relate the electron and positron densities with the molecular size. The exact solutions of the ρ_+ and ρ_- densities are intractable quantum mechanical problem. Nevertheless an equation can be derived for the hole size by considering a simple model in which the Ps particle resides in a spherical well with a radius having an infinite potential barrier [15]. This semi–empirical approach assumes a homogeneous electron layer with a thickness of $\Delta R = R_0-R$ inside the wall, where R_0 is the infinite spherical potential radius and R is the hole radius. This leads to the following analytical relationship between the o-Ps lifetime τ_3 and the free volume radius:

$$\tau_3^{-1} = 2\left[1 - \frac{R}{R_0} + \frac{1}{2\pi}\sin\left(\frac{2\pi R}{R_0}\right)\right] \quad (ns^{-1}) \tag{3}$$

where $R_0 = R + \Delta R$, and ΔR is an empirical parameter. Nakanishi et. al. [16] determined ΔR by fitting the observed lifetimes with the known hole and cavity sizes in molecular substrates. The best fitted value of ΔR for all known data is found to be 1.66 Å. The correlation between τ_3 and free volume is shown in Figure 10.2. The line fitted in the Figure 10.2 is that of Eq. 3. Equation 3 and Figure 10.2 are the foundation for the determination of the mean size of free-volume holes using positron annihilation lifetime spectroscopy. A good correlation is found for hole sizes up to about a radius of 1 nm. Recently Ito et al [17] and Dull et al [18] extended this equation for hole sizes larger than 1nm or o-Ps lifetimes longer than 20 ns. It is assumed that in the larger pores, the o-Ps behaves more like a quantum particle, bouncing back and forth between the energy barriers as the potential well becomes large.

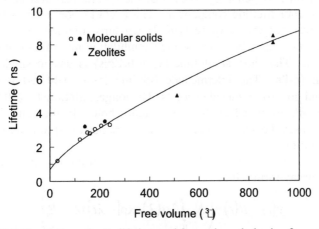

Figure 10.2 Correlation of o-Ps lifetime and free volume hole size for molecular solids and zeolites [16].

10.2.1 Data analysis of PAL spectra in polymer

The analysis of the positron annihilation lifetime spectra is a very important aspect of using the PAL techniques to analyze polymers. Without proper data analysis interpretation of data might be misleading and important scientific information will be lost. In PAL studies of polymers the PAL spectrum can be analyzed in two ways; (1) a finite lifetime analysis or (2) continuous lifetime analysis. In the finite lifetime analysis the PAL spectra is resolved into a finite number of negative exponentials decays. The experimental data *y(t)* is expressed as a convoluted expression (by a symbol *) of the instrument resolution function *R(t)* and a finite number (n) of negative exponentials:

$$y(t) = R(t) * \left(N_t \sum_{i=1}^{n} \alpha_i \lambda_i \exp[-\lambda_i t] + B \right) \qquad (4)$$

where N_t is the normalized counts, B is the background. λ_i is the inverse of the i-th lifetime component (τ_i) and $\alpha_i \lambda_i$ is its intensity.

In polymers it is generally found that the spectra can best be resolved into three or four lifetime components. The PATFIT computer program is often employed for this purpose [19]. When three lifetimes are resolved, each lifetime corresponds to the average annihilation rate of a positron in a different state. The shortest lifetime ($\tau_1 \approx 0.12$ns) is due to singlet para-positronium (p-Ps). The intermediate lifetime ($\tau_2 \approx 0.40$ ns) is due to positrons and positron-molecule species. The longest lifetime ($\tau_3 \geq 0.5$ ns) is due to the o-Ps localized in the free-volumes holes. In the finite lifetime analysis one uses the single longest lifetime parameter (τ_3) to determine the mean free volume hole size via Eq. 3.

In continuous lifetime analysis, a PAL spectrum is expressed in a continuous decay form [20]:

$$y(t) = R(t) * \left(N_t \int_0^\infty \lambda \alpha(\lambda) \exp[-\lambda t] d\lambda + B \right) \qquad (5)$$

in which the annihilation decay integral function is simply a Laplace transformation of the decay probability density function (pdf) $\lambda \alpha(\lambda)$. The exact solution of $\alpha(\lambda)$ and λ in the above equation is a very difficult mathematical problem since the resolution function *R(t)* is not known exactly. However, the solution can be obtained if one measures a reference spectrum from a sample with known positron lifetimes and uses this to deconvolute the unknown spectrum. For this purpose, extra high purity and defect free single crystals, such as Cu (τ_r = 122 ps), or Ni (τ_r = 102 ps) can be used. The PAL spectrum for the reference must be obtained under the same experimental conditions and configurations as used in the samples in order to preserve the same instrument resolution.

In practice the lifetime distributions are usually obtained using a computer program such as the MELT [21] or CONTIN [22, 23] programs. The reliablity of these programs for measurring the o-PS lifetime distribution in polymers was shown by Cao et al [24]. A detailed description of these methods of data analysis is presented in Chapter 4. The advantage of the continuous lifetime analysis is that one can obtain free volume hole distributions rather that the average values obtained in the finite analysis.

10.2.2 Free volume fractions

Besides determining the hole size of polymers, it is useful to determine changes in fractional free volumes (f_v) as this parameter is directly related to the mechanical properties of a polymer. In terms of the free-volume holes, the fractional free volume can be thought of as the product of the average hole size and the hole concentration. It has been suggested that the total fraction of o-Ps formed (I_3) in the polymer is related to the number of free volume holes in the matrix. The probability of o-Ps formation is assumed to be proportional to the number of regions of low electron density in which the o-Ps can locate. Accordingly the fractional free volume, f_v, can be determined using the following equation [25, 26]

$$f_v = CI_3 \langle v_f(\tau_3) \rangle \qquad (6)$$

where $\langle v_f(\tau_3) \rangle$ (in Å^3) is the mean hole volume determined using Eq. 3 and assuming a spherical cavity and C is a empirical scaling constant that reflects the probability of o-Ps formation that do not depend on the free volume and is usually determined from specific pressure-volume-temperatures (PVT) measurements above and below glass transition temperature (T_g). The validity of Eq. 6 has been confirmed by comparing PAL data against independent free-volume values estimated from bulk volume temperature data via appropriate theory, using the scaling constant [25, 26, 27]. Care needs to be taken when using Eq. 6 to determine the free volume. This equation will only be valid for polymers that do not contain groups, which inhibit positronium formation. Exposure of the sample to the radiation source may also lead to a decrease/increase in the I_3 values and need to be taken into consideration when using Eq. 4 (see sec. 10.4)

Alternative ways of determining the free volume fraction without using I_3 have also been proposed by Dlubek et al [28], as well as, Brandzuch et al [29]. Dlubek et al used the coefficient of thermal expansion of the amorphous regions and hole volume determined from positron data to determine the number density of the free volume holes. Brandzuch et. al. used the coefficient of thermal expansion just above and just below the T_g to estimate the fractional free volumes. This model is based on the assumption that the expansion of the holes of the free volume, as seen by positrons, reflects the expansion of the total volume of the material.

10.3 Examples of PAL studies in polymers

10.3.1 Phase transition phenomena in polymers

Amorphous polymers exhibit widely different physical and mechanical behaviours, depending on the temperature and structure of the polymer [30]. The temperature dependence of the polymers properties constitutes a very important part of polymer studies. The most important thermal transition in a polymer is the glass transition temperature. At low temperatures amorphous polymers are glassy, hard and brittle. As the temperature is raised, they go through the so-called glass-rubber transition and the properties change drastically and are best described as being "leathery"[30]. The glass transition temperature (T_g) is defined as the temperature at which the polymer "softens" because of the onset of long-range co-ordinated molecular motion. This transition is accompanied by a change in the free volume properties in the polymer.

Many PAL temperature related studies have been carried out on polymers in order to correlate the changes in the free volume properties with phase transition phenomena (particularly T_g) [31-37] Generally, as might be expected, there is a significant change in the slope of the τ_3-temperature curve as the polymers pass through the glass transition temperature, with an increase in the slope above T_g. Figure 10. 3 shows a typical τ_3 –temperature curve for polymethylmethacrylate (PMMA). The T_g is observed as an inflection point in the curve. Data reported on epoxy resins [12], polystyrene [27], methacrylates [35], cis-1,4-poly(butadiene) and polyisoprene [33]show a sharp change of slope of τ_3 versus temperature at T_g. Figure 10.4 shows the hole size distribution for the PMMA samples at varoius temperatures determined using the CONTIN program.

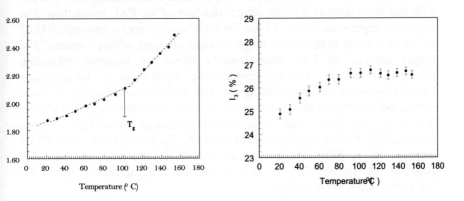

Figure 10.3 Typical τ_3 and I_3 temperature curve for PMMA showing the inflection point, which corresponds to the T_g of the polymer.

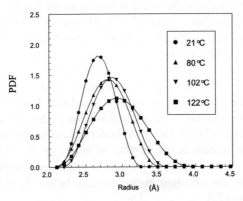

Figure 10.4 Hole size distribution curves (determined using the CONTIN program) for PMMA at various temperatures.

The increase in the slope of the τ_3-temperature curve can be correlated with the greater rate of increase of the free volume with temperature as the molecular motions intensify and a growing number of conformational states for the polymer chains become populated. Not all these states allow for a dense packing of the polymers chains, with the resultant increase in free volume [38]. The T_g values determined from PAL data tend to be lower than those determined from mechanical or viscoelastic measurements or by DSC. This has been attributed to the longer duration of the PAL measurement at any one temperature (about 1 hour for each temperature) compared to DSC where the entire temperature range is usual scanned within minutes. The longer duration of PAL measurements allows for longtime relaxation processes to occur. The temperature-time dependence of the transition kinetics in molecular materials means that methods requiring longer time scales indicate transitions at lower temperatures than methods with shorter time scales [39]. Interestingly it has been demonstrated that the hole volumes measured by PAL at the location of the DSC T_g, increase linearly with an increase in T_g [40]. Srithawatpong et al [41] showed that data for polyisoprene, polyvinylactete, polystyrene and polycarbonates all fit the following simple relationship

$$\langle v_f \rangle = aT_g$$

where $\langle v_f \rangle$ is the hole volume determined from τ_3 using Eq. 3, T_g is the T_g determined by DSC and a has a value of 3.27×10^{-4} nm^3/K. One of the interesting consequences of this is that the T_g is not an iso-free-volume state; that is the value of the fractional free volume at T_g increases with an increasing T_g.

The variation in the o-Ps intensity (I_3) with temperature is more complex than that of the o-Ps lifetime. A number of authors have reported that the I_3 values increases below the T_g of the polymers and then flattens out above the T_g [25, 42, 27] This has lead to the suggestion that in the melt (above T_g) the increase in total free volume with temperature involves the growth in hole size rather than an increase in the number of holes. On the other hand since τ_3 increases only slightly below T_g while I_3 increases significantly, it has been suggested that the variation in free volume below T_g primarily results from hole formation [25].

A number of discontinuities in the τ_3-temperature and I_3-temperature curves have been observed for polymers at temperatures well below the T_g of the polymer. [35, 43]. In the case if polyethylene these discontinuities have been interpreted as being associated with the secondary thermal transitions in the polymer [35]. Lin et al observed an inflection point in the τ_3 vs. temperature curve at 160K (about 80K below the Tg) and identified this as being associated with the gamma transition in the polymer. Uedono et al [44] also demonstrated that the β and γ transitions could be observed as changes in the slope of the τ_3 curves for polyethylene. The variation of the lifetime parameters as a function of temperature for polyethylene is shown in Fig. 10.5. The temperature ranges corresponding to the mechanical relaxation processes, α, β and γ are shown in the figure. Similar transitions have been observed in temperature studies of polyether-based network polymers [34]. The γ transition is related to the small segment motions of the polymer molecules (the so-called crankshaft motion), while the β transitions are associated with longer range cooperative motion of the chains. The ability to detect these transitions illustrates the sensitivity of the Ps probe in detecting very small changes in the free volume properties of the polymer. The detection of these secondary transitions is extremely difficult with many conventional techniques such as dillatometry and DSC.

As the temperature increases to well above the glass transition and the polymer chains begin to relax, the lifetime behavior exhibited by these materials becomes similar to that observed in organic liquids [45]. In these instances it is thought that the surface tension of the materials becomes small enough to allow the repulsive forces between the Ps and the surrounding molecules to form a "bubble" around the Ps atom [43]. As the temperature is increased further, the effect of surface tension decreases until the size of the bubble is determined solely by the repulsive action of the Ps atom. Thus above a certain temperature the bubble no longer expands since its dimensions have reached an equilibrium level. This is observed as a flattening of the τ_3-temperature curves, which reaches a constant value above this threshold temperature. Typically this plateau is observed at about 60-100K above the observed T_g in the polymer [33, 34, 41]. At this point equation 3 will no longer be valid since the o-Ps lifetime is no longer correlated to the hole size.

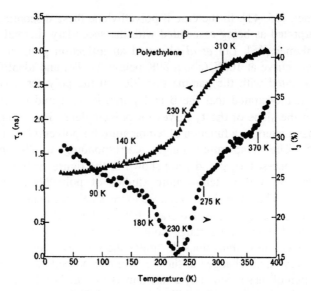

Figure 10. 5 The o-Ps lifetime (τ_3) and intensity (I_3) for polyethylene as a function of temperature. The temperature ranges corresponding to mechanical relaxation processes (α, β and γ) are shown in the figure [44].

10.3.2 Studies in thermoplastics

A large number of PAL studies on thermoplastic polymers have been carried out over the past 35 year. Once again this section will not attempt to review all the relevant literature. Instead an overview will be presented to illustrate the types of information, which can be obtained from PAS. An comprehnsive review, by Pethrick [45] can be found for the various studies conducted on a varity of thermoplastics. In addition a comphrehensive list of published positron data for a large number of polymers is included as an appendix for this book.

Many of the early studies done on polymers were carried out on commercial materials. Unfortunately in many cases these materials where not fully characterize in terms of the chemical structure and purity. The result is that data presented by different workers on the same polymers, apparently show a large deviation in both the lifetimes and intensities reported. These discrepancies are most probably a result of differences in the sample preparation, molecular mass, residual catalyst content, processing additives and the thermal histories of the samples [45]. More recently studies have been carried out on well defined and characterized polymer samples.

10.3.2.1 Studies in semi-crystalline thermoplastics

The morphology of semi-crystalline polymers is far more complex than that of completely amorphous polymers. The degree or amount of crystallinity in the polymer depends to a large degree on the chemical structure of the polymers, as well as, its thermal history. It is, for example, possible to influence the crystallinity and crystal structures by varying the temperature at which crystallization takes place [30]. In addition the crystallites that grow during the crystallization process are not clearly defined entities with sharp boundaries or crystal faces separating them from the surrounding amorphous material. The crystallites are small volume elements in which chain segments have a crystalline arrangement. These are connected to the surrounding amorphous material via chain segments in which the crystalline order is partially maintained. The crystallites themselves also contain many defects and dislocations arising from the extensive chain entanglements of molecules in the bulk state. There is, therefore, not a sharp discontinuity in the value of the physical parameters between the crystalline and amorphous regions. This phenomena presents unique challenges in interpreting PAL data. These factors are at least partially responsible for the fact that there is no clear consensus in the literature as to how many lifetimes can be resolved in semi-crystalline polymers.

A number of authors have found that the lifetime spectra of semi-crystalline polymers are best resolved into three components. In a study of PEEK (poly(ether ether ketone)), Nakanishi *et al* [14] found that a three component fit was best. They observed that the o-Ps lifetime (τ_3) did not change with an increasing amounts of crystallinity, but the o-Ps yield (I_3) decreases linearly with an increasing amount of crystallininty. It was also demonstrated that I_3 extrapolates to 0 at 100% crystallinity. No o-Ps lifetime was observed that might be attributed to the annihilation within the crystalline regions of the polymer. Lind *et al* [43] found a similar result for polypropylene where the τ_3 component changed very little with the amount of crystallinity, but the I_3 value decreased with an increasing amount of crystallinity. This was once again interpreted as being indicative of the fact that o-Ps annihilation occurs only in the amorphous region of the polymer. From this it was concluded that due to the dense packing, no o-Ps is formed in the PEEK or polypropylene crystals. On the other hand, four lifetime components have been found for studies in polyethylene [46, 47], polytetrafluororthylene [28], polyamides [48] and a series of poly(a-olefins) [49].

In the case of a four components fit, there is strong evidence to suggest that the shorter τ_3 ($\approx$ 1 ns) component corresponds to o-Ps annihilation in the crystalline regions of the polymers and the longer τ_4 (2-4 ns) component corresponds to o-Ps annihilation in the amorphous regions. Serna *et al* [50] have shown that the intensity of the third component increases while the intensity of the fourth component decreases with increasing crystallinity in polyethylene. It has also been shown that in the case of polyethylene the τ_3 value increases between 80 and 300K, as would be expected from the thermal expansion of the PE crystal unit cell [28]. A recent article by Dlubek *et al* [47], however illustrates that this simplistic interpretation of the data may not be entirely valid. As it is pointed out, this interpretation assumes a simple two-phase model (crystalline and amorphous), when at the very least a three-phase model would be expected where the crystal/amorphous interface is included. There is also evidence to suggest that significant diffusion of o-Ps from the crystalline to amorphous phase of the polymer exists. It has also recently been demonstrated how artifacts in the spectrum analysis caused be the broad distribution in the o-Ps lifetimes can cause τ_3 and τ_4 analyzed from the spectrum to be too large whereas the intensities appear too high (I_3) or too low (I_4) when compared to the true (or simulated) values [47]. It should also be pointed out that several other possibilities have been proposed for the occurrence of four lifetime components in semi-crystalline polyethylene. Stevens and Lichtenberger proposed that the τ_3 component is due to free positrons annihilating in the less ordered regions between the folded chains [51]. Balta Calleja *et al* [52] did a PAL study on lamellar polyethylene with chain defects and deduced that the τ_3 component may be ascribed to Ps trapped at the crystal interface. Kindle *et al* [53] conducted PAL measurements on branched polyethylene. The lifetime spectra were decomposed into four components. They showed that both the lifetime and intensity of the third component was constant over the entire temperature range even through the T_g. They concluded that this component should be ascribed to the annihilation from two different states, which are attributed to an attachment of the Ps atom to the molecule and relate to the trapping sites in crystalline regions or at the crystalline – amorphous interface.

The above discussion illustrates a very important aspect of PAL studies in polymers. In order for PAL studies to be meaningful there needs to be a sound theoretical basis for the interpretation of the data and results. The discussion also illustrates the increasing complexity of the underlying theory as more complex systems are studied. It is clear that in order for PAL to be developed as a tool for the characterization of these systems there needs to be simultaneous development and investigation of the underlying theory applicable to these systems.

10.3.2.2 The influence of other structural properties on PAL measurement

The effect of the molecular weight of polystyrene samples on the free volume of the polymers was demonstrated by Yu et al [27]. They showed that for a series of polystyrene with molecular weights ranging from 4000 to 400 000, the free volumes below T_g were not significantly different, however, above T_g the lower molecular weight compound has a significantly higher fractional free volume at any given temperature. The end-group contribution to the free volume in these polystyrene samples is clearly evident from the o-Ps annihilation results that . The molecular chain length not only affects the T_g, but also the o-Ps lifetimes and intensity. As expected, the polystyrene with longer molecular chains has a higher T_g (determined from PAL data) and smaller average cavity size in the melt. Porto et al [54] found a similar result in a series of poly(ethylene oxides), where once again, the contributions from the end groups to the free volume was demonstrated. It was, however, found that above a molecular weight of about 6000 the PAL properties become less sensitive to the growth of the chains.

Dlubek et al [49] studied a series of metallocene-catalyzed poly α-olefins) with progressively longer chains as the pendant side groups from polypropylene to poly-1-eicosene (20 carbons). Their results show an interesting relationship between the o-Ps lifetime and intensity in the amorphous phase for this series of polymers. They found that the average hole size and o-Ps intensity from the amorphous phase, decreased from polyethylene to polypropylene, followed by a slight increase to poly-1-butene. There was a rapid rise in hole size and intensity to poly-1-dodecene.

This was once again followed by a gradual decrease up till poly-1-eicosene. They were able to show a good correlation between the hole size at room temperature and the T_g of the samples as measured by DSC. They presented a novel way of using PAL data to estimate the T_g of the higher α-olefins, which were not able to be detected using DSC due to the crystallinity of the samples. A plot of T_g (from DSC measurements) versus hole volume in the amorphous phase (from PAL data at room tempature) for the lower α-olefins was extrapolated to the higher α-olefins. Qi et al [55] found that in a series of styrene-methly acrylate copolymers and styrene-butyl methacrylate copolymers there is an increase in the number of side groups of larger volume will cause an increase in the measured τ_3 values.

The affect of polymer stereoregularity in the chains on the PAL data has also been studied. Hamielec et al [56] found what appears to be an increased lifetime (hole size) with increased randomness of the chain configuration in a series of polyvinlychloride (PVC) polymers, despite the large degree of scatter in the sample (probably due to the fact that a series of commercially available products were used.). They however found little correlation with tacticity in polypropylene. More recently a PAL study on a series of very well characterized polystyrene and poly(p-methlystyrene) samples of differing tacticity [57] was performed. In addition to finding that the polystyrene samples have smaller free volume holes than the poly(p-methylstyrene) samples, they found that the syndiotactic samples had broader hole distributions than the attactic samples.

10.3.3 Correlation with mechanical and other properties

10.3.3.1 Gas permeation

The diffusion of gases through a polymer matrix is determined by the mobility of gas molecules through the matrix. The diffusion coefficient is therefore, at least partially determined by the free volume size of the polymer. It has been shown, for example, that there is a correlation between the free volume measured by PAL and the diffusivity of carbon dioxide in a seriers of polycarbonates [58]. In a study of poly (trimethylsilyl propyne) (PTMSP), which has an extremely high gas permeability and diffusion coefficients, it was found that the lifetime data could be resolved into four components [59]. The longest lifetime component (τ_4) had a lifetime of about 6.67 ns, which corresponds to a hole size with a radius of about 6 Å. It was concluded that these exceptionally large holes for a glassy polymer were responsible for the extremely high gas permeability. A similar study of poly(2,2-bistrifluoro 4,5 difluoro 1,3 dioxole) which has a similarly very high permeability, found that the lifetime data could also be resolved into four components [60]. Once again the longest lifetime component was found to be around 6.07 ns. The high permeability was once again attributed to the presence of the exceptional large holes. Table 10.1 shows the lifetime data for these materials, as well as, the lifetime data for some common glassy polymers as a comparison of the extremely large hole size found in these polymers. V_f is the average hole volume calculated via equation 3 and assuming a spherical hole.

Table 10.1 Positron Lifetime data for some glassy polymers.

Polymer	τ_{o-Ps} (ns)	I_{o-Ps} (%)	V_f (Å^3)
Polystryene[a]	2.100	48.11	106
Polycarbonate[a]	2.154	18.70	110
PTMSP[b]	6.668	30.00	860
Poly(2,2 bis trifluoro 4,5 di-fluoro 1,3 dioxole)[c]	5.26-6.07	23.7-23.3	479-591

[a] data from reference [61]
[b] data from reference [59]
[c] data from reference [60]

12.3.3.2 Mechanical properties

A number of PAL studies on polymers have attempted to correlate PAL parameters with macroscopic mechanical properties. The value of PAL in providing microscopic structural information for the observed macroscopic properties in polymer systems is illustrated by the examples bellow.

Hsieh et al [62] found that for a range of thermotropic liquid crystalline polymers, the greater the free volume measured by PAL the greater the chain mobility at Tg and the higher the value of tan δ (damping strength) measured by dynamic mechanical analysis.

Wang et al [63] used PAL to study the effects of deformation on the microstructure of polytetrafluoroethylene (PTFE). They found that the PAL parameters could be correlated with three stages of deformation, namely; the elastic region at low deformation, the plastic flow region and the strain stregthning region. In each region it was found that the o-Ps intensity remained constant, but there is a steep increase at the transition points between each region. This illustrates the sensitivity of PAL to probe microstructual changes of polymer during deformation.

The effects of plasticizers has also been studied by PAL [64, 65]. The addition of a plasticizer to polymers generally has the effect of lowering the T_g, however in some cases an anti-plasticization can occur. Borek et al [65] have shown that the fraction of free volume in PVC polymers could be fit with a fourth order polynomial as a function of plasticizer concentration. The decrease in th T_g with increasing amount of plasticizer is attributed to this increase in the free volume of the polymers.

Antiplasticization effects have also been studied by PAL. Anderson et al [64] found that in polystyrene/mineral oil blends, with small amounts of mineral oil, the antiplasticization effect was dependent on the molecular mass of the polymer. The high molecular weight polymers (270 000) showed only plasticizations effects, whereas the low molecular weight polymers (40 000) exhibited an antiplasticization at low mineral oil content and a plasticization effect at higher mineral oil content. In the low molecular weight polymers it was found that the fractional free volume showed a 10% decrease up to 6% mineral oil content. This correlates to a two-fold increase in the flexural moduli and flexural strength in these polymers. It was concluded that the antiplasticization is a phenomenon that can be attributed to the chain ends. At low concentrations, the mineral oil fills the small holes at the chain ends, thus restricting their mobility and decreasing the fractional free volume. It was further shown that the antiplasticization occurs when the average diameter of the mineral oil domains approximates the average size of the free volume voids. At higher oil concentrations, above the solubility limit of the mineral oil, where the mineral oils domain sizes are significantly larger than the average free volume hole diameter, phase separation occurs and plasticization is dominant.

10.3.4 Chemical sensitivity

One of the recent interesting developments in the application of PAS to polymers has been the development of techniques to probe the chemical environment of the free volume holes in polymers [66]. The technique is similar to the technique used extensively for probing the chemical environment of voids in other materials. It makes use of the Doppler broadened annihilation energy line or alternatively the angular correlation of the annihilation radiation (ACAR). These spectra contain information on the momentum density of the radiation and are a superposition of a narrow component due to self-annihilation of p-Ps and a much wider distribution arising from the annihilation of free positrons and the o-Ps pick-off annihilation. The narrow component reflects the localization momentum of the p-Ps inside the holes and its full width at half maximum may be related to the hole dimensions. In contrast the broader components of the momentum distribution contain information about the electronic configurations of the surrounding molecules. Since it is believed that the free positrons and positronium annihilate within the free volumes holes, the broad components of the momentum distributions are expected to reflect the chemical surroundings of the annihilation sites in the hole. The techniques involves the subtraction of the narrow component obtained from a three Gaussian fit from the Doppler broadened spectra. The remaining broad distribution can then be related to the nature and density of the chemical species in the vicinity of the holes.

10.3.5 Polymer blends

PAL has been used to study both miscible and immiscible polymer blends [41, 61, 67-70]. PAL results have shown both positive and negative deviations from additivity of free volume with blend composition. In the case of multi phase systems, PAL data analysis is complicated by the fact that Ps may diffuse between the different blend phases.

10.4 Potential problems in polymer studies

There are a number of artifacts of the PAL technique that need to be considered proir to meaningful analysis of PAL data.

10.4.1 Exposure to the positron source

One of the main factors which needs to be considered in PAL analysis of polymers, is the affect which prolonged exposure to the positron source has on the lifetime parameters. It has been found that on prolonged exposure to a positron source, the o-Ps lifetimes are largely unchanged, but that there are significant variations in the o-Ps intensities for some polymers. Examples of these effects for a wide variety of polymers can be found; polypropylene (PP), polyethylene (PE) [71], polystyrene [72], polycarbonates [73] poly(α-olefins) [49], poly(vinlyacetate) [74], poly(methyl methacrylate) [74] and a number of copolymers [75].

The effects of e^+-irradiation are strongly dependent on time, temperature and the nature of the polymer. In general, at moderate temperatures, it has been found that there is a decrease in the I_3 intensity with increasing exposure time. It has also been found that the decrease in I_3 on exposure only occurs in non-polar or weakly polar polymers and not in strongly polar polymers at moderate temperatures [75]. This decrease in I_3 on prolonged exposure has been attributed to the build-up of an electric field inside the polymers during prolonged PAL measurements [73, 75]. In terms of the spur model for positronium formation, this field effect can lead to positrons and electrons diffusing out of the terminal spur and therefore inhibit Ps formation. In the case of polar polymers reactions of the polar group, trapping of positrons and electrons and the recombination of electrons with the positive ions is faster than the positrons and electron diffusing out of the spur under the influence of the electric field. The effect of the build up of charge in the polymer was demonstrated by Qi *et al* [76]. They found that the presence of a very thin (20-40 nm) conductive film on the polymer surface decreased the rate of decease of the I_3 value with time. In addition it was found that the rate of decrease of I_3 was less with films with increasing conductive properties, most probably as a result of the reduction of the electric field at the surface. A similar result is obtained when the samples are grounded. Alternatively the observed decrease have been explained in terms of the formation of radicals in the sample, which in principle can inhibit Ps formation [72].

The effects of temperature have also been demonstrated. Fig. 10.6 shows the influence of temperature on the observed changes in o-Ps intensities in polystyrene on prolonged exposure to e^+ irradiation. Peng et al [67] showed that in high vinly polybutadiene (HVBD)/cis polyisoprene (CPI) blends the o-Ps intensity decreased more rapidly in the melts as the temperature approaches the T_g and then again more slowly in the glassy. It was also found that at room temperature (about 70K above the blend T_g), no decrease in I_3 with e^+ exposure time was found. When the temperature was decrease to only 10 above the T_g a decrease in the I_3 was once again observed. It was found that this decrease was completely reversible after annealing the sample at the higher temperature. It has also been demonstrated that the decrease in the I_3 values with e^+ irradiation in polystyrene samples at temperatures below T_g, is also completely reversible after annealing the samples at higher temperatures [72]. It was, however found that in the case of polypropylene, there was no recovery of the I_3 value after annealing at higher temperatures [72]. This lack of recovery is explained in terms of the fact that in polypropylene the free radicals formed at the low temperatures lead to cross-linking between the chains and the permanent decrease in the I_3 value is in fact due to permanent structural changes produced in the vicinity of the radiation spur as a result of cross-linking of the matrix.

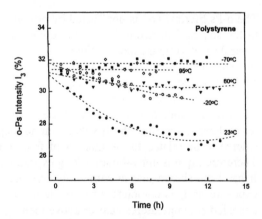

Figure 10.6 The dependence on exposure time to a positron source of the o-Ps intensity in polystyrene at different temperatures [72].

Dlubek *et al* [49] also showed an interesting relationship between the observed decrease in the o-Ps intensity and the temperature of the measurement in relation to the T_g for a series of poly (α-olefins). They found that the rate of decrease of the o-Ps decreased, but the saturation levels increased, the more the T_g is below the temperature of the measurements.

At very low temperatures (typically below 150K) the o-Ps intensity increases as a function of exposure time for a large number of polymers. This increase in the o-Ps yields at very low temperature is explained by the reaction of free positrons with trapped electrons produced by the previously injected positrons [77].

The variation of I_3 on e^+ exposure illustrates that after prolonged exposure the I_3 values can no longer be considered to be related to the number of free volume holes and therefore equation 6 cannot be used. These effects have also lead a number of authors to completely discount the reliability of equation 6 as a measure of the free volume fractions [49, 75, 78, 79]. Other authors have suggested that Eq. 6 may still be used provided the samples are rejuvenated at high temperatures between each measurement [67].

In summary the effects of exposure to the e^+ source have several very important consequences for polymer studies, particularly in those where the polymer is exposed to the source for extended periods (for examples in temperature dependent studies):

(1) The o-Ps lifetime (τ_3) is unaffected by prolonged exposure to the source. Therefore Eq. 3 can be considered reliable regardless of sample exposure time.

(2) In experiments where samples are exposed to the source for long periods, the effects of this exposure on I_3 need to be investigated and considered.

(3) The decrease/increase in I_3 on exposure illustrates that after prolonged exposure the I_3 values can no longer be considered to be related to the number of free volume holes. Therefore Eq. 6 is not valid after long exposure times.

(4) It has been shown that in some polymers the effect of e^+ exposure on I_3 is completely reversible if the samples are annealed at temperatures near or above their T_g's. This implies that in thermal studies, methods should be developed

to rejuvenate the samples (temperature cycling) to minimize anomalous decreases in o-Ps intensities resulting from prolonged exposure to e^+ radiation (for examples see reference [67, 72].

10.4.2 Inhibition and quenching

In polymer systems, positron chemistry can occur if the e^+ is able to attach itself to a particular atom or group. Halogenated polymers, radicals and certain electron rich unsaturated structures have the capability of forming stable species of the form (e^+,R^-), with free positrons. These processes can lead to quenching and inhibition of Ps formation.

When inhibition of positronium formation occurs, it is seen as a decrease in the o-Ps intensity, while quenching of o-Ps, as a result of chemical reactions, will shorten the o-Ps lifetime (τ_3). While quenching and inhibition of positronuim have been extensively studied in solution (see Chapter 5) it is a factor that has often been overlooked in polymer studies. There are numerous examples of inhibition and quenching effects in polymers [42, 80-82]

In the case of quenching the reduction (shortening of the o-Ps lifetime) is explained by the chemical reaction of positronium and the quenching species (which in many cases are additives present in the polymer matrix). The chemical rate constant for the reaction between the Ps and quenching species can be expressed in terms of the concentration of the quenching species [M], and it is found that it can be described by pseudo-first order kinetics as [42, 80]:

$$\frac{1}{\tau_3} = \frac{1}{\tau_3^0} + k_{Ps}[M] \qquad (7)$$

where τ_3 is o-Ps lifetime in the polymer containing the quenching species, τ_3^0 is the o-Ps lifetime in the pure polymer and k_{Ps} is the Ps quenching rate constant. Generally the k_{Ps} values are 1-2 orders of magnitude smaller than the k_{Ps} of the same quenching species in liquids [42, 80]. This lower chemical reactivity can be expected by the lower mobility of the quenching species in the polymer matrix in comparision to liquids. Assuming that the reaction between the Ps and the quenching species is completely diffusion controlled, the k_{Ps} value allows for the determination of the Ps diffusion coeffcient (D_{Ps}) in the polymer [42, 80]

Inhibition or the reduction in the o-Ps yield is most probaly due to the inhibiting species scavaning precursors of Ps such as electrons and/or hot Ps atoms in the terminal positron spur [80]. It is also found that the inhibition effect is dependent on the concentration of the inhibiting species. Figure 10.7 shows the dependence of the o-Ps intensity (I_3) versus the concentration of a dopping chromophore in a PMMA polymer at two temperatures. With increasing amounts of the chormophore there is a decrease in the I_3 value. This data can be fitted with the following relationship [42]:

$$I_3 = \frac{I_0}{(1 + K_{inhib} M)} \quad (8)$$

where I_0 is the I_3 in the PMMA polymer without doping and K_{inhib} is called the Ps inhibition constant. A simlar realtionship was found for number of inhibiting species in a large number of polymers [42, 80-82].

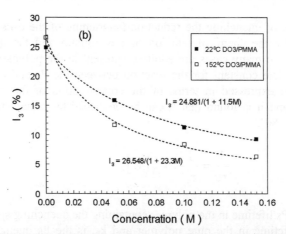

Figure 10.7 o-Ps intensity (I_3) versus the concentration of a doping chromophore (DO3) in PMMA. The lines in the figure are fitted according to Eq. (7) [42]. The values of I_3^0 and K_{inhib} are indicated in the figure for both temperatures.

Clearly in any polymer studies the possibility of positronium chemistry influencing the lifetime parameters needs to be considered. Once again the interpretation of the PAL data can be misleading if quenching and inhibition effects are present in the polymer system.

10.5 Concluding remarks

PAL has become well established as an analytical tool in polymer characterization. The results of PAL studies on polymers have led to many important insights into the relationship between microscopic structure at a nanoscale and bulk properties of many polymers. The usefullness of the technique is seen in the fact that a large number of polymer systems have been studied in recent years.

As more complex polymers systems have been studied it is evident that subtle effects in the interpretation of data and the underlining theory have come to the fore. Advances in polymerization chemistry, as well as, the enormous advances in techniques used in polymer characterization, has meant that PAL polymer studies can now be conducted on well characterized and controlled systems. There is no doubt that PAL will continue to be a valuble tool for the polymer scientist. The application of monoenergitic slow positron beams to polymers surfaces will be discussed in the next chapter.

Acknowledgements
The assistance and support from Prof. Y.C Jean is gratefully acknowledged. Financial support from the University of Stellenbosch is also acknowledged. C. Greyling for assistance in preparing this chapter.

References

[1]   H. Eyring, *J. Chem. Phys.* **4**, 283 (1936)
[2]   T.G. Fox and P.J. Flory, *J. Appl. Phys.*, **21**, 21 (1951)
[3]   A. K. Doolittle, *J. Appl. Phys.*, **22** ,1471 (1951)
[4]   R.J. Roe and J.G. Curro, *Macromolecules*, **16**, 428 (1983)
[5]   J.G. Curro and R. J. Roe, *Polymer*, **25**, 1424 (1985)
[6]   S. Nojima, R. J. Roe, D. Rigby and C.C. Han, *Macromolecules*, **23**, 4305 (1990)
[7]   W.C. Yu, C.S.P. Sung and R.E. Robertson, *Macromolecules*, **21**, 355 (1988).
[8]   W.C. Yu and C.S.P Sung, Macromolecules, 21, 365 (1988)
[9]   J.G. Victor and J. M. Torkelson, *Macromolecules*, **20**, 2241 (1987)
[10] J.G. Victor and J. M. Torkelson, *Macromolecules*, **21**, 3490 (1998)
[11] J.S. Royal, J.G. Victor and J.M. Torkelson, *Macromolecules*, **25**, 4792 (1992)

[12] Y.C. Jean, T.C. Sandreczki and D.P. Ames, *J. Polym. Sci. Part B: Polym. Phys.*, **24**, 1247 (1986)
[13] Q. Deng, C.S. Sundar and Y.C. Jean, *J. Phys. Chem.*, **96**, 492 (1992)
[14] H. Nakanishi, Y.C. Jean, E.G. Smith and T.C Sandreczki, *J. Polym. Sci. Part B: Polym. Phys.*, **27**, 1419 (1989)
[15] S. J. Tao, *J. Chem. Phys.*, **56**, 5499 (1972)
[16] H. Nakanishi, S. J. Wang and Y.C. Jean, in *Positron Annihilation Studies of Fluids*, S.C. Sharma ed., World Scientific, Singapore 1988, p. 292.
[17] K. Ito, H. Nakanishi and Y. Ujihira, *J. Phys. Chem B*, **103(21)**, 4555 (1999)
[18] T.L. Dull, W.E. Frieze, D.W. Gidley, J.N. Sun and A.F. Yee, *J. Phys. Chem. B*, **105**, 4657 (2001)
[19] PATFIT package (1989), Riso National Laboratory, Denmark
[20] Q. Debg, F. Zandiehnadem and Y.C. Jean, *Macromolecules*, **25**, 1090 (1992)
[21] A. Shukla, M. Peter and L. Hoffmann, *Nucl. Instrum. Methods A*, **335**, 310 and 335 (1993)
[22] S.W. Provencher: CONTIN program, EMBL Technical Report DA 05, European Biology Laboratory, Germany 1982; *Comput. Phys. Commum.*, **27**, 229 (1982)
[23] R.B. Gregory, Y. Zhu, *Nucl. Instrum. Methods Phys. Res. A*, **290**, 172 (1990)
[24] H. Cao, G.H. Dai, J.P. Yuan, Y.C. Jean, *Mat. Sci. Form.*, **255-257**, 238 (1997)
[25] Y. Kobayashi, W. Zheng, E.F. Meyer, J.D. McGervey, A.M. Jamieson and R. Simha, Macromolecules, 22, 2302 (1989)
[26] Y.Y. Wang, H. Nakanishi, Y.C. Jean and T.C. Sandreczki, *J. Poly. Sci. Part B: Polym. Phys.*, **28**, 1431 (1990)
[27] Z. Yu, U. Yahsi, J. McGervey, A.M Jamieson and R. Simha, *J. Polym. Sci. Part B: Polym. Phys.*, **32**, 2637 (1994)
[28] G. Dlubek, K. Saarinen and H.M. Fretwell, *J. Polym. Sci. Part B: Polym. Phys.*, **36**, 1513 (1998)
[29] P. Bandzuch, J. Kristiak, O. Suasa and J. Zrubcova, *Phys. Rev. B*, **61(13)**, 8784 (2000)
[30] For examples see L.H Sperling *Introduction to Physical Polymer Science*, John Wiley & Sons, New York, 1986
[31] H.A. Hristov, B. Bolan, A.F. Yee, L. Xie and D.W. Gidley, *Macromolecules*, **29**, 8507 (1996)
[32] B. Djermouni and H.J. Ache, *Macromolecules*, **13**, 168 (1980)
[33] J. Bartos, P. Bandzuch, O. Sausa, K. Kristiakova, J. Kristiak, T. Kanaya and W. Jenninger, *Macromolecules*, **30**, 6906 (1997)
[34] A. Uedono, T. Tanigawa, M. Watanabe and N. Nihimoto, *J. Polym. Sci. Part B: Polym. Phys.*, **36**, 1919 (1998)
[35] B. Malhotra and R. Pethrick, *Macromolecules*, **16**, 1175 (1983)
[36] D. Bamford, G. Dlubek, A. Reiche, M.A. Alam, W. Meyer, P. Galvosas and F. Rittig, *J. Chem. Phys.*, **115(15)**, 7260 (2001)

Application to Polymers 279

[37] D. Lin and S.J. Wang, *J. Phys. Condens. Matter*, **4**, 3331 (1992)
[38] G. Strobl *The Physics of Polymers* 2nd ed. Springer, Berlin (1997)
[39] J.D. Ferry, *Viscoelastic Properties of Polymers*, Wiley, New York (1970)
[40] J. E. Kluin, Z. Yu, S. Vleeshouwers, J. D. McGevrvey, A.M. Jamieson, R. Simha and K. Sommer, *Macromolecules*, **26**, 1853 (1993)
[41] R. Srithawatpong, Z. L. Peng, B.G. Olson, A.M. Jamieson, R. Simha, J. D. McGervey, T.R. Maier, A. F. Halasa and H. Ishida, *J. Polym. Sci. Part B: Polym. Phys.*, **37**, 2754 (1999)
[42] C.M. Huang, E. W. Hellmuth and Y.C. Jean, *J. Phys. Chem B*, **102(14)**, 2474 (1998)
[43] J.H. Lind, P. L. Jones and G.W. Pearsall, *J. Polym. Sci. Part A: Polym. Chem.*, **24**, 3033 (1986)
[44] A. Uedono, R. Suzuki, T. Ohdaira, T. Uozumi, M. Ban, M. Kyoto, S. Tangawa, T. Mikado, *J. Polym. Sci. Part B: Polym. Phys.*, **36**, 2597 (1998)
[45] R. A. Pethrick, *Prog. Polym. Sci.*, **22**, 1 (1997)
[46] J. Borek and W. Osoba, *Polymer*, **42**, 2901 (2001)
[47]] G. Dlubek, J. Stejny, T.H. Lupke, D. Bamford, K. Petters, C.H. Hubner, M.A. Alam, M.J. Hill, *J. Polym.Sci. Part B: Polym. Phys.*, **40**, 65 (2002)
[48] G. Dlubek, M. Stolp, C Nagel, H.M Fretwell, M.A. Alam and H.J. Radusch, *J. Phys. Condens. Matte*, **10**, 10443 (1998)
[49] G. Dlubek, D. Bamford, O. Henschke, J. Knorr, M.A. Alam, M. Arnold, T. Lupke, *Polymer*, **42**, 5381 (2001)
[50] J. Serna, J. C. Abbe and G. Duplatre, *Phys. Stat. Sol. A*, **115**, 289 (1989)
[51] J.R Stevens and P.C. Lichtenberger, *Phys. Rev. Lett.*, **29**, 166 (1972)
[52] F.J. Balta Calleja, J. Serna, J. Vicente, *J. App. Phys.*, **58**, 253 (1985)
[53] P. Kindl and G. Reiter, *Phys. Stat. Sol. A*, **104**, 707 (1987)
[54] A.O. Porto, G. Silva and W.F Magalhaes, *J. Polym. Sci. Part B: Polym. Phys.* **37**, 219 (1999)
[55] C. Qi, S. Zhang, Y. Wu, H. Li, G. Wang, G. Huai, T. Wang, J. Ma, *J. Polym. Sci. Part B: Polym. Phys.*, **37**, 2476 (1999)
[56] A.E. Hamielec, M. Eldrup, O. Mogensen and J. Jansen, *J. Macromol. Sci. Revs. Macromol. Chem.*, **C9(2)**, 305 (1973)
[57] R. M. Dammert, S. L. Maunu, F. H. J. Maurer, I. M. Neelov, S. Niemela, F. Sundholm and C. Wastlund, *Macromolecules*, **32**, 1930 (1999)
[58] Y.C. Jean J.P. Yuan, J. Liu, Q. Deng and H. Yang, *J. Polym. Sci. Part B: Polym. Phys.*, **33**, 2365 (1995)
[59] Y.P. Yampolskii, V.P. Shantorovich, F.P. Chernyakovskii, A.I. Kornilov and N.A. Plate, *J. App. Polym. Sci.*, **47**, 85 (1993)
[60] W.J. Davis and R. A. Pethrick, *Eur. Polym. J.*, **30(11)**, 1289 (1994)
[61] J. Liu, Y.C. Jean and H. Yang, *Macromolecules*, **28**, 5774 (1995)
[62] T.T. Hsieh, C. Tiu, G. P. Simon, J. App. Polym. Sci., 82, 2252 (2001)

[63] C.L. Wang, B. Wang, S. Q. Li and S. J. Wang, *J. Phys. Condens. Matter*, **5**, 7515 (1993)
[64] S.L. Anderson, E.A. Grulke, P.T. DeLassus, P.B. Smith, C.W. Kocher, B.G. Landes, *Macromolecules*, **28**, 2944 (1995)
[65] J. Borek, W. Osoba, *J. Polym. Sci. Part B: Polym. Phys.*, **36**, 1839 (1998)
[66] G. Dlubek, H.M. Fretwell and M.A. Alam, Macromolecules, 33 187 (2000)
[67]] Z.L. Peng, B.G. Olson, R. Srithawatpong, J.D. McGervey, A.M. Jamieson, H. Ishida, T.M. Meier, A.F. Halasa, *J. Polym. Sci. Part B: Polym. Phys.*, **36**, 861 (1998)
[68] C.M. McCullagh, Z. Yu, A.M. Jamieson, J. Blackwell and J.D. McGervey, *Macromolecules*, **28**, 6100 (1995)
[69] G.P. Simon, M.D. Zipper and A.J. Hill, *J. App. Polym. Sci*, **52**, 1191 (1994)
[70] G. Dlubeik, M.A Alam, M. Stolp, H.J. Radusch, *J. Polym. Sci. Part B: Polym. Phys.*, **37**, 1749 (1999)
[71] T. Suzuki, T. Miura, Y. Oki, M. Numajiri, K. Kondo and Y. Ito, *Radiat. Phys. Chem.*, **45(4)**, 657 (1995)
[72] Z.L. Peng, B.G. Olson, J.D. McGervey, A.M. Jamieson, *Polymer*, **40**, 3033 (1999)
[73] X.S. Li and M.C. Boyce, *J. Polym. Sci. Part B: Polym. Phys.*, **31**, 869 (1993)
[74] C.L. Wang, T. Hirade, F.H.J. Maurer, M. Eldrup and N.J. Pedersen, J. Chem. Phys., **108(11)**, (1998)
[75] C. Qi, W. Wei, Y. Wu, S. Zhang, W. Haijun, H. Li, T. Wang, F. Yan, *J. Polym. Sci. Part B: Polym. Phys.*, **38**, 435 (2000)
[76] C. Qi, D. Ma, Y. Hu, F. Yan, H. Gao, H. Yang, X. Zhou, T. Wang, *J. Polym. Sci. Part B: Polym. Phys.*, **39**, 332 (2001)
[77] T. Hirade, F.H.J. Maurer, M. Eldrup, *Radiat. Phys. Chem.*, **58**, 465 (2000)
[78] F. H. J. Maurer, M. Schmidt, *Radiat. Phys. Chem.*, **58**, 509 (2000)
[79] M. Schmidt, F.H.J. Maurer, *Polymer*, **41**, 8419 (2000)
[80] K. Hirata, Y. Kobayashi, Y. Ujihira, *J. Chem. Soc. Faraday Trans.*, **92(6)**, 985 (1996)
[81] P.E. Mallon, C.M. Huang, H. Chen, R. Zhang, M.H.S. Gradwell, Y.C. Jean, Mat. Sci. For. , **363-365**, 281 (2001)
[82] A. Baranowski, M. Debowska, K. Jerie, G. Mirkiewicz, J. Rudzinska-Girulska, R. Tadeusz Sikorski, *J. De Physique IV*, **C4 3**, 225 (1993)

Chapter 11

Applications of Slow Positrons to Polymeric Surfaces and Coatings

Y.C. Jean[1], P.E. Mallon[2], Renwu Zhang[1], Hongmin Chen[1], Y.C. Wu[1], Ying Li[1], and Junjie Zhang[1]

Department of Chemistry, University of Missouri-Kansas City, USA[1]

Division of Polymer Science, Department of Chemistry, University of Stellenbosch, South Africa[2]

11.1 Introduction

The development of monoenergetic positron beams in recent years has enabled studies of surface properties of materials, in particular studies of atomic-scale defects at the surface and near-surface regions. Variable positron implantation energy beams have been extensively used to study metal and semiconductor surfaces [1-3]. Recently, variable mono-energetic positron beams have also been used to study polymer surfaces [4-10]. In the case of polymers, these beams allow for the determination of local free-volume properties of the polymers as a function of the depth from the surface. Since the Positronium (Ps) work function is negative, the positron data using a variable-energy positron beam contains information about surface chemical states, morphology and Ps dynamics near the surface.

In the case of coating, which contains a composite of polymer (typically 80% wt) and pigments (typically 20% wt, such as TiO_2), the obtained positron data strongly depend on chemical compositions and structures of coatings. In such heterogeneous and complicated chemical systems, scientific curiosity has focused on applications of slow positrons to both fundamental and applied research in recent years. The early stage of research [11-22] uses positrons emitted from ^{22}Na radioisotope, which penetrates coatings to a depth of about 100 μm. The information is an integrated bulk property. Recently, the use of slow positrons has opened a new area of research. The positrons have been used to probe the surface, interfaces, multilayers, and bulk properties of coating systems [23-36]. It is fascinating in nature that slow positrons are capable of specifically probing the change in local free-volume properties that control the applicability of coatings for protecting structural materials.

Two main types of positron experiments have been made in polymers using a positron beam, namely, Doppler broadening of energy spectra (DBES) and positron annihilation lifetime measurements (PAL). Angular correlation of annihilation radiation (ACAR) measurements using fast positrons have also been performed [37-40], but will not be discussed in this chapter. PAL measurements generally provide more quantitative information (such as free-volume size and distribution) than DBES studies, but in monoenergetic positrons beams PAL are more complicated than DBES measurements due to the challenges of detecting the start signal for the positron lifetimes. In DBES studies the momentum distribution around the 511 keV energy peak is used. The DBES spectra is characterized by an S parameter, defined as the ratio of integrated counts in the central part of the peak to the total counts after the background is properly subtracted. Although o-Ps (ortho-positronium, a triplet state) has a longer lifetime in polymers than p-Ps (para-positronium, a singlet state), neither its three-photon nor two-photon pick-off annihilation contribute much to the S parameter of DBES spectra. Therefore, only the p-Ps part of Ps annihilation contributes to the S parameter in DBES, while the o-Ps part of annihilation contributes to the long lifetime components, which are used to measure free-volume size, distribution and content in PAL. In PAL, one uses the well-established relationship between o-Ps pick-off lifetime (τ_3 in ns) and spherical hole radius R (in Å) as [41]:

$$\tau_3^{-1} = 2\left[1 - \frac{R}{R_0} + \frac{1}{2\pi}\sin\left(\frac{2\pi R}{R_0}\right)\right] \quad (ns^{-1}), \tag{1}$$

where $R_0 = R + \Delta R$, and ΔR is an empirical parameter. Nakanishi et al. [42] determined ΔR by fitting the observed lifetimes with the known hole and cavity sizes in molecular substrates. The best-fit value of ΔR for all known data is found to be 1.66 Å, while the free-volume fraction f_v is usually determined by fitting empirical data with thermal expansion data of polymers. A useful equation to estimate f_v (in %) based on the Williams-Landel-Ferry (WFL) equation [43] is given as [41]:

$$f_v = 0.0018 I_3 \langle v_f(\tau_3) \rangle \quad , \tag{2}$$

where I_3 is the o-Ps intensity (in %) of pick-off annihilation lifetime (τ_3) and v_f is the free-volume hole volume calculated from eq. (1) in Å. In the Appendix, we listed a table of free-volume hole size (Å) and f_v (%) as calculated from Eqs. (1) and (2) from the o-Ps data for polymeric materials.

For DBES data three main factors contribute to the S parameter in polymers: (1) free-volume content, (2) free-volume size, and (3) chemical composition. First, larger free-volume content contributes to a larger S value. DBES measures radiation near 511 keV where a major contribution comes from p-Ps. This p-Ps contribution is only 1/3 the o-Ps intensity as that in I_3 of PAL data. Second, when p-Ps is localized in a defect with a dimension Δx, the momentum Δp has a dispersion according to the Heisenburg uncertainty principle $\Delta x \Delta p \geq h/4\pi$. The S parameter from DBES spectra is a direct measure of the quantity of momentum dispersion. In a larger size hole where Ps is localized, there will be a larger S parameter due to smaller momentum uncertainty. Therefore, in a system with defects or voids, such as polymers, the S parameter is a qualitative measure of the defect size and defect concentration. The value of the S parameter also depends on the momentum of the valence electrons, which annihilate with the positrons. The absolute value of the S parameter therefore, may differ from polymer to polymer. Third, the S parameter depends on the electron momentum of the elements. As the atomic number of the elements increases, the electron momentum increases, and thus the S parameter decreases. Fortunately, in chemicals of the same composition, the elemental effect on S could be small, and S data reflect mainly free-volume properties in polymers.

11.1.1 Depth profiles in polymeric materials

When a positron with a well-defined energy is injected from a vacuum into a polymer, it is either reflected back to the surface or it penetrates into the polymer. The fraction of positrons that enter the polymers increases rapidly as a function of the positron energy. As the positrons enter the polymer, inelastic collisions between the positron and molecules slow down the positrons by ionization, excitation and phonon processes. The implantation—stopping profile $P(z,E)$ of the positrons varies as a function of depth as [1, 2]:

$$p(z,E) = -d\left[\exp-(z/z_0)^2\right]/dz , \quad (3)$$

where z_0 is related to the mean implantation depth Z_0 through:

$$z_0 = 2Z_0/\sqrt{\pi} . \quad (4)$$

Z_0 depends on the incident energy as

$$Z_0(E) = (400/\rho)E^{1.6} , \quad (5)$$

where Z_0 is expressed in Å and ρ is the density in g/cm^3. Figure 11.1 shows the mean stopping distance and stopping profile of positrons as a function of incident energy for a typical polymer.

As shown in Figure 11.1, the depth of the polymer from the surface could be defined by applying a well-defined positron incident energy from a slow beam [24]. A university-based slow beam could be operated at an energy up to about 30 keV (which is equivalent to about 10 μm) from the topmost layer of the surface of a few eV of positron energy [28]. The depth resolution is better when the energy is low. A typical estimated depth resolution is about 10% of the mean stopping distance z_0. For example, a 100 eV beam gives a mean depth of 36 Å with a depth resolution of about 3 Å, while a 1000 eV beam gives a depth resolution of about 40 Å at a mean depth of 360 Å in a polymer.

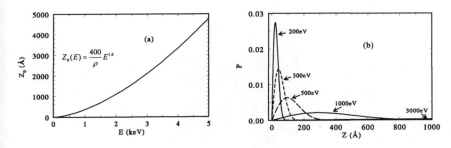

Figure 11.1 Mean stopping distance and stopping profiles for positrons as a function of incident energy [24].

11.1.2 Positron and positronium dynamics near the surface

For positrons and Ps in polymeric surfaces, one needs to consider three additional important effects in addition to the bulk: (1) the diffusion of the positron and Ps back to the surface, (2) the formation of Ps from the positrons by abstracting the surface electron, and (3) the Ps emission to the vacuum from the surface or the sticking of Ps on the polymeric surfaces. The dynamic behavior of the positron and Ps near the surface is schematically shown in Figure 11.2 below. The lifetimes of the positron and of Ps are different among those three types in addition to that of the bulk. If each has one distinct lifetime, a typical PAL lifetime spectrum could contain eight lifetimes for a complete analysis. This is beyond the current resolving power of the PAL data analysis method, either discrete or continuous. A practical approach is to invoke some good theoretical models before one applies the conventional data analysis method to a PAL spectrum near the surface for polymeric materials.

After losing their kinetic energy the penetrated positrons may either directly annihilate with surrounding electrons into two gamma rays, or combine with an electron to form a Ps atom. Although both positrons and Ps are known to localize within the free volumes, a certain fraction of them may diffuse back to the surface and escape to the vacuum. The probability of positrons and Ps annihilating in the polymer depends on their diffusion coefficients.

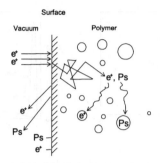

Figure 11.2 Schematic diagram of the positron and Ps states near the surface of polymers [10].

The mean diffusion length of the positron or Ps undergoing one-dimensional Brownian motion can be expressed as [44]:

$$l_{Ps} = \sqrt{\tau D_{Ps}}, \qquad (6)$$

where D_{Ps} and τ are Ps diffusion coefficients and lifetime, respectively. For a polymer containing a large fraction of defects, i.e., holes, free volumes, and interfaces, the D_{Ps} for Ps is much smaller than that for the positron due to its localization, neutral charge and high polarizability. Typical values of D_{Ps} in polymers for Ps are 10^{-6} cm^2/s, while the corresponding values for the positron are 10^{-3} cm^2/s. In the case of polystyrene, for example, the D_{Ps} value has been reported as 3×10^{-6} cm^2/s and $\tau = 2$ to 3×10^{-9}s [45]. The Ps diffusion length is therefore as short as 10 Å. On the other hand, the diffusion length of the positron in amorphous polymers is on the order of 50 to 100 Å [41]. In the case of DBES experiments of polymer surfaces, the main effect of Ps diffusion is very near the surface, while that of the positron spans 100 Å. Figure 11.3 (left) shows a typical S parameter vs positron incident energy curve for a polystyrene. There is an initial increase in the S parameter vs energy curve as the fraction of both positrons and Ps annihilating in the polymers increases. Increasing S parameters, as shown in Figure 11.3, are typical for slow positron data near the surface of amorphous polymers, which contain a large fraction of sub-nanometer defects. On the other hand, the S parameter decreases in solids which do not contain a large fraction of defects, such as oxides or semiconductors [1-3]. The similarity between S and I_3 variation with respect to the depth strongly supports the theory that S probes p-Ps, which is directly related to the same free-volume properties that PAL measures. However, a sharper increase of I_3 vs energy is observed than that of S near the surface. This indicates that the S parameter is contributed not only from p-Ps but also positrons, which have a longer diffusion length than Ps in polymers

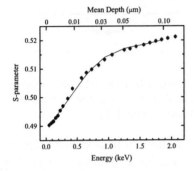
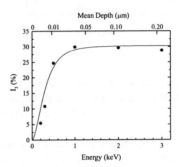

Figure 11.3. S parameter from DBES (left) and the o-Ps intensity (I_3) from PAL vs positron incident energy of polystyrene thin film on Cu substrate [44, 10]. The line was fit to a simple diffusion model from the VEPFIT program [51].

The o-Ps lifetime also varies as a function of the depth of the polymer from the surface. In polymers near the surface, at least two o-Ps states could be resolved from a PAL spectrum, o-Ps in the polymer and o-Ps diffusing to the vacuum from the surface. The lifetime of diffused o-Ps is close to 142 ns (free o-Ps), although a shorter o-Ps lifetime (from 142 to about 10 ns) has been reported due to o-Ps collision with the beam chamber [6]. In our four-lifetime analyses two o-Ps are identified, with the longest lifetime fixed at 142 ns and the sum of the two o-Ps intensities nearly constant, as shown in Figure 11.4 in polystyrene [11]. This indicates that all o-Ps are formed in the bulk of polymers, specifically in free volumes and holes, in which inner surfaces serve as effective sites for supplying electrons for the positron to form Ps. This also supports the notion that Ps is formed in free volumes and holes in polymers, as shown schematically in Figure 11.5 [10]. Therefore, the formed Ps near the surface is dynamically diffusing under competition between the surface (as an infinitely large hole, Figure 11.2) and inner surfaces (from free volumes and holes, Figure 11.5).

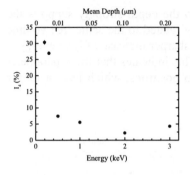 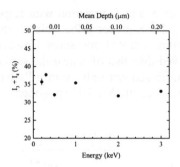

Figure 11.4. The free o-Ps intensity (I_4, with lifetime 142 ns) and the sum of the free o-Ps (I_4) and o-Ps in the bulk (I_3, as shown in Figure 11.3) near the surface of polystyrene vs positron incident energy [11].

Since both positrons and Ps could be localized in free-volume holes, the data of positron lifetime (τ_2) and o-Ps bulk lifetime (τ_3) provide information about the size and distribution of free-volume size as a function of the depth near the surface. Figure 11.6 shows the variation of positron lifetime and o-Ps bulk lifetime vs the depth. A significant increase of lifetimes near the surface shows a larger size of free volumes near the surface than in the bulk. Similar variations vs the positron energy indicates that both positrons and Ps are localized in free volumes and holes. Figure 11.7 shows the distributions of hole size in the polymer from the data of o-Ps lifetime distribution. Near the surface, not only the size is larger than the bulk, the distribution is significantly wider [10].

The larger free volume and distribution also indicates a larger fraction of free volume near the surface than in the bulk. According to the WFL theory [43], a larger free volume leads to a lower T_g. Indeed a significant T_g depression (as much as 70 °C) has been reported in the surface of polystyrene by using PAL method [10]. Other studies of polymer surfaces have shown that the size of the free volume holes near the surface of polyethylene [47] and polypropylene [48] are larger than the bulk.

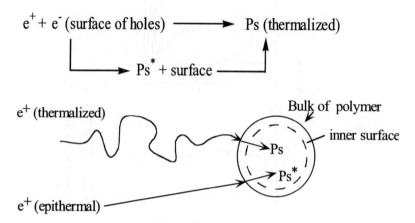

Figure 11.5 Schemes for Ps formation in free volumes and holes of polymers [10].

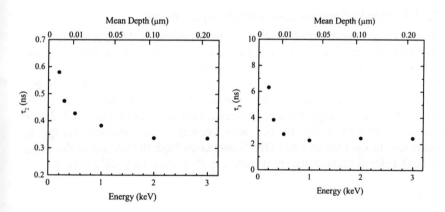

Figure 11.6 The positron lifetime (left) and the o-Ps lifetime in the bulk of polystyrene near the surface vs positron incident energy [10].

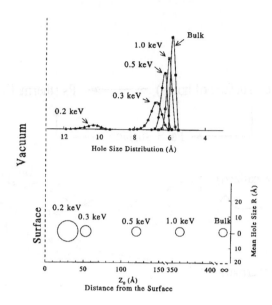

Figure 11.7 Hole-size distribution and mean size near the surface of a polystyrene film on a Cu substrate [10].

11.1.3 Chemical composition and the S parameter

As mentioned previously the value of the S parameter depends on both the size and amount of holes as well as the momentum of the electrons of the substrate with which the positrons annihilate. The variation in the S parameter may therefore, differ depending on the chemical nature of the polymer. Figure 11.8 shows the measured S parameter as a function of positron incident energy for a few selected polymers [28] Figure 11.8 shows that a variety of S dependence with respect to the incident energy is observed. In the case of TEFLON, S decreases from the surface to the bulk; in KAPTON, S remains nearly constant; in PVC (polyvinyl chloride), it increases at energies below 2 keV and then deceases and in polyurethane (and in most amorphous polymers) S increases sharply from the surface. A more comprehensive study on the chemical effect on S parameter has been reported as a Ph.D. dissertation [48].

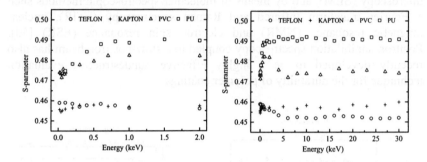

Figure 11.8 S parameter versus positron incident energy in different polymers. PVC-polyvinyl chloride, PU-polyurethane. TEFLON—polytetrafluoroethylene, KAPTON—polyimide [28].

11.2 Applications to coatings

11.2.1 Depth profile of coatings

Durability is a primary concern for coating systems. Environmental degradation from moisture, light, weather, and fluids reduce the service life of the coating. There is limited understanding of the causes of low durability in most coating systems. Existing methods of assessing durability and degradation are chiefly macroscopic approaches, measuring mechanical properties such as adhesion, hardness, pulling strength etc. Most of the knowledge of coating degradation and failure is based on these evaluations of performance [49]. However, the origin, mechanisms and progression of degradations are not yet fully ascertained for coating systems. Characterization of physical and chemical changes during degradation requires a multi-technique approach The most powerful approach is to monitor the physico-chemical changes at the molecular level by means of atomic probes such as atomic force microscopy (AFM), scanning electron microscopy (SEM), and by means of molecular spectroscopic methods such a Fourier transform infrared and Raman spectroscopy (FTIR), nuclear magnetic resonance (NMR) and electron spin resonance (ESR) [50]. Positron annihilation spectroscopy coupled to a slow-positron beam has also recently developed to be a highly effective nondestructive evaluation technique for the durability of polymer coatings.

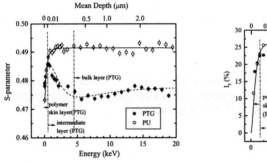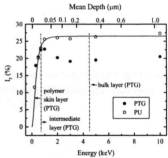

Figure 11.9 S parameter (left) [24] and o-Ps intensity (right) [52] in the bulk, as a function of positron incident energy for pure polyurethane (PU) and a polyurethane topcoat containing TiO_2 pigments (topcoat).

Figure 11.9 shows an S parameter and o-Ps intensity (I_3) vs energy curve for a typical polyurethane (PU) and a polyurethane topcoat paint (PTG) containing a TiO_2 pigment. A distinct difference in the depth profile is observed.

In both PU and PTG samples, there is an initial increase in the S parameter to a maximum value. In the coating containing the pigment, there is a decrease in the S parameter at a depth greater than about 8 nm. This decrease is attributed to the effect of the TiO_2 pigment, which contains electrons with high momentum and is indicative of the fact that the coating contains a thin layer of pure polymers near the surface. The S value for TiO_2

powders has been measured to be 0.405 in the same slow-positron beam. The dashed lines in the figure indicate the location of the layer boundaries for the paint sample determined using the VEPFIT program [51]. From the depth data it has been concluded that in coating systems containing TiO_2, (1) there exists approximately 10 nm of polymer skin from the surface of the paint, (2) inside the polymer skin there exists an intermediate transition layer (approximately 1μm) where TiO_2 starts to disperse with the polymer and other additives, and (3) inside the transition layer there exists the bulk layer of paint.

11.2.2 Exposure to UV and degradation of coatings

On exposure to various weathering and aging techniques, significant deceases in the S parameter have been observed. Slow-positron beam techniques are very successful in monitoring these changes in the very early stages of degradation of polymer coatings caused by accelerated weathering [24-29] as well as under natural weathering conditions [30-36]. The great value of the technique has been the depth-profiling ability, which allows for the determination of the kinetic data of the degradation process as a function of the depth from the surface.

Figure 11.10 shows the decrease in the S parameter as a function of exposure time due to natural weathering for a polyurethane coating (no pigment) [30]. Figure 11.11 shows the magnitude of the change in the S parameters as a function of exposure time as:

$$-\Delta S = S_t - S_0 , \qquad (7)$$

where S_t is the S parameter at a certain depth after exposure time t, and S_0 is the S parameter at that depth of the unexposed sample. These figures clearly illustrate that there is a decrease in the S parameter as a function of exposure time.

It has been suggested that the decrease in the S parameter on exposure time is a result of the decrease in the defect (free volume) size or concentration due to irradiation [30]. This is a result of a net increase in the cross-linking density of the samples as a result of photodegradation. It has been demonstrated that there is a direct relationship between the decrease in the S parameter (loss of free volume) and an increase in the cross-linking density due to degradation for polyurethane coatings [32]. We can postulate that the —ΔS is a measure of the product concentration of the degradation process (which appears reasonable in light of the relationship between —ΔS and cross-linking density). The —ΔS vs time plots can be fit to an exponential function [24-30]:

$$-\Delta S = -\Delta S_{max}\left(1 - e^{-kt}\right), \tag{8}$$

where t is the time of exposure and k is a rate constant. A good fit of this exponential function indicates a first-order kinetics of the degradation process, at least very near the surface. The half life of the degradation can also be determined from the first-order kinetics using the relationship [30]:

$$t_{1/2} = \ln 2/k \,. \tag{9}$$

This $t_{1/2}$ value is a characteristic parameter representing the durability of the coating with respect to weathering or photodegradation [30].

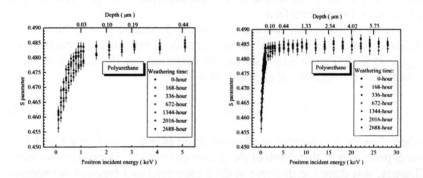

Figure 11.10 Parameter vs energy and depth in polyurethane coatings at different periods of natural weathering [30].

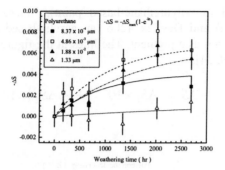

Figure 11.11 -ΔS vs weathering time at different depths3 from the surface [30]. ΔS is calculated according to eq. 9 from data shown in Figure 11.11. Lines are fit to Eq. 8.

It can also be seen from Figures 11.10-11.11 that effects of weathering are more pronounced near the surface and decrease progressively into the bulk. This is not surprising and the decrease in the magnitude of the change of the S parameter as a function depth can be explained by the attenuation of the light intensity with increasing depth from the surface. Figure 11.12 shows a plot of the —ΔS vs depth for two exposure times of the samples shown in Figure 11.X. The variation of —ΔS vs depth (d) can be expressed by an exponential decay function according to the following equation:

$$-\Delta S = -\Delta S_0 10^{-\varepsilon d}, \qquad (10)$$

where -ΔS$_0$ is the fitted -ΔS at d = 0 and ε represents a parameter with a property similar to the extinction coefficient for UV absorption.

11.2.3 Durability of commercial coatings and paints

In commercial coatings containing TiO_2 the —ΔS vs depth curves are best fitted with a two-exponential decay function [34 to 36]. Figure 11.13 shows the —ΔS vs depth plot for three commercially available coatings. The three polymeric coatings shown are produced under the commercial names of

Carboline 133HB (a polyurethane-based paint), Macro Epoxy 646 (an epoxy-based paint) and Devflex 4206 (an acrylic-based paint). All these samples contain TiO_2 as a pigment. The lines in the figure are fit according to the equation [29, 34]:

$$-\Delta S = -\Delta S_1 \times 10^{-\varepsilon_1 d} + (-\Delta S_2) \times 10^{-\varepsilon_2 d}, \qquad (11)$$

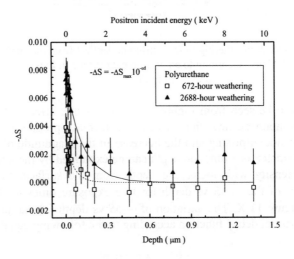

Figure 11.12 -ΔS vs positron incident energy and depth for polyurethane coating exposed to natural weathering for 672 and 2688 h [30]. The line in the figure is fitted according to Eq. 10.

where d is the depth and ε_1 and ε_2 are the first and second extinction coefficients. The fact that a two-exponential decay function best fits the data for the commercial coatings is indicative of the fact that the coatings consist of a mulitlayer with a polymer skin, followed by a transition layer and the bulk.

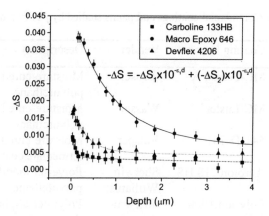

Figure 11.13 —ΔS vs positron implantation depth for commercial coatings after 2688 h of natural weathering. Lines in the Figure are fitted with eq. 11.

Positron data of S parameters have been used to characterize the durability of a variety of commercial coatings and paint products as listed in Table 11.1 and the resulting loss of S (-ΔS) is shown in Figure 11.14, below [36]. A larger -ΔS indicates a poorer durability against weathering.

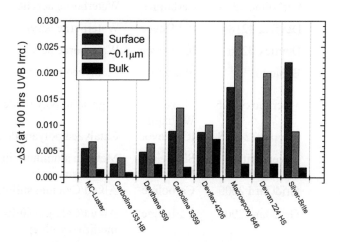

Figure 11.14 Comparison of -ΔS in a varity of commercial paints [34].

Table 11.1 Commercial coatings and paints studied by positron method [34].

Polymer base	Coating Name	Vender	Description
Polyurethane	MC-Ferrox A	Wasser	Moisture-cured aliphatic polyurethane
Polyurethane	MC-Luster	Wasser	Moisture-cured aliphatic polyurethane
Polyurethane	541-D-101	Valspar	Moisture-cured urethane intermediate coat
Polyurethane	Acrolon 218 HS	Sherwin-Williams,	Polyester modified acrylic polyurethane
Polyurethane	Poly-lon 1900	Sherwin-Williams	Polyester-aliphatic polyurethane
Polyurethane	Carboline 133 HB	Carboline	Aliphatic polyurethane
Polyurethane	Carbothane 134 HG	Carboline	Acrylic aliphatic polyurethane
Polyurethane	Devthane 359	ICI Devoe	Acrylic aliphatic polyurethane
Polyurethane	PU	Bayer	N-100 polyisocyanate +631A-75 polyol
Acrylic	Carboline 3359	Carboline	Waterborne acrylic
Acrylic	Devflex 4218	ICI Devoe	Waterborne acrylic
Acrylic	Devflex 4206	ICI Devoe	Waterborne acrylic
Epoxy	Epolon II	Sherwin-Williams	Catalyzed polyamide epoxy
Epoxy	Macropoxy 646	Sherwin-Williams	High solid polyamide epoxy
Epoxy	Devran 224 HS	ICI Devoe	Catalyzed polyamide epoxy
Alkyd	Silver-Brite	Sherwin-Williams	Metallic aluminum in petroleum resin
Alkyd	GridGard 2600	Carboline	Alkyd/Calcium sulfonate
Alkyd	GridGard 2901	Carboline	Alkyd/Calcium sulfonate modified with epoxy/polyester

11. 3 Correlations with other properties

11.3.1 ESR

The changes in the S parameter on exposure to weathering have been correlated to many other chemical and physical changes in the coatings. It has been shown, for example, that there is a direct correlation between the — ΔS value and radical formation measured by electron spin resonance (ESR) in the coating [26]. Figure 11.15 shows the correlation between the intensity of the ESR signal and the change in the S parameter for both a catalyzed and noncatalyzed polyurethane coating due to *ex-situ* UV irradiation. This figure illustrates the role of free radicals in the microstructural changes that take place in the coatings.

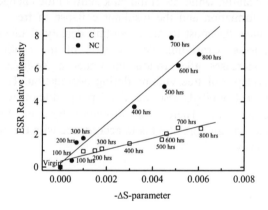

Figure 11.15 Correlation between intensity of ESR signal and decrease in the S parameter of PAS data due to *ex situ* UV-irradiation for catalyzed (C) and non-catalyzed (NC) polyurethane [26]. The —ΔS values are the average of the bulk data.

Correlations between the change in the S parameter and changes in other physical structures of the coating surface have also been demonstrated. It has been shown, for example, that there is a direct relationship between the change in the gloss (a measure of the amount of light reflected at a certain angle from the polymer surface) of the coating and the change in the S parameter [26, 27]. A correlation between the change in the S parameter and the mean surface roughness, measured by AFM, has also been demonstrated. These observed correlations support the usefulness of applying PAS to detect the early stages of degradation.

11.3.2 Cross-link density

The change of cross-link density indicates some new bonds are formed between the polymer chains. The formation of these new bonds and the increase in cros-link density will decrease the free volume in the polymer. The ΔS parameter describes the change of free volume in polymeric material. Figure 11.16 expresses the correlation of ΔS and ΔX_c at different depths of polyurethane (PU) [28]. ΔX_c is the overall change of cross-link density for the whole sample. ΔS on the surface and near the surface increases significantly, while ΔS in the bulk shows little change. This means that cross-link formation and the resultant collapse in free volume at the surface contributes the most to the cross-link density changes in the early stages of exposure. Cross-link density (X_c) of the coating material—polyurethane—increases after irradiation under a Xe-light source. This indicates the collapse of free volume during degradation. The S parameter from DBES and I_3 from PAL describes the depth profile of free volume from the surface to the bulk. There is a correlation between ΔX_c and ΔS at different depths. The interpretation is in accordance with the observation we obtained previously.

Applications of Slow Positrons to Polymeric Surfaces and Coatings 301

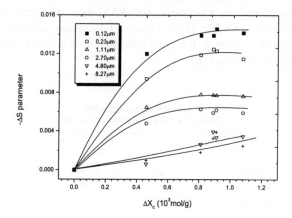

Correlation between ΔX_c and average $-\Delta S$ parameter

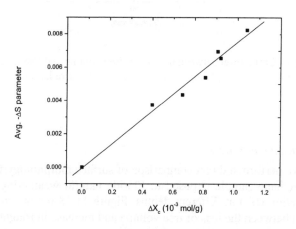

Figure 11.16 Correlation between the loss of free volume ($-\Delta S$) and increase of cross-link density. The top is for all depths and the bottom figure is an average of all depths as Xe-lamp light penetrates through the PU samples [28].

11.3.3 Gloss

One physical property regarded by the coating industry as important is gloss, which is a measure of surface roughness. Since PAS has a capability to measure atomic-scale defects from the surface to the bulk, we conducted parallel experiments on glossiness and PAS measurements in a model polyurethane coating (Table 11.1). Figure 11.17 shows S-defect parameter vs positron incident energy for polyurethane under different durations of UVB (313 nm) irradiation [30].

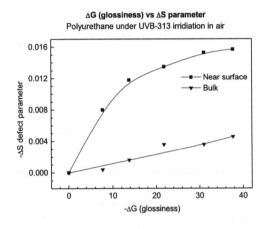

Figure 11.17 Correlation between the loss of free volume and loss of gloss in a polyurethane sample exposed to UVB (313 nm) irradiation [30].

11.3.4 Surface roughness

Recently, we perform a direct comparison of surface morphology from AFM in a polyurethane sample exposed to Florida natural weathering and free-volume fraction (ffv) at different depths. Figure 11.18 below, shows direct correlations between the loss of free volume and increase in roughness [30].

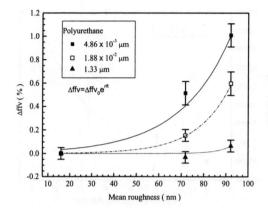

Figure 11.18 Correlations between the loss of free-volume fraction (ffv) from PAL vs mean roughness measured by AFM [30].

11.3.5 Mechanical loading

The mechanical durability of coatings has been tested by applying the cyclic loading until the coating is failed visibly. A parallel experiment of positron annihilation studies on the same samples have been performed [35]. A correlation between the loss of free volume ($-\Delta S$) and the loading cycle has been observed in a commercial coating, as shown in Figure 11.19 below. It is interesting to observe a direct link between an engineering parameter and a molecular physical parameter obtained by positron annihilation method.

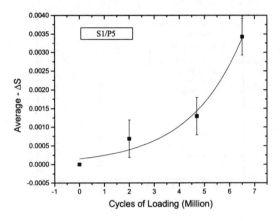

Figure 11.19 A direct correlation between the loss of free volume (-ΔS) at the molecular level and cyclic loading in a sulfonate coating [35].

Conclusion

Positron annihilation in polymeric coatings provides new information about, (1) early detection of deterioration on the order of a few days of natural weathering, (2) depth profile information from the surface to the depth, (3) molecular level information. The existing data shows great promise to develop positron annihilation as a new sensitive tool for the detection of coating degradation and improving coating durability. Real world applications as a field tool require additional systematic studies and engineering collaboration in the future

Acknowledgment

This research has been supported by the AFOSR and the NSF of the USA. Fruitful collaborations with Professor T.C. Sandreczki, J. R. Richardson, D-M. Zhu, Dr. R. Suzuki, and T. Ohdaira are acknowledged

References

[1] P.J. Schultz and K.G. Lynn *Rev. Mod. Phys.* **1989**, 60, 701.
[2] P. G. Coleman, Ed., *Positron Beam and Their Applications*, Singapore: World Sci. Pub. (2000).
[3] R. Krause-Rehberg, H.S. Leipner, *Positron Annihilation in Semiconductors*, Berlin: Springer (1999).
[4] Y.C. Jean, G.H. Dai, H. Shi, R. Suzuki, and Y. Kobayashi, in *Positron Beams for Solids and Surfaces, AIP Conference Proceedings*, E. Ottewitte, A.H. Weiss, Eds., 303, New York: Academic Press (1994), p. 129.
[5] L. Xie, G.B. DeMaggio, W.E. Frieze, J. DeVries, D.W. Gidley, H. Hristov, and A.F. Yee *Phys. Rev. Lett.* **1995**, 74, 4944.
[6] D.G. Gidley, D.N. McKirrsey, P.W. Zitzewitz *J. Appl. Phys.* **1995**, 28, 1406.
[7] Y.C. Jean, H. Cao, G.H. Dai, R. Suzuki, T. Ohdaira, and Y. Kobayashi, *Appl. Surf. Sci.* **1997**, 116, 251.
[8] Y. Kobayashi, I. Kojima, S. Hishita, R. Suzuki, E. Asari, and M. Kitajima *Phys. Rev. B: Condens. Matter* **1995**, 52, 823.
[9] G.B. DeMaggio, W.E. Frieze, D.W. Gidely, M. Zhu, H. Hristov, and A.F. Yee Phys. Rev. Lett. 1997, 78, 1524.
[10] H. Cao, R. Zhang, J.-P. Yuan, C.-M. Huang, Y.C. Jean, R. Suzuki, T. Ohdaira, and B. Nielsen *J. Phys.: Condens. Matter* **1998**, 10, 10429.
[11] H. Leidheiser, Jr., Cs. Szeles, A. Vertes, *Nucl. Instrum. Methods Phys. Res. Sect. A* **1987**, 255, 606.
[12] Cs. Szeles, K. Suvegh, A. Vertes, M.L. White, H. Leidheister, Jr. *J. Coat. Tech.* **1988**, 60, 47.
[13] K. Suvegh, Cs. Szeles, M.L. White, A. Vertes, and H. Leidheister, Jr. *Cryst. Res. Technol.* **1988**, 23, 285.
[14] Cs. Szeles, A. Vertes, M.L. White, H. Leidheiser, Jr., *Nucl.Instrum. Methods Phys. Res. Sect. A.* **1988**, 271, 688.
[15] R.C. MacQueen and R.D. Granata *Mater. Sci. Forum* **1992**, 105-110, 1649.
[16] R.C. MacQueen and R.D. Granata, in *Positronium Chemistry*, Y.C. Jean, Ed., World Sci. Pub. Singapore (1990), p. 82.
[17] R.C. MacQueen and R.D. Granata *J. Polym. Sci. Part B: Polym. Phys.* **1993**, 31, 97.
[18] J. Audreas, F. Anwain, B.J. Carlosso, M. DiLorenzo, S. Grossman, C.J. Knauss, J. McCarthy, B. Mysza *J. Coat. Tech.* **1994**, 66, 49.
[19] R.C. MacQueen and R.D. Granata *Prog. Org. Coat.* **1996**, 28, 97.
[20] M.M. Madani, R.D. Granata *J. Appl. Phys.* **1996**, 80, 2555.
[21] M.M. Madani, H.L. Vedage, and R.D. Granata *J. Electrochem. Soc.* **1997**, 144, 3298.
[22] A.J. Hill, M.R. Tant, R.C. McGill, P.P. Shang, R.C. Stochi, D.L. Murrey, J.P. Cloyd *J. Coat. Tech.* **2001**, 73, 115.
[23] L.D. Hulett, Jr., S. Wallace, J. Xu, B. Nielsen, Cs Scales, K.G. Lynn, and J. Pfau, A. Schaub *Appl. Surf. Sci.* **1995**, 85, 234.

[24] H. Cao, R. Zhang, C.S. Sundar, J.P. Yuan, Y. He, T. C. Sandreczki, and Y.C. Jean *Macromolecules* **1998**, 31, 6627.
[25] H. Cao, J.P. Yuan, R. Zhang, C.M. Huang, Y. He, T.C. Sandreczki, Y.C. Jean, B. Nielsen, R. Suzuki, and T. Ohdaira *Macromolecules* **1999**, 32, 5925.
[26] H.Cao, Y. He, R. Zhang. J. P. Yuan, T.C. Sandreczki, Y.C. Jean, and B. Nielsen *J. Polym. Sci. Part B: Polym. Phys.* **1999**, 37, 1289.
[27] H.Cao, R. Zhang, H.M. Chen, P. Mallon, C.M. Huang, Y. He, T.C. Sandreczki, Y.C. Jean, B. Nielsen, T. Friessnegg, R. Suzuki, and T. Ohdaira *Radiat. Phys. Chem.* **2000**, 58, 639.
[28] R. Zhang, H. Cao, H.M. Chen, P. Mallon, T.C. Sandreczki, J. R. Richardson, Y.C. Jean, B. Nielsen, R. Suzuki, and T. Ohdaira *Radiat. Phys. Chem.* **2000**, 58, 639.
[29] R. Zhang, P.E. Mallon, H. Chen, C.M Huang, J. Zhang, Y. Li, Y. Wu, T.C. Sandreczki, and Y.C. Jean *Prog. Org. Coatings* **2001**, 42, 244.
[30] Y. C. Wu, C.M. Huang, Y. Li, R. Zhang, H. Chen, P.E. Mallon, J. Zhang, T.C. Sandreczki, D.M. Zhu, Y.C. Jean, R. Suzuki, and T. Ohdaira *J. Polym. Sci. Part B: Polym. Phys.* **2001**, 39, 2290.
[31] R.E. Galindo, A. van Veen, A. Gracia, H. Schut, and J.T. De Mosson *Mater. Sci. Forum* **2001**, 363-365, 502.
[32] R. Zhang, P. Mallon, H.M. Chen, C.-M. Huang, Junjie Zhang, Ying Li, H. Cao, Q. Peng, J.R. Richardson, Y.Y. Huang, T.C. Sandreczki, and Y.C. Jean, B. Nielsen *Mater. Sci. Forum* **2001**, 363-365, 505.
[33] H.M. Weng, Y.M. Fan, B.J. Yie, X.Y. Zhou, J.F. Du, R.D. Han, C.G. Ma, C.C. Ling *Mater. Sci. Forum* **2001**, 363-365, 511.
[34] Y.C. Jean, H. Cao, R. Zhang, H. Chen, P. Mallon, Y. Huang, T.C. Sandreczki, R.J. Richardson, J.J. Calcara, and Q. Peng, Amer. Chem. Soc. Symp. Ser. 805, 299-315 (2002), J.W. Martin and D.R. Bauer, Eds., Amer. Chem. Soc. D.C.
[35] H. Chen, Q. Peng, Y.Y. Huang, R. Zhang, P.E. Mallon, J. Zhang, Y. Li, Y. Wu, J.R. Richardson, T.C. Sandreczki, Y.C. Jean, R. Suzuki, and T. Ohdaira *Appl. Surf. Sci.* (in press, 2002).
[36] P.E. Mallon, Y.Li, H. Chen, R. Zhang, J. Zhang, Y. Wu, Y.C. Jean, R. Suzuki, and T. Ohdaira *Appl. Surf. Sci.* (in press, 2002).
[37] Y.C. Jean, H. Nakanishi, L.Y. Hao, and T.C. Sandreczki *Phys. Rev. B: Condens. Matter* **1990**, 42, 9705.
[38] Y.C. Jean, Y. Rhee, Y. Lou, H.L. Yen, H. Cao, K. Cheong, and Y. Gu *Phys. Rev. B: Condens. Matter* **1996**, 54, 1785.
[39] Y.C. Jean, Y. Rhee, Y. Lou, D. Shelby, and G.L. Wilkes *J. Polym. Sci. Part B: Polym. Phys.* **1996**, 34, 2979.
[40] P.E. Mijnarends, C.V. Falub, S.W.H. Eijt, and A. van Veen *Mater. Sci. Forum*, **2001**, 303-305, 332.
[41] For positron reviews in polymers, see: Y.C. Jean, in *Positron Spectroscopy of Solids*, A. Dupasquier and A.P. Mills, Jr., Eds., ISO Press: Amsterdam (1995), p.503, and Y.C. Jean *Microchem. J.* **1990**, 42, 72.

[42] H. Nakanishi, S. J. Wang and Y.C. Jean in *Positron Annihilation Studies of Fluids*, S.C. Sharma, Ed., World Sci. Pub., Singapore (1988), p. 292.
[43] M.L. Williams, R.F. Landel, and J.D. Ferry *J. Am. Chem. Soc.* **1955**, 77, 3701.
[44] Y.C. Jean, R. Zhang, H. Cao, J.-P. Yuan, C.-M. Huang, B. Nielsen, and P. Asoka-Kumar *Phys. Rev. B: Condens. Matter* **1997**, 56, R8459.
[45] K. Hirata, Y. Kobayashi, and Y. Ujihira *J. Chem. Soc. Faraday Trans.* **1996**, 92, 985.
[46] A. Uedono, T. Tanigawa, M. Watanabe and N. Nihimoto *J. Polym. Sci. Part B: Polym. Phys.* **1998**, 36, 1919.
[47] A. Uedono, R. Suzuki, T. Ohdaira, T. Uozumi, M. Ban, M. Kyoto, S. Tanigawa, T. Mikado *J. Polym. Sci,. Part B: Polym. Phys.* 1998, 36, 2597.
[48] Renwu Zhang, PhD. Dissertation at the University of Missouri-Kansas City, May, 2002, "Studies of Coating Degradation and Polymer Chemical Environment Using Positron Annihilation Spectroscopy."
[49] For examples see Industrial Coatings: Properties, Applications Quality, and Environmental Compliance. Proceedings of the Advances in Coating Technology Conference, Chicago, Illinois, November 1992: ASM International: Materials Park, OH (1992).
[50] For example see *Polymer Characterization*, B. J Hunt, M.I. James, Eds., Polymer Characterization, Blackie Academic and Professional, London (1993).
[51] A. Van Veen, H. Schul, J. de Vries, I. Hakroot, M.R. Ijpmr *Am Inst. Phys Proc* **1990** 218 117.
[52] H. Cao, Ph.D dissertation at the University of Missouri-Kansas City, Dec., 1999, "Degradation of Polymeric Coatings Studied by Positron Annihilation Spectroscopy."

Chapter 12

Positron Annihilation Induced Auger Spectroscopy

Shafaq Amdani, Anat Eshed, Nail Fazleev, and Alex Weiss

Physics Department, The University of Texas at Arlington

12.1 Introduction

Positron annihilation induced Auger Electron Spectroscopy, (PAES), is a spectroscopic technique that has important applications in the study of chemical composition and structure of surfaces and thin films. In addition PAES provides a powerful means of studying positron interactions at surfaces and the momentum distribution of core electrons. PAES makes use of matter-antimatter annihilation to create the core holes that result in Auger electron emission [1]. This unique excitation mechanism gives PAES a number of significant advantages over conventional Auger methods. The use of very low (~20eV) positron beam energies permits PAES spectra to be obtained which are free [2] of the large secondary electron background which makes it impossible to unambiguously determine Auger line shapes using conventional Auger techniques [3]. In addition, PAES is extremely surface-selective. The PAES signal originates almost exclusively from the topmost atomic layer due to the trapping of positrons in an image potential at the surface before annihilation.

In this chapter we will provide a brief introduction to the PAES technique including a discusion of the PAES mechanism, the advantages and applications of PAES in surface analysis, and an outline of the theory underlying calculation of PAES intensities. In addition, this chapter will include the first atlas of PAES spectra of the elements in which the intensities have been plotted using a comon normalization method (based on a new analysis of original data) so as to permit direct comparison of relative PAES intensities. The reader is refered to previous PAES reviews for additional PAES examples, applications, and references [4-7].

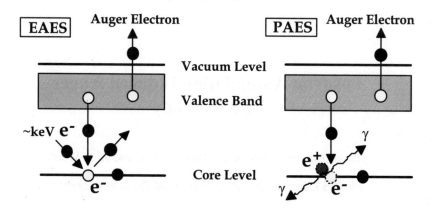

Figure 12.1 Comparison of PAES and EAES mechanisms.

12.2 PAES Mechanism

The Auger process is a non-radiative transition in which an atom with an inner shell hole relaxes by filling this hole with a less tightly bound electron while simultaneously emitting another electron (the Auger electron), which carries off the excess energy. The energy of the Auger electron is given to a first approximation by the equation, $E_{XYZ} = E_X - E_Y - E_Z^*$ where E_X and E_Y are the binding energies of the electron removed to form the original inner shell hole and the electron that fills the hole respectively, and E_Z^* is the binding energy of the outgoing electron where the binding energy has been calculated for an atom with a preexisting hole in the Y level. Because the energy levels of different elements are in general unique, the elemental identity of an atom may be deduced from the energies of the Auger electrons emitted as a result of core hole excitations. This fact along with the short escape depth of low energy electrons has been exploited in the widely used surface analysis tool, Auger Electron Spectroscopy (AES) [8].

Electron induced Auger Electron Spectroscopy (EAES) makes use of high-energy electrons to remove core electrons via impact-ionization. However, in many instances the utility of EAES is limited by problems associated with the large secondary electron background and the lack of surface specificity inherent in the EAES excitation process [2, 8].

The PAES mechanism, first demonstrated in 1987 [1], can be outlined as follows: (1). A positron implanted at low energy diffuses to and gets trapped at the surface. (2). A few percent of the trapped positrons annihilate with core electrons leaving atom in excited state. (3). The atom relaxes via emission of an Auger electron. The PAES mechanism is contrasted with that of electron induced Auger Spectroscopy (EAES) in Figure 12.1.

Due to the fact that the core holes are created by matter-antimatter annihilation and not impact-ionization the large secondary electron that plagues conventional AES can be eliminated by using a incident beam energy below the energy of the Auger electron energies. Secondary electrons cannot be created through impact-processes with energies in excess of an energy E_k (which we term the kinematic edge) given by:

$$E_{sec} \leq E_k = E_p.\text{-}\phi^- + \phi$$

where E_{sec} is the kinetic energy with which the secondary electrons leave the surface of the sample, E_p is the kinetic energy of the primary beam at the sample surface, and ϕ^- and ϕ are the positron and electron work functions respectively [2]. Thus the impact-excited secondary electron background in the energy range of the Auger electron can be eliminated by using incident beam energies less than the Auger electron energy.

The enhanced surface selectivity of PAES stems from the fact that positrons implanted into a metal or semiconductor at low energies have a high probability of diffusing to the surface and becoming trapped in an "image-correlation" well before they annihilate [3, 9]. The positrons in this well are localized at the surface and annihilate almost exclusively with atoms at the surface. As a result almost all of the Auger electrons originate from the top-most atomic layer [3]. This is in contrast to conventional Auger techniques in which the Auger electrons originate from an excitation volume which extends hundreds of atomic layers below the surface, limiting the surface selectivity to the 4-20Å escape depth of the Auger electron [8].

A comparison between EAES and PAES spectra is shown in Figure 12.2. This comparison demonstrates the ability of PAES to eliminate the large secondary electron background associated with electron and photon excited Auger Electron emission [2]. The EAES signal [Figure 12.2(a)] can be seen to be sitting on top of a large secondary electron background. This background is almost entirely absent in the Positron annihilation Induced Auger spectra shown in Figs. 4B and 4D. Note the rapid increase in the number of detected electrons below the energy 25 eV due to the onset of impact-induced secondary electrons.

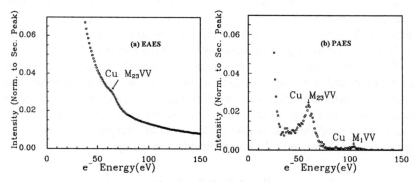

Figure 12.2 Comparison of spectra obtained using EAES and PAES.

An important result of the PAES measurements has been an indication that the assumption, used in most background subtraction routines, that the Auger line shape quickly goes to zero is incorrect. The ability of PAES to eliminate the secondary electron background should be particularly useful in determining the line shape of the low energy Auger lines (those most surface sensitive). In addition, the large improvement in signal to background permits PAES measurements to be made with energy doses ~5 orders of magnitude lower than doses required for to EAES [2]. The large reductions in charge and especially energy dose may make PAES useful for determining the elemental content of insulators and fragile adsorbate systems damaged by conventional AES methods.

The enhanced surface selectivity of PAES was first demonstrated by a series of measurements performed on single crystal Cu with varying coverage of S [3]. The presence of a half monolayer of S caused a four-fold reduction in the PAES intensity. This is in contrast to EAES where only a 15% reduction in the Cu intensity was observed. The reduction in the PAES signal is consistent with the fact that the S overlayer pushes positron wave function away from the surface, thereby reducing its overlap with the top layer of Cu atoms. Detailed calculations of the probability of core-hole formation due to positrons trapped at clean and covered, Cu surfaces predict decreases in the Cu PAES signal in agreement with the measurements. A similar sensitivity of PAES to overlayers of H_2 on Pd has been demonstrated by Lee et al. [10]. The utility of PAES in determining the elemental content of the topmost layer is currently being exploited in studies of ultra-thin metal layers on metal and semiconductor substrates.

12.3 Theory

The calculation of PAES intensities largely reduces to the calculation core annihilation probabilities for positrons in the surface state [11]. This follows from the fact that almost all of the core hole excitations of the outer cores relax via Auger emission and that almost all of the positrons incident at low energies become trapped in a surface state before annihilation. First-principles calculations of the positron states and positron annihilation characteristics at metal and semiconductor surfaces are based on a treatment of a positron as a single charged particle trapped in a "correlation well" in the proximity of surface atoms. The calculations were performed within a modified superimposed-atom method using the corrugated-mirror model of Nieminen and Puska [12].

The positron potential $V^+(r)$ at a semiconductor surface contains an electrostatic Hartree-Coulomb potential $V_H(r)$, and a correlation potential $V_{corr}(r)$,

$$V^+(\mathbf{r}) = V_H(\mathbf{r}) + V_{corr}(\mathbf{r}). \quad (1)$$

The Hartree potential $V_H(r)$ is constructed as a superposition of the atomic Coulomb potentials $V_{Coul}^{at}(|\mathbf{r} - \mathbf{R}|)$, where $\mathbf{R}$ defines the positions of the host nuclei. The method of Wienert and Watson[2] is used to correct for misrepresentation of the surface dipole in the superimposed-atom model [13, 14].

We perform atomic calculations within the local-spin-density approximation [15] using the exchange-correlation atomic configurations from [16] and [17], respectively. The positron-electron correlation potential V_{corr} *(r)* reflects the response of the electron charge density to the presence of the positron by taking into account many body effects. The correlation component of the positron potential in the region where the electron density is high (deep inside) and far outside the metal surface where the electron density is assumed to be zero is described well by the local density approximation (LDA) and the image potential, respectively. We then divide the space into two regions, namely, the bulk and image potential regions, where the two models are applied. The border between these regions is chosen to pass through the cross-over point of the bulk and image potentials, located immediately outside the surface.

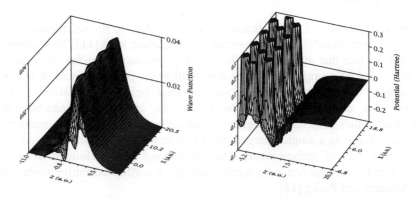

Figure 12.3 The positron-surface-state wave function (right panel) calculated for Cu(100) using the potential calculated for Cu(100) shown in (left panel).

The local density correlation potential is joined to the image potential by taking $V_{Corr}(r)$ to be the larger of the two at each point outside the surface of the solid. In the LDA $V_{Corr}(r)$ is obtained at a given position by considering the positron to be embedded in a homogeneous electron gas with an electron density n corresponding to the electron density at that particular point, i.e.:

$$V_{corr}(\mathbf{r}) = V_{corr}^{LDA}(r;n_-) = V_{corr}^{EG}(n_-)[\ f(n_-,\varepsilon_g)]^{1/3}, \quad (2)$$

Where, $V_{corr}^{EG}(n_-)$ is the correlation energy for a positron in a homogeneous electron gas of density n_- [18]. The term, $f(n_-,\varepsilon_g)$ is a reduction factor that is introduced for semiconductors to account for the diminished screening response of semiconductors to charged particles due to the existence of a band gap, in comparison to metals for which $f = 1$. $\varepsilon_g = E_g/E_F$, where E_g is the energy gap and E_F is the Fermi energy. Brandt and Reinheimer [19] obtained a functional form for the reduction factor f, from a fit to their numerical data, obtained from screening calculations for point charges in a model semiconductor:

$$f(n,\varepsilon_g) = 1 - \frac{0.37\varepsilon_g}{1+0.18r_s}, \quad (3)$$

with $r_s = (3/4\pi n_-)^{1/3}$.

The positron-electron correlation potential outside the metal or semiconductor surface described by the image potential can be written as

$$V_{image}(\mathbf{r}) = -\frac{e^2}{4\pi\varepsilon_0}\chi\frac{1}{4\left[Z_{eff}(n_-(r)) - Z_0\right]}, \quad (4)$$

where χ is equal 1 or $\dfrac{(\varepsilon_\infty - 1)}{(\varepsilon_\infty + 1)}$ for metal or semiconductor surfaces respectively, e is the charge of a positron, $\tilde{\varepsilon}$ is the vacuum permittivity, ε_∞ is the dielectric constant, $Z_{eff}(n_-(r))$ is the effective distance from the surface, represented as a function of the total electron density at the surface, $n(r)$, and Z_0

defines the effective image-plane position on the vacuum side of the top layer of atoms.

The image potential is constructed to have the same corrugations as the total electron density, $n\_(r)$. The assumption is made that at large distances the corrugations in the image potential are negligible and Z_{eff} is equal to the coordinate perpendicular to the surface. The joining of the image potential to the local density correlation potential is done by taking V_{corr} to be the larger of the two at each point outside the surface.

The single particle Schrödinger equation

$$-\frac{\hbar^2}{2m}\nabla^2 \psi_i^+(r) + \left[V_H(r) + V_{Corr}(r) \right] \psi_i^+(r) = E_i \psi_i^+(r) \quad (5)$$

is solved numerically using a finite difference relaxation technique. Calculations are performed assuming that the positron is in the ground state in the image correlation well at the surface and is delocalized in the plane of the surface. The positron binding energy and the positron wave function are found through iteratively solving for the energy, and then correcting the wave function based on the energy, the positron potential, and the surrounding values of the wave function. In numerical calculations of the positron binding energy the mesh density is doubled repeatedly until the calculated energy converges.

The positron annihilation rate $\lambda_{n,l}$ is computed from the overlap of the positron and electron densities:

$$\lambda_{n,l} = \pi r_0^2 c \int d^3 r \left| \psi^+(r) \right|^2 \left[\sum_i \left| \psi_{n,l}^i(r) \right|^2 \right] \quad (6)$$

where r_0 is the classical electron radius, c is the speed of light, Ψ^+ is the positron wave function, and $\Psi^i{}_{n,l}$ is the wave function of the core electron described by quantum numbers n and l. The summation is over all electron states in the atomic level defined by quantum numbers n and l. Comparison of experimental positron lifetimes with calculations of positron annihilation rates in metals and semiconductors [20] and angular correlation of annihilation radiation (ACAR) results [21], indicate that it is necessary to multiply the result for $\lambda_{n,l}$ by an enhancement factor (of the order of 1-2 for core electrons) to account approximately for the electron-positron correlation effects.

The total annihilation rate, λ, of the surface trapped positrons is found from the following equation:

$$\lambda = \frac{\pi r_o^2 c}{e^2} \int d^3r \; n^+(r) \; n(r) \; \Gamma(n(r)) \tag{7}$$

where $n^+(r)$ is the positron charge density, $n(r)$ is the electron density, and $\Gamma(n(r))$ is the short-range annihilation enhancement factor in an electron gas of density $n(r)$. Outside the metal or semiconductor surface the LDA must break down since the positron correlation potential is no longer related to the electron density at the position of the positron, but is due to the presence of electrons located on the surface. Far from the surface, the positron correlation potential must approach the image potential.

To correct for the inherent inconsistency with the modeling of the positron correlation potential the LDA result for λ was modified by assuming the factor $\Gamma(n(r))$ to be non-zero for all r inside the bulk region, and to be zero for all r inside the image-potential region, assuming the local annihilation rate in this region is zero. Due to the existence of a band the positron screening in semiconductors is not as effective as in metals. This has to be taken into account in calculations of the total annihilation rate. For the bulk region of a we use the following expression for the enhancement factor $\Gamma(n(r))$, obtained for the positron in the homogeneous electron gas [22]

$$\Gamma(n(r)) = 1 + 1.23 r_s + 0.8295 r_s^{3/2} - 1.26 r_s^2 + 0.3286 r_s^{5/2} + \Lambda r_s^3/6, \tag{8}$$

where $(4\pi/3) r_s^3 n = 1$, r_s is the usual electron density parameter of a homogeneous electron gas, Λ is equal 1 for a metal and $(1 - 1/\varepsilon_\infty)$ for semiconductor, and ε_∞ is the high-frequency dielectric constant.

We calculate λ over the bulk region defined by the cut-off point. The core annihilation probabilities $p_{n,l}$ with the specific core electron shells, described by quantum numbers n and l, can be obtained by dividing the partial positron annihilation rate $\lambda_{n,l}$ with the different core shells by the total positron annihilation rate λ: $p_{n,l} = \lambda_{n,l}/\lambda$. Calculated annihilation probabilities for core levels relevant to PAES for a number of elements are tabulated in Table 12.1.

12.4 Catalog of PAES spectra of slected elements

As noted in the introduction, the energy distribution of Auger electrons from each element is in general unique and provide a way of "finger printing" the elements at the surface. However, quantitative analysis requires knowledge of the relative sensitivity of PAES for each element. While PAES spectra have been collected for more than a 16 elements, up till now, the published data were lacking in a uniform method of normalization making it difficult or impossible to determine the relative elemental sensitivities. In Figs. 12.4-12.7 we present a set of PAES spectra for 13 elements, all normalized using a common method so as to permit direct comparison of the PAES intensities.

The common normalization is based on an analysis using original data sets as archived in lab books and computer files [23]. The chosen data are representative of the best spectra currently available (i.e. highest statistics from the cleanest samples). Only data from the magnetically guided PAES system were chosen in order to maximize the uniformity of experimental conditions. Note that while PAES spectra have been obtained from additional elements including C [24-26], O [27] and Cl [28], and higher resolution spectra were obtained for Cu [29], Ag [30], GaAs [31] and Ge [32] these spectra were not included in this catalog because the differences in the response functions for the various spectrometers used to make the measurements made it difficult to make quantitative intensity comparisons with the spectra presented here. Normalization was accomplished as follows. The data sets to be compared were first normalized by dividing each spectra by a term proportional to the number of incident positrons hitting the sample (the number of gamma-counts in a NaI detector located adjacent to the sample). The height of the Cu $M_{2,3}M_{4,5}M_{4,5}$ Auger peak (at ~ 54 EV) was arbitrarily fixed at 1.0. The spectra of the other elements were then normalized by multiplying the data by a factor which resulted in an Auger peak height which was in correct proportion to that of a Cu spectra taken under nearly identical conditions, close to the time of the acquisition of the spectra for the other element (denoted as a "contemporaneous spectra"). In a few cases, no contemporaneous Cu spectra were available for comparison and comparisons were made instead to contemporaneous spectra obtained from elements that could be directly compared to Cu (using other spectra). *No correction* was made for positronium (Ps) emission (Ps emission reduces the number of positrons in the surface state available to annihilate with core electrons and thereby reduces PAES intensities). All spectra were acquired at room temperature with the exception of that of Rh which was acquired at 173K

12.5 Applications of PAES

The ability of PAES to determine the elemental content of the topmost atomic layer has been exploited in the study of the initial growth and stability of ultrathin films [9, 33-40], nano-structures on surfaces [35, 41] and the intitial stages of gas adsorption and oxidation [41-45]. The strongly dependence of PAES intensities on the positron density distribution at the surface has been exploited in studies of the nature of the positron surface state [33, 34, 46, 47]. PAES's ability to eliminate the secondary electron background has been exploited in studies of the line shape of low energy Auger transitions [29, 30, 48]. Most recently, PAES measurements have been made in coincidence with Doppler broadening measurements to obtain the annihilation gamma spectra associated with individual core levels [49].

One interesting example of how PAES can be used in studies of surface chemistry is a measurements carried out on a model system (O/Au/Si) which demonstrated (see Figs. 12.8 and 12.9) that PAES is able to identify sites of preferential adsorption, a crucial starting point in the study of catalytic reactivity and selectivity. The preferential adsorption of molecules to a particular type of site may be identified by the sharp decrease in the PAES intensity of that element due to the displacement of the positron wave function by the adsorbate [41]. Future experiments can be expected to elucidate mechanisms leading to poisoning and promotion.

Table 12.1

Surface	Fraction of Positrons in Surface State Annihilating with Electrons in Indicated Core levels (%)				
Cu(100)	3s 0.83	3p 3.03			
Cu(110)	3s 0.78	3p 2.86			
Cu(111)	3s 0.77	3p 2.85			
Ag(100)	3s 0.75	3p 3.08			
Al(100)	2s 0.39	2p 1.24			
Ni(100)	3s 0.95	3p 3.42			
Si(100)	2s 0.32	2p 0.94			
Ge(100)	3s 0.25	3p 0.88	3d 3.00		
Li(100)	1s 3.34				
Na(100)	1s 0.22	2s 3.17	2p 4.15		
K(100)	2p 0.05	3s 1.05	3p 4.37		
Rb(100)	3s 0.02	3p 0.06	4s 0.90	3d 0.14	4p 4.18
Cs(100)	4s 0.02	4p 0.06	4d 0.51	5s 0.52	5p 3.88
Au(100)	4p 0.04	4d 0.080	4f 0.32	5s 0.72	5p 3.34
GaAs(100) As on top	As 3p 0.87	As 3d 3.40	Ga 3p 0.15	Ga 3d 0.46	
GaAs(100) Ga on top	As 3p 0.28	As 3d 1.09	Ga 3p 0.47	Ga 3d 1.45	

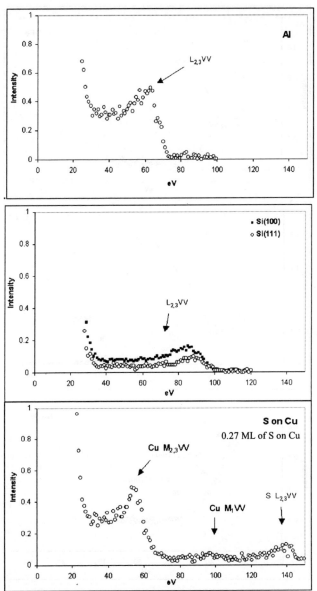

Figure 12.4 PAES spectra of Al, Si and S ($S_{0.27}Cu_{0.63}$) normalized to clean Cu.

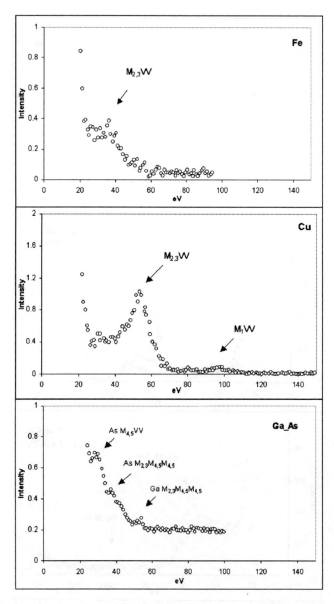

Figure 12.5 PAES spectra of Fe, Cu, and GaAs normalized to clean Cu.

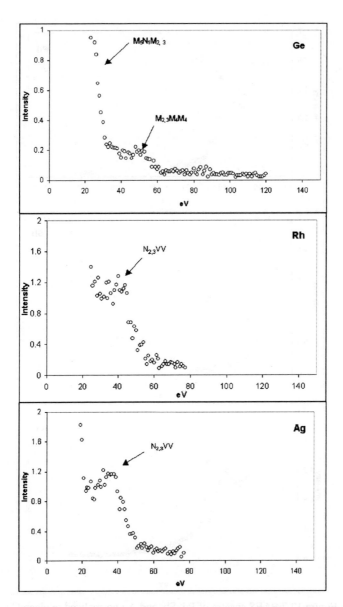

Figure 12.6 PAES spectra of Ge, Rh, and Ag normalized to clean Cu.

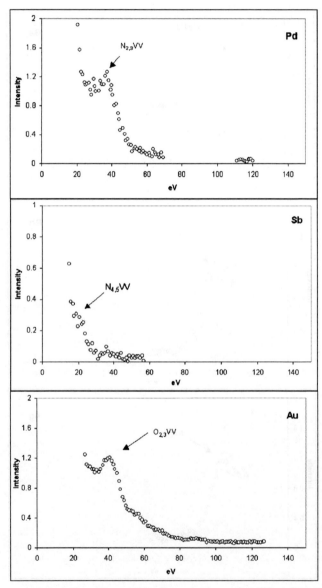

Figure 12.7 PAES spectra of Pd, Sb, and Au normalized to clean Cu.

Positron Annihilation Induced Auger Spectroscopy 325

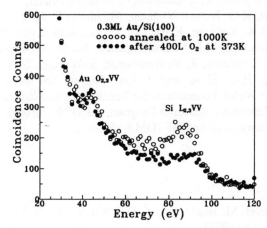

Figure 12.8 PAES spectra for a Si(100) surface covered with 0.3 ML of Au (a) before and (b) after exposure to 400 Langmuir of O_2. The oxygen exposure caused a decreas in the Si peak but not the Au peak. This indicates that the positrons are being pushed away from the Si due to selective adsorption of oxygen at exposed Si sites. A schematic drawing indicating this mechanism is shown in Figure 12.9 below.

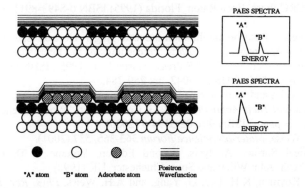

Figure 12.9 Schematic Diagram illustrating how active adsorption sites may be identified due to "blocking." The presence of the overlayer atoms will push the positron wave function away from the group of ["B"] atoms (the adsorption site) leading to a large reduction in the PAES signal from the element(s) ["B" atoms] present at that site.

Acknowledgements

The results reported in this chapter represents the work of many researchers. Contributions are gratefully acknowledged from: G. Brauer, W.C. Chen, J.L. Fry, F. Gotwald, K.O. Jensen, N. Jiang, M. Jibaly, E. Jung, J. Kaiser, J.H. Kim, S. Kim, A.R. Koymen, K.-H. Lee, C. Lei, K.G. Lynn, R. Mayer, D. Mehl, G. A. Mulhollan, M.P. Nadesalingam, S. Naidu, A. Nangia, R. Nayak, S. Rajalakshmi, S. Starnes, R. Venkataraman, S. Wheeler, S. Xie, J. Yan, G. Yang, S. Yang, H.Q. Zhou, and J. Zhu. Support for this research was provided by: the Welch Foundation, the Texas Advanced Research Program, the NATO Collaborative Research Program, the DOE (through Brookhaven National Laboratory) and the NSF (DMR-981262).

References

A.Weiss, R. Mayer, M. Jibaly, C. Lei, D. Mehl, and K. G. Lynn, *Phys. Rev. Lett.* **61**, 2245-2248 (1988).

[1] Alex Weiss, David Mehl, Ali R. Koymen, K. H. Lee and Chun Lei, *J. Vac. Sci. Technol* **A8**, 2517-2520 (1990).
[2] David Mehl, A. R. Koymen, Kjeld O. Jensen, Fred Gotwald, and Alex Weiss, *Phys. Rev. B (R.C.)* **41**, 799-802 (1990).
[3] A.H. Weiss, Positron Annihilation Induced Auger Spectroscopy, Chapt. 45 in *The Handbook of Surface Imaging and Visualization*, Arthur T. Hubbard, Ed., CRC Press, Boca Raton, Florida (1995) ISBN 0-8493-8911-9, pp. 617—633.
[4] A. Weiss, Positron-annihilation-induced Auger electron spectroscopy, in *Positron Spectroscopy of Solids*, Proceedings of the International School of Physics "Enrico Fermi," IOS Press, Amsterdam, (1995), ISBN 90 5199 203 3 and ISBN 4 274 90028 2 C3042, pp. 259-284.
[5] Alex Weiss and Paul Coleman, "Surface Science with Positrons," in *Positron Beams and Their Applications*, P. G. Coleman, Ed., World Scientific, Singapore, (59 pages) ISBN 981-02-3394-9 (2000)
[6] A.H. Weiss, *Materials Science Forum* **363-365**, 537 (2001).
[7] Practical Surface Analysis, Second Edition, Volume 1, D. Briggs and M.P.Seah, John Wiley and Sons, Chichester, UK (1990).
[8] A.R. Koymen, K.H. Lee, G. Yang, and A.H. Weiss, *Phys. Rev. B (RC)* **48**, 2020-2023, (1993).
[9] K.H. Lee, Ali R. Koymen, David Mehl, K.O. Jensen, and A. Weiss, *Surf. Sci.* **264**, 127-134 (1992).

[10] Kjeld Jensen and Alex Weiss, *Phys. Rev. B.*, **41**, 3928-3936, (1990).
[11] R. M. Nieminen and M. J. Puska: *Phys. Rev. Lett.* **50**, 281 (1983).
[12] M. Weinert and R. E. Watson, *Phys. Rev. B* **29**, 3001 (1984).
[13] N. G. Fazleev, J. L. Fry, K. Kuttler, A. R. Koymen, and A. H. Weiss, *Phys. Rev. B* **52**, 5351 (1995).
[14] O. Gunnarsson and B. I. Lundqvist, *Phys. Rev. B* **13**, 4276 (1976).
[15] D. M. Ceperly and B. J. Adler, *Phys. Rev. Lett.* **45**, 566 (1980).
[16] F. Herman and S. Skillman: Atomic Structure Calculations (Englewood Cliffs, Prentice-Hall, N.J., 1963).
[17] J. Arponen and E. Pajanne, *Ann. Phys. (N.Y.)* **121**, 343 (1979).
[18] W. Brandt and J. Reinheimer, *Phys. Rev. B* **2**, 3104 (1970).
[19] E. Bonderup, J. U. Andersen, and D. N. Lowy, *Phys. Rev. B* **20**, 883 (1979); B. Chakraborty, in Positron Annihilation, edited by P. C. Coleman, S. C. Sharma, and L. M. Diana, 207 (North-Holland, Amsterdam, 1982).
[20] M. Sob, *Solid State Comm.* **53**, 255 (1985).
[21] K. O. Jensen, *J. Phys.: Cond. Matt.* **1**, 10595 (1989).
[22] These data were acquired as part of the Ph.D. research of D. Mehl, K.-H. Lee, G. Yang and W.-C. Chen.
[23] E. Soininen, A. Schwab, and K.G. Lynn, *Phys. Rev. B*, **43**, 10051-61, (1991).
[24] R. Suzuki, Y. Kobayashi, T. Mikado, H. Ohgaki, M. Chiwaki, and T.Yamazaki, *Hyperfine Interactions* **84**, 345 (1994).
[25] R. Suzuki, T. Ohdaira, T. Midado, H. Ohgaki, M. Chiwaki, and T. Yamazaki, *Appl. Surf. Sci.* **100-101**, 297 (1996).
[26] T. Ohdaira, R. Suzuki, T. Mikado, H. Ohgaki, M. Chiwaki and T. Yamazaki, *Appl. Surf. Sci.* **100-101**, 73, (1996).
[27] Positron Surface States, K.G. Lynn and Y. Kong, in Positron at Metallic Surfaces, A. Ishii, Editor, *Solid State Phenomena* Vol. **28 & 29**, 275-292, Trans Tech Publications, Aedermannsdorf, Switzerland (1992/93).
[28] A.H. Weiss, S. Yang, H.Q. Zhou, E. Jung, and S. Wheeler, *J. Elect. Spec. and Rel. Phenom.***72**, 305-309, (1995).
[29] E. Jung, H.Q. Zhou, J.H. Kim, S. Starnes, R. Venkataram, and A.H. Weiss, *Appl. Surf. Sci.* **116**, 318-323, (1997).
[30] W.-C. Chen, *et al.*, to be published.
[31] S. Starnes *et al.*, to be published.
[32] A.R. Koymen, K.H. Lee, D. Mehl, Alex Weiss, and K.O. Jensen, *Phys. Rev. Lett.* **68**, 2378-2381, (1992).
[33] N.G. Fazleev, J.L. Fry, J.H. Kaiser, A.R. Koymen, K.H. Lee, T.D. Niedzwiecki, and A.H. Weiss, *Phys. Rev. B* **49**, 10577-10584,(1993)
[34] K.H. Lee, Gimo Yang, A.R. Koymen, K.O. Jensen, and A.H. Weiss, *Phys. Rev. Lett.* **72**,,1866-1869.
[35] G. Yang, S. Yang, J.H. Kim, K.H. Lee, A.R. Koymen, G. A. Mulhollan, and A.H. Weiss, *J. Vac. Sci. Technol.* **A12**, 411-417, (1993).

[36] G. Yang, J. H. Kim, S. Yang, and A.H., *Surf. Sci.* **367**, 45-55, (1996).
[37] J.H. Kim, S. Wheeler, A. Nangia,E. Jung, and A.H. Weiss, *Appl. Surf. Sci.*, **116**, 324-329, (1997).
[38] W.-C. Chen, N.G. Fazleev, J.L. Fry, and A.H. Weiss, *Mat. Sci. Forum*, **363-365**, 621-623 (2001).
[39] Kim JH, Lee KH, Yang G, Koymen AR, Weiss A.H, *Appl. Surf. Sci.*, **173**, 2001, pp.203-7(2001).
[40] G. Yang, J.H. Kim, S. Yang, and A.H. Weiss, *Appl. Surf. Sci.* **85**, 77-81, (1995).
[41] J.H. Kim, G. Yang, and A.H. Weiss, *Surf. Sci.*, **396**, 388-393 (1998).
[42] Fazleev, N. G.; Fry, J. L.; Weiss, A. *H Radiation Physics and Chemistry*, **58**, 659-665 (2000).
[43] T. Ohdaira, R. Suzuk, T. Mikado, H. Ohgaki, M. Chiwaki, T. Yamazak, M. Hasegawa, *Appl. Surf. Sci.* **100/101**, 73 (1996).
[44] T. Ohdaira, R. Suzuki, and T. Mikado, *Appl. Surf. Sci.*,**149**, 260 (1999).
[45] R. Mayer, A. Schwab, and Alex Weiss, *Phys. Rev. B* **42**, 1881-1884, (1990).
[46] N.G. Fazleev, J.L. Fry, K.H. Kuttler, A.R. Koymen, and A. H. Weiss, *Phys. Rev. B.*, **52**, 5351-5363, (1995).
[47] S. Yang, H. Q. Zhou, E. Jung, and A. H. Weiss, *Rev. Sci. Instrum.* **68**, 3893-7 (1997).
[48] A. Eshed *et al.*, to be published.

Chapter 13

Characterization of Nanoparticle and Nanopore Materials

Jun Xu

Oak Ridge National Laboratory P. O. Box 2008, Oak Ridge, TN 37831

13.1 Nanoparticle materials

Nanoparticle materials are important because they exhibit unique properties due to size effects, quantum tunneling, and quantum confinement. As sizes of embedded particles are reduced to the nanometer scale, the surface-to-bulk ratio increases significantly. Therefore, surface effects can dominate bulk properties and an understanding of nanosurfaces becomes important. In this chapter, we discuss characterization of vacancy clusters that reside on surfaces of embedded nanoparticles as well as studies on the correlation of surface vacancy clusters to the properties of the nanomaterials.

Gold nanoparticles embedded in MgO are of importance because of their unique optical and electronic properties [1-3]. A long-standing goal is to produce a size-tight network of metallic nanoparticles in an oxide matrix with minimal imperfections. Ion implantation and sequential annealing are being used as a means to achieve this goal [4]. Noticeably, creation of vacancy clusters and their various combinations are often associated with the nanoparticle formation processes. Also, unavoidably, these vacancy clusters are expected to interact with embedded nanoparticles and alter the properties of nano systems.

A new spectroscopic method for the characterization of surface vacancy clusters is a combination of positron lifetime spectroscopy, which determines the size of vacancy clusters, and coincidence Doppler broadening of annihilation radiation, which gives information on where vacancy clusters are located [5, 6]. If these clusters are located on the surface of gold nanoparticles, namely the interface between the particle and host matrix, the surroundings of the clusters should include both particle atoms and the matrix atoms. Doppler broadening of annihilation radiation (DBAR) with two-detector coincidence should be able to reveal these atomic constituents, and therefore elucidate the location of vacancy clusters.

13.1.1 Vacancy clusters on the surface of gold nanoparticles embedded in MgO

Generation of embedded nanoparticles involves the following two steps: (1) MgO (100) single crystals were implanted with 1.1 MeV gold ions at a dose of 6×10^{16} Au ions/cm^2. The depth profiles of the gold concentration were measured by Rutherford back-scattering spectroscopy (RBS), which shows that the Au implants are primarily located at a depth of 0.16—0.4 µm, while MgO remains crystalline after implantation. (2) The implanted MgO crystals were annealed at 1200 °C in an Ar + O$_2$ (5%) or Ar + H$_2$ (5%) atmosphere for up to 10 h. Cross-section tunneling electron microscopy (XTEM) of these samples shows that rectangular Au—metal nanoparticles are distributed in a size range of 0.5 to 7 nm with the most probable size of 1.2 nm, as shown in Figure 13.1.

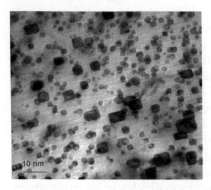

Figure 13.1. Cross-section TEM image of Au nanoparticles embedded in MgO.

Insert of Figure 13.2 shows the positron lifetime spectra for MgO (open circles), Au-implanted MgO (crosses) and Au nanoparticles embedded in MgO (solid circles). These spectra were deconvoluted using Laplace inversion [CONTIN, 7] into the probability density functions (pdf) as a function of vacancy size. Figure 13.2 shows the pdf spectra for the MgO samples accordingly. The positron lifetime components obtained for the MgO layer are 0.22 ± 0.04 ns with 89 ± 3% contribution and 0.59 ± 0.07 ns with 11 ± 3% contribution. For the Au-implanted sample without annealing, the major lifetime component is at 0.32 ns. For the Au nanoparticle-embedded MgO, lifetime components are 0.41 ± 0.08 ns at 90% and 1.8 ± 0.3 ns at 7%.

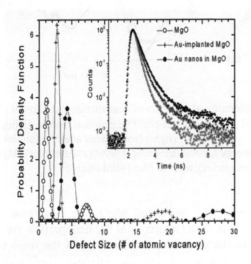

Figure 13.2 Probability density functions (pdf) as a function of defect size, resulting from Laplace inversion (CONTIN) of lifetime spectra (insert) for MgO (open circles), Au-implanted MgO (crosses) and Au nanoparticles embedded in MgO (solid circles) [5, 6].

The positron lifetimes for different defects in MgO are calculated using the insulator model of Puska and co-workers. In this model, the annihilation rates are determined by the positron density overlapping with the enhanced electron density that is proportional to the atomic polarizability of MgO [8, 9]. Based on comparison between experimental and calculated values [5, 6], the positron lifetime of the embedded Au nanoparticle layer, 0.41 ns, suggests that positrons are predominantly trapped in clusters consisting of two Mg vacancies and two O vacancies. These four vacancy clusters are referred to as "v_4". It is believed that the disturbance of the possible Au atoms near vacancy clusters on the electron density of the clusters is small and therefore was not considered in this calculation. The main component of the MgO layer, 0.22 ns, is due to positrons trapped at single vacancies of MgO (v_1).

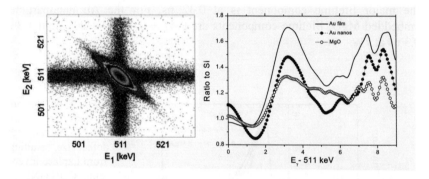

Figure 13.3 (left) Two-dimensional spectrum of annihilation radiation of positrons injected into a p-Si (100), with 8 Ω-cm. The diagonal feature indicates the condition of $E_1 + E_2 = 1.022$ MeV; (right) Normalized annihilation lines as a function of photon energy for Au nanoparticle layer (solid circles), MgO layer (open circles), and Au film (solid line) [6].

After detecting v_4 sites in the implanted MgO, the next task is to determine whether these sites lie in the matrix or at the nanoparticle surfaces by obtaining information on the elemental surroundings of the traps using two-detector coincidence Doppler broadening (2D-DBAR). 2D-DBAR has been measured using a system similar to that developed by Asoka-Kumar and Lynn [10]. Figure 13.3 (left) shows the two-dimensional data, while Figure 13.3 (right) shows the normalized ratio as a function of photon energy for Au nanoparticle (solid circles) and MgO layers (open circles). In the high-energy range (> 513 keV), the spectrum for the Au nanoparticle layer shows two enhanced bands at 514.3 keV and 518.5 keV. These bands are dramatically different from that of the MgO layers, where the normalized intensity is flat above the silicon. The difference is largely attributed to the surrounding Au atoms of the defects since the Au nanoparticle is different for the two environments. To confirm that the difference is due to the Au atoms, we measure the Doppler broadening spectrum for a pure Au film, shown as the solid line in Figure 13.3 (right). The Si-normalized spectrum for the Au film also shows two main enhanced bands: 514.3 keV and 518.5 keV. This is consistent with the features of Au films observed in the work of Myler and co-workers [11] in which the Au annihilation line was also normalized by that of perfect Si. There is a high degree of similarity between the Au film structures and the Au nanoparticle layer. This supports the interpretation that the positrons, trapped at the v_4 sites (as we know from the lifetime spectra), annihilate with the electrons associated with Au nanoparticles. Of course, the defect environment also includes the MgO matrix, which contributes to the energy deviation spectrum. It is concluded that vacancy sites where positrons are trapped are located on the Au nano surfaces.

13.1.2 Optical properties affected by surface vacancies

Recently it has been reported that the surface plasmon resonance wavelength of Au nanoparticles embedded in magnesia is shifted from 524 nm, which results from H_2 annealing atmosphere, to 560 nm when the nanoparticles are generated with an O_2 annealing atmosphere, as shown in Figure 13.4 [12]. The mechanism for such a shift is not known. Based on positron data described below and [13], the "red shift" was found to correlate with the presence of vacancy clusters on the nanoparticle surfaces. Electrons are expected to transfer from the nanoparticles to the vacancy clusters, leading to a decrease in electron density of the metallic nanoparticles; therefore, the surface plasma resonance frequency changes accordingly.

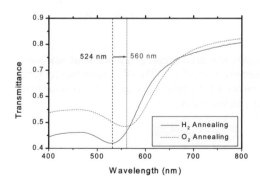

Figure 13.4 Optical transmittance spectra for Au nanoparticles formed by annealing Au-implanted MgO in H_2 (solid line) and O_2 (dashed line) atmospheres.

Figure 13.5 (left) compares the positron lifetime spectra of positrons injected into layers of Au-nanoparticles formed by annealing at ~1200 °C for 10 h in Ar + 5% H_2 or in Ar + 5% O_2 with that for the MgO layer. The figure on the right shows the resulting pdf spectra. For the lifetime spectrum of H_2 annealing, a fast lifetime component, 0.22 ± 0.05 ns, which contributes 93 ± 3% of events, is about the same as the MgO reference lifetime, 0.22 ± 0.04. The slow lifetime component is 1.9 ± 0.4 ns at 7% contribution. For post-implantation O_2 annealing, the fast component is 0.41 ± 0.08 ns at 90% contribution, which was found to be due to clusters of four atomic vacancies, v_4, on the surface of nanoparticles. The O_2 annealed sample does not exhibit a peak at 0.22 ns, which is characteristic of trapping in the MgO reference layer, because all positrons are trapped at v_4 sites associated with the 0.41-ns lifetime. The slow component is 1.8 ± 0.3 ns at 7%, which is approximately the same as that for the H_2 annealed sample. This component is attributed to larger pores. It is seen that the lifetime component for the surface v_4 clusters that appears for O_2 anealing, as described above, disappears if Au nanoparticles are generated under H_2 annealing conditions.

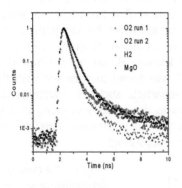
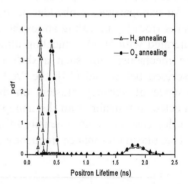

Figure 13.5 Probability density functions (right) as a function of positron me, resulting from Laplace inversion (CONTIN) of lifetime spectra (left) Au nanoparticle layers generated in H_2 (triangles) and O_2 (circles) annealing atmospheres [13].

Depth profiles of vacancy clusters were evaluated by measuring the energy distribution of annihilation photons as a function of positron incident energy. The annihilation photon peak is centered at 511 keV, and broadened due to the Doppler effect induced by the electron momentum. Doppler broadening can be characterized by a shape parameter, S, defined as the ratio of the counts appearing in the central region to the total counts in the annihilation photon peak. The S parameter mainly represents the contribution from the vacancy concentration and size. Figure 13.6 shows the S parameter, measured as a function of positron incident energy corresponding to an implantation depth shown in the top scale, for MgO samples embedded with Au nanoparticles that were annealed at 1200 °C in Ar + 5% H_2 and Ar + 5% O_2, respectively. For O_2 annealing, the S parameter (open circles) is much larger than that of H_2 annealing. Therefore, the increased S parameter for O_2 annealing shows a higher vacancy concentration than that for H_2, which is consistent with the findings of positron lifetime measurements, i. e. excess vacancy clusters are found for O_2 annealing compared to H_2 annealing.

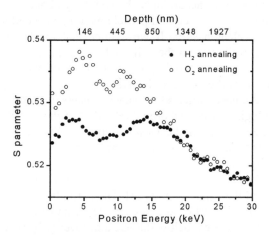

Figure 13.6 S parameters as a function of positron energy for MgO samples that are Au-ion implanted and annealed at ~1200 °C for 10 h in Ar + 5% H_2 (solid circles) and Ar + 5% O_2 (open circles), respectively.

13.1.3 Quantum dot—anti dot coupling

Subnanometer vacancy clusters are a special group of quantum anti-dots that have a localized electronic structure. A great deal of literature has been accumulated on "anti-dot" or quantum void systems in semiconductor or metal—oxide thin films [14]. The positron data described above may provide information on electronic structure of anti-dots, i. e. vacancy clusters. The presence of vacancy clusters, which is observed after annealing in O_2, is correlated to the red shift of the surface plasmon resonance frequency, which is also observed under O_2-annealing conditions. This correlation between the presence of vacancy clusters and the red shift observed as a result of annealing in O_2 can be explained by the following mechanism. Electrons in Au particles are transferred to a localized state of the vacancy clusters, reducing the electron density, n. The surface plasmon resonant frequency, ω_{sp}, is determined in the first order by electron density [15]:

$$\omega_{sp} = \sqrt{\frac{ne^2}{\varepsilon_0 m}} \quad (1)$$

where e and m are the electron charge and mass, and ε_0 is the permittivity of a vacuum. Therefore, with a decrease of the electron density in the nanoparticles the frequency decreases accordingly. Indeed, only about a 6% decrease in the number of electrons on a colloid particle would be needed to account for the observed wavelength shift. If we assume that the largest particles make the greatest contribution to the shift, a 1.2-nm size particle, which consists of 72 atoms per particle, would produce such a shift with the transfer of about 5 electrons to nearby vacancy clusters.

Electronic transfer from Au nanoparticles to vacancy clusters depends on the relative positions of the highest occupied orbital (HOMO) of the metallic nanoparticles and the localized states of the vacancy clusters. Figure 13.7 schematically shows the density of states diagram for the interface between a metallic nanoparticle and MgO matrix in the presence of vacancy clusters. The bottom of the MgO conduction band is 1.3 eV below the vacuum level with a 7.8-eV gap to the top of the valance band [16]. The partially occupied $6sp$ orbital of Au metal is 4.6 eV below the vacuum level and the fully occupied band of $5d$ electrons is 6.3 eV below the vacuum level [17]. If Au is present as nanoparticles, these energy levels are found by photon emission spectroscopy to depend on the size of the particle [18]. For 1.2-nm Au cubic particles, the $6sp$ electron energy is shifted by 0.9 eV toward the vacuum level and $5d$ electrons shift 0.4 eV in a similar way, as shown in the figure. For electron transfer from the Au particles to the v_4 clusters, our data suggest that the v_4 localized state should be 2.6 eV higher than that of an F center, an oxygen vacancy occupied by two electrons. This suggestion is qualitatively consistent with an earlier finding that F_n centers produce an energy level higher than that of F centers [19]. Such a transfer mechanism is also plausible considering that vacancy clusters are adjacent to the Au nanoparticles [6]. For H_2 annealing, there are no vacancy clusters observed. Consistently, no pol is observed either. It is clear that additional measurements and theoretical calculations are needed to evaluate localized states of the vacancy clusters and Au nanoparticle energy levels as a function of particle size.

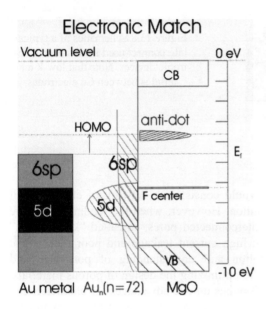

Figure 13.7 Schematic diagram of the energy band structure for the interface between Au nanoparticles and MgO with various defects. For comparison, the energy diagram for Au metal is also shown.

13.2 Nanopore materials—Ultra-low k

Typical nanopore materials are porous oxides or polymers that have an ultra-low dielectric constant (k < 2). Ultra-low k is significant in producing high-speed electronic devices such as the interconnect structure shown in Figure 13.8. SiO_2, which has a dielectric constant of about 4, is currently used as a dielectric material between interconnects in most microelectronic devices. When the packing density between multilevel interconnects increases, a low dielectric constant material is needed to minimize RC delay. By taking advantage of the low dielectric constant of air (~1) and introducing pores into a dielectric material, the dielectric constant can be effectively reduced to less than 2.0. This approach has initiated a major effort in semiconductor interconnect research and development [20, 21]. Porous methylsilsesquioxane (MSQ) films are some of these new low-k materials that have been actively studied by the microelectronics industry. For example, it has been shown that porous MSQ films can have a dielectric constant as low as k = 1.5 with 50% polymer loadings and still have high breakdown voltages [22].

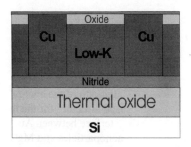

Figure 13.8 Schematics of a typical interconnect used in advanced electronic devices. Note that low-k material is between Cu electrodes.

An ideal porous material would consist of a network of closed, small pores with narrow size distribution. However, when the loading of a pore generator is high, large and interconnected pores, so-called "killer pores" may be generated, resulting in high current leakage and poor mechanical strength. Clearly, characterization and understanding of pore size and interconnectivity are important to optimizing the design of porous materials. Positron annihilation spectroscopy has unique advantages for characterizing pores because ortho-positronium (o-Ps), a positron—electron pair with spin state of 1, is preferentially formed and trapped in the pores. Its subsequent annihilation photons carry characteristics that are associated with pore structures. Recently, positron annihilation spectroscopy has been used to characterize pore structures of low-k films [23—29]. In this chapter, I discuss studies of porosity and interconnectivity of pores in porous MSQ films by measuring o-Ps 3γ emission and o-Ps lifetimes. The results clearly show that pore percolation in the films strongly depends on the characteristics of the polymers and provides a direction for developing the ideal low-k films.

Porosity in the films is introduced by mixing MSQ with a triblock copolymer (porogen), followed by spin coating on a Si (100) substrate and subsequent thermal curing and decomposition of the porogen. Various poly(ethylene oxide-b-propylene oxide-b-ethylene oxide) (PEO-b-PPO-b-PEO) triblock copolymers obtained from BASF (Pluronic® P103, P105, F38, and F88) were used as the porogen [30]. The polymers were in the form of either a paste or a solid with various relative molecular mass and ethylene oxide mass fractions, as described in Table 13.1. The loading of the porogen was 30% by weight. The thickness of each film was approximately 0.6 μm. The mixture was cured by heating at 120 °C after spin coating. The porogen was decomposed during subsequent annealing at a temperature of 500 °C, leaving pores in the MSQ matrix. The dielectric constants of these films were around 2.0.

13.2.1 Ps out diffusion—interconnectivity of pores

When monoenergetic positrons inject into a porous MSQ film, they are quickly thermalized and distributed around a depth determined by the positron incident energy. Both thermal and nonthermal positrons diffuse in films. Some of the positrons form positronium, which is trapped in the pores. If pores in the MSQ films are interconnected, an open channel results. If the open channel ends at the surface, o-Ps can diffuse from the film into the vacuum through the channel. The deeper the Ps is formed the longer its diffusion path, the greater the chance it will annihilate into 2γ via collision with the channel wall, and the lower the yield of the 3γ annihilation events. Such a decrease in 3γ yield as a function of depth constitutes a means for measuring the diffusion length and therefore to elucidate the pore interconnectivity. A larger 3γ yield implies a larger o-Ps fraction.

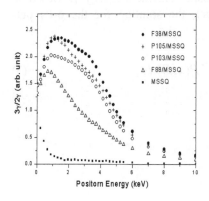

Figure 13.9 The profiles of relative 3γ o-Ps fraction as a function of positron beam energy, representing a mean depth from the surface, for porous MSQ films [28].

Figure 13.9 shows relative 3γ o-Ps fraction as a function of positron energy, or depth from the surface. At the surface, the 3γ o-Ps fraction was approximately 1.5 for all films, which is attributed to o-Ps formation at the MSQ surface. In the porous layers, represented by positron energies below ~4.5 keV, large 3γ/2γ ratios are generally found to be dependent on the porogen molecular weight. At higher incident energies, the o-Ps emission yield is zero because the positrons are implanted deeper into the Si substrate, where there are no pores. Two main features are displayed in all profiles: the first is the increase of 3γ/2γ ratio as the depth increases in the near surface region. This feature is believed to be due to the profile of the o-Ps formed initially. Originally, the near surface positrons may re-emit into the vacuum and so the resulting o-Ps is low there. The second feature is the decrease of o-Ps fraction as a function of depth. This fits into the Ps out- diffusion model. The depth profiles of these 3γ o-Ps fractions depend on the type of porogens initially used for pore generation.

These 3γ o-Ps fraction profiles were analyzed in a manner similar to the models used for Ps diffusion in ice [31], and for Ps diffusion in low-k films [26]. In the previous models, the initial positron deposition profile is an exponential function with respect to the depth. Although an analog expression of the out diffusion probability can be given by the exponential, the deposition profile generally is not believed to be exponential. In the analysis decribed below, a Makhovian distribution was used to represent the initial positron deposition profile. Let F to be the fraction of o-Ps particles that diffuse out into the vacuum, proportional to the 3g/2g ratio presented in Figure 9. It is given by the initial positronium yield, f_0, multiplied by the diffusion probability, $J(E)$:

$$F(E) = f_0(E)J(E). \qquad (2)$$

Both $f_0(E)$ and $J(E)$ depend on the positron incident energy, E. Based on the one-dimensional out-diffusion model, the solution for the o-Ps out-diffusion probability is

$$J(E) = \int_0^\infty e^{-\beta x} P(x,E)dx, \qquad (3)$$

where β is the reciprocal of the diffusion length and $P(x, E)$ is the positron deposition profile with incident positron energy, E. In this work, we used a Makhovian distribution for the deposition profile [32]:

$$P(x) = \frac{mx^{m-1}}{x_0^m} e^{-(x/x_0)^m}, \qquad (4)$$

where x denotes the depth into the solid from the surface, m is known as a shape parameter, and x_0 depends on the positron incident energy and relates to the mean implantation depth, x_{mean}, by

$$x_0 = \frac{x_{mean}}{\Gamma(\frac{1}{m}+1)}, \qquad (5)$$

where Γ is the gamma function and m is chosen to be 2 for MSQ oxide. The mean depth, x_{mean}, in nm, is related to positron incident energy, E, in keV,

$$x_{mean} = \frac{40}{\rho} E^{1.6}, \qquad (6)$$

where ρ is the density of films in g/cm^3. The initial profile of positronium formation, f_0, is chosen as an empirical function of the positron incident energy:

$$f_0(E) = f_{max} - f_{re} e^{-(E/E_0)^{1.6}}, \qquad (7)$$

where f_{max} is the saturation yield of Ps formation and $(f_{max}-f_{re})$ is the Ps formation yield at $E = 0$. E_0 is a parameter that can be related to the out-diffusion of the precursor positrons.

In this analysis, E_0, f_{max}, f_{re}, and β are obtained by fitting the experimental data. In fact, fitting is quite straightforward: the decay as a function of E determines the diffusion length, β; the increase region determines E_0; f_{max} is the saturation Ps yield; and $(f_{max}-f_{re})$ is determined by the surface Ps yield. Neither f_{max} nor f_{re} are affected by the profiles. The data for energy < 3.0 keV were only considered in the calculation in order to avoid the effect of positrons overlapping with the substrate.

Figure 13.10 shows the fittings for the low-k films. The diffusion lengths of o-Ps, listed in Table 13.1, depend on molecular weights of porogens. For F88/MSQ film, the diffusion length is the shortest among the group, suggesting a low degree of interconnectivity between pores for this film. For other films, diffusion lengths are long, causing many Ps atoms to escape into the vacuum. We infer that the interconnectivity of pores is high for these films. It is noted that the parameters obtained by the fitting may vary with the models used. However, their relative variation with molecular weight of porogens is still valid.

13.2.2 Positronium lifetime spectroscopy

o-Ps trapped in pores annihilate mainly via collisions with pore walls, in which case the o-Ps lifetime reveals the collision frequency and the pore size. If o-Ps escape into the vacuum via interconnected open pores, the lifetime is the characteristic o-Ps vacuum lifetime, 142 ns. Therefore, the intensity of this component povides the degree of interconnectivity. When the porous films are capped with a thin layer on the surfaces, those open pores are closed and act like additional o-Ps traps. o-Ps lifetimes are expected to be longer for capped than uncapped films. In this section, the o-Ps lifetimes measured with and without a thin SiO_2 cap on the porous MSQ films will be discussed.

The porous MSQ films (without the cap) were evaluated using positronium lifetime spectroscopy. Figure 13.11 shows positron lifetime spectra for porous MSQ films with positron energy of 3.3 keV. This energy corresponds to a mean depth of approximately 340 nm, which is roughly the center of the films. Fitting of the lifetime spectra shows that the pore component of the lifetime spectra varies with the type of porogen, as shown in Table 13.1. In addition, a long-lived component with lifetime nearly equal to the vacuum o-Ps lifetime, 142 ns, was observed for porogen F88 with a fraction of about 4%. The true fraction of this component is believed to be larger than the measured value because the gamma-ray collimator may block the photons emitted from the o-Ps in the vacuum. For comparison, Figure 13.11 also shows a positron lifetime spectrum of MSQ film without pores generated. The main positron component is 4.7 ns and is attributed to inherent free volumes of the nonporous MSQ film.

For subnanometer free volumes, the Tao-Eldrup model [33] is conventionally used to relate positron lifetime to free-volume size. For nanometer pores as studied here, Gidley's model [23, 24] was used to relate the positron lifetimes to pore sizes. The 47-ns lifetime for the F88 copolymer-generated porous film yields a diameter pore size of 3.7 nm if the pores are assumed to be a closed sphere, while the 54-ns lifetime for the P103 copolymer-generated film corresponds to a diameter pore size of 4.3 nm. It is pointed out that future work is needed to relate positronium lifetimes and pore sizes, especially for uncapped films, since positronium lifetimes of those samples include contributions from both closed and open pores.

The porous MSQ films were capped with a thin SiO_2 layer and evaluated with positronium lifetime spectroscopy, conducted by Gidley and his co-workers. The pore components for the porous MSQ films are listed in Table 13.1. Indeed, the lifetimes with capped films are longer than those for their corresponding uncapped films. The uncapped film of F88/MSQ shows a similar lifetime, 47 ns, to that of the capped film, 49 ns, indicating a low degree of interconnectivity for this film. The lifetimes with other capped films are longer than uncapped films by 6 to 10 ns. These increases are because the capping layers prevent o-Ps from escaping into the vacuum. The original open interconnected pores serve as o-Ps traps after capping. These lifetime results suggest that the interconnectivity of pores for the F88/MSQ film is lower than for the other films. This conclusion is consistent with that obtained from the diffusion-length measurements.

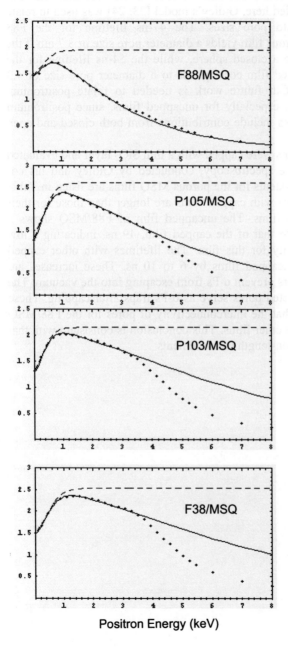

Figure 13.10 Theoretical model accounting for Ps formation (dashed lines) and Ps out-diffusion (solid lines) for 3γ profiles of porous low-k films. In this analysis, the fits are only extended to 3 keV, where the nonporous substrate effects are minimized. [29]

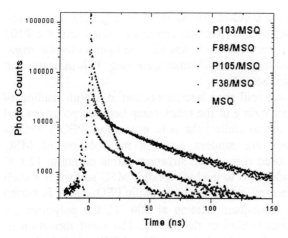

Figure 13.11 Positron annihilation lifetime spectra for porous MSQ films generated by mixing MSQ film with different triblock copolymers: P103, F88, P105, and F38. A spectrum for non-porous MSQ is listed for comparison [28].

Table 13.1 Molecular mass, PEO fraction, o-Ps diffusion length (L_{ps}), pore lifetime component (uncapped), pore diameter (D), and pore lifetime component (capped) for porous MSQ films templated with various porogens.

Porogen	Molecular Mass	PEO mass fraction (%)	L_{Ps} (nm)	$\tau_{o\text{-}Ps}$ (ns) Uncapped	D (nm)	$\tau_{o\text{-}Ps}$ (ns) Capped
F88	11400	80	330	47	3.7	49
P105	6500	50	750	53	4.2	60
P103	4952	30	1200	54	4.3	63
F38	4700	80	1330	54	4.3	63

13.2.3 Dependence on molecular weight of porogen

For the porous MSQ film templated by the F88 triblock copolymer porogen, which has the largest molecular mass in the samples studied, the positronium diffusion length was found to be the shortest and the pore size the smallest.

This result suggests that the pores in porous MSQ films generated with this larger molecular-mass polymer are smaller and more closed. For the P103 porogen, which has a low ethylene oxide fraction and light molecular mass, both the positron diffusion length and lifetimes were long. This indicates that pores are large and interconnected.

Since it is difficult to avoid pore interconnection at a high loading of pore generators, an understanding of the relationship between polymers and pore structures is important to control the pore generation. PEO-b-PPO-b-PEO triblock copolymers are random coils in the solution of MSQ precursors and form a closed core—shell structure during curing at 120 °C [22]. The PPO block resides next to the hydrophobic MSQ surface as a shell to minimize interfacial energy, while the hydrophilic PEOs block is buried inside as a core. After subsequent heating at 500 °C, the polymers are decomposed, leaving pores behind in the matrix. The rapid formation of core—shell structure depends on the chain mobility of PPO and PEO blocks which, in turn, determines the pore size and its percolation. Shorter polymer chains, such as F38, are more mobile and therefore likely to form larger core—shell domains and be percolated. The larger pore size and longer o-Ps diffusion length for the F38/MSQ film compared to those of the F88/MSQ film are consistent with this assumption. Solid state NMR shows that PEO blocks are predominantly crystallized in the bulk block copolymers and less mobile than the PPO block [22, 34]. Polymers with higher PEO wt% are more restricted in forming larger core—shell domains; thus each domain is more isolated, similar to the effect of higher molecular weight. The larger interconnectivity in the P103/MSQ films seems to be related to both the smaller molecular weight and the smaller PEO wt%.

Acknowledgments

The Author thanks Professor Allen P. Mills, Jr., of the University of California at Riverside; Drs. R. Suzuki and T. Ohdaira of the National Institute of Advanced Industrial Science and Technology, Japan; Dr. Shu Yang of Bell Labs; Dr. J. Maxam of Oak Ridge National Laboratories; and Professor D. Gidley of the University of Michigan; for their contribution and helpful discussions. This research was sponsored by the Division of Chemical Sciences, Geosciences, and Biosciences, BES, U.S. DOE, under

Contract DE-AC05-00OR22725 with ORNL, managed and operated by UT—Batelle, LLC.

References

[1] J. D. Budai, C. W. White, S. P. Withrow, M. F. Chisholm, J. Zhu, R. A. Zuhr, *Nature*, **390**, 384 (1997).
[2] W. C. W. Chan, S. Nie, *Science*, **281**, 2016 (1998).
[3] S. Chen, R. S. Ingram, M. J. Hostetler, J. J. Pietron, R. W. Murray, T. G. Schaff, J. T. Khoury, M. M. Avarez, R. L. Whetten, *Science*, **280**, 2098 (1998).
[4] C. W. White, et al., *Mater. Sci. Reports*, **4**, 43 (1989).
[5] Jun Xu, A. P. Mills, Jr., A. Ueda, D. Henderson, R. Suzuki, S. Ishibashi, *Phys. Rev. Lett.*, **83**, 4586(1999)
[6] Jun Xu, J. Moxom, B. Somieski, C. W. White, A. P. Mills, Jr., R. Suzuki, S. Ishibashi, *Phys. Rev. B*, **64**, In press, (2001).
[7] R. B. Gregory, *Nucl. Instrum. Methods*, **A302**, 496 (1991).
[8] M. J. Puska, S Makinen, M. Manninen, and R. M. Nieminen, *Phys. Rev. B* **39**, 7666 (1989).
[9] M. J Puska, in Positron annihilation, L. Dorikens-Vanpraet, M. Dorikens, D. Degers, Eds. (World Scientific, New Jersey, 1988), P101.
[10] P. Asoka-Kumar, M. Alatalo, V. J. Ghosh, A. C. Kruseman, B. Nielsen, K. G. Lynn, *Phys. Rev. Lett.*, **77**, 2097 (1996).
[11] U. Myler, R. D. Goldberg, A. P. Knights, D. W. lawther, P. J. Simpson, *Appl. Phys. Lett.*, **69**, 333(1996).
   A. Ueda, et al., *Mater. Sci. Forum.*, **239-241** (1997) 675.
[12] Jun Xu, J. Moxom, S. H. Overbury, C. W. White, A. P. Mills, Jr., and R. Suzuki, *Phys. Rev. Lett. Vol88, In press, (2002).*
[13] Maasilta I. J. & Goldman, V., *J. Phys. Rev. Lett.*, **84**, 1776, 2000; and references thereafter.
[14] Fukumi, K., Chayahara, A., Kadono, K., Sakaguchi, T., and Hirono, Y., *J. Appl. Phys. 75, 3075 (1994).*
[15] L. H. Tjeng, A. R. Vos, G. A. Sawatzky, *Surf. Sci.*, **235**, 269 (1990).
[16] W. DeHeer, *Rev. Mod. Phys.*, **65**, 611 (1993).
[17] Taylor, Cheshnovsky, Smalley, *J. Chem. Phys.*, **96**, 3319 (1992).
[18] B. D. Evans, J. Comas, P. R. Malmberg, *Phys. Rev. B*, **6**, 2453 (1972).
[19] Semiconductor Res. Corp., International Technology Roadmap for Semiconductors 1999 Edition.
[20] R. D. Miller, *Science* **286**, (1999) 421.
[21] Shu Yang, Peter A. Mirau, Chien-Shing Pai, Omkaram Nalamasu, Elsa Reichmanis, Eric K. Lin, Hae-Jeong Lee, David W. Gidley, Jianing Sun, *Chem. Mater.* **13**, (2001) in press.

[22] D.W. Gidley, W.E. Frieze, T.L. Dull, A.F. Yee, E.T. Ryan, H.-M. Ho, *Phys. Rev. B* **60**, (1999) R5157.
[23] D.W. Gidley, W.E. Frieze, T.L. Dull, J. Sun, A.F. Yee, C.V. Nguyen, D.Y. Yoon, *Appl. Phys. Lett.* **76**, (2000) 1282.
[24] M.P. Petkov, M.H. Weber, K.G. Lynn, K.P. Rodbell, submitted to *Appl. Phys. Lett.* (2000).
[25] K.P. Rodbell, M.P. Petkov, M.H. Weber, K.G. Lynn, W. Volksen, R. D. Miller, *Mater. Sci. Forum* **15**, (2001) 363-365
[26] R. Suzuki, T. Ohdaira, Y. Shioya, T. Ishimaru, *Jpn. J. Appl. Phys.*, **40**, (2001) L414.
[27] Jun Xu, Jeremy Moxo, Shu Yang, R. Suzuki, T. Ohdaira, *Appl. Surf. Sci.*, (2002), in press.
[28] Jun Xu, Jeremy Moxo, Shu Yang, R. Suzuki, and T. Ohdaira, *Appl. Phys. Lett.*, Submitted (2002).
[29] http://www.basf.com/static/OpenMarket/Xcelerate/Preview_cid-982931200019_pubid-974236729499_c-Article.html.
[30] M. Eldrup, A. Vehanen, P.J. Shultz, K.G. Lynn, *Phys. Rev. B* **32**, (1985) 7048.
[31] P. Asoka-Kumar, K. G. Lynn, D. O. Welch, *J. Appl. Phys.* **76** (1994) 4935.
[32] S. J. Tao, J. Chem. Phys. **56**, (1972) 5499-5510.
[33] P. A. Mirau, S. Yang, *Chem. Mater.* **14**, (2002) 249.

Chapter 14

AMOC in Positron and Positronium Chemistry

Hermann Stoll, Petra Castellaz and Andreas Siegle

Max-Planck-Institut für Metallforschung, D-70569 Stuttgart, Germany

14.1 Introduction

The two quantities which can be observed when an individual positron annihilates in condensed matter are the positron age τ, which is the time interval between implantation and annihilation of the positron, and the momentum p of the annihilating positron-electron pair. Time-resolved information on the evolution of positron states is obtained by correlated measurements of the individual positron lifetime (= positron age) and the momentum of the annihilating positron-electron pair (*Age-Momentum Correlation*, AMOC). AMOC measurements are an extremely powerful tool for the study of reactions involving positrons. It not only provides the information obtainable from the two constituent measurements but allows us to follow directly, in the time domain, changes in the e^+e^- momentum distribution of a positron state (cf. Sect. 1).

In the field of positronium chemistry AMOC combines the positron-lifetime analysis which is especially sensitive to detect long-lived ortho-positronium (o-Ps) with the Doppler-broadening-measurement technique which is particularly suitable for the observation of the para-positronium (p-Ps) state with its characteristic narrow momentum distribution. In addition,

AMOC allows time-dependent observations of the occupations and transitions of different positron states tagged by their characteristic Doppler broadening. Chemical reactions of positronium have been studied by beam-based AMOC as well as bound states between positrons (e^+) and halide ions (cf. Sect. 2).

For all positronium forming solids and liquids investigated so far epithermal positronium and its slowing-down could be observed, indicating that no special sites are required for positronium formation (cf. Sect. 3). Evidence for the existence of two kinds of o-Ps in liquid rare gases and for irreversible transitions between them is given in Sect. 4.

14.2 Beam-based age-momentum correlation (AMOC)

Speed and performance of positron age-momentum-correlation (AMOC) measurements have been improved significantly by using an MeV positrons beam and by taking advantage of the "beam-based" $\beta^+\gamma\Delta E_\gamma$-coincidence technique [1-5].

By letting the relativistic positrons pass through a about 5 mm thick plastic scintillator before implantation into the sample, a start signal with nearly 100% detection efficiency is generated [6-8]. Compared with efficiencies of less than 10% for prompt γ detectors in conventional "source-based" measurements, the unity detection efficiency of the β^+ start detector permits better statistics and/or drastically reduced measuring time. Relativistic positrons are essential in order to avoid jitter in the time of flight between the β^+ start detector and the sample [9]. Detection of pile-up pulses in the β^+ start detector or of short intervals between two pulses in the β^+ start detector can be used efficiently to discriminate against random coincidences due to positrons arriving in the sample almost simultaneously. In this way spectra with very large peak-to-background ratios are obtained even at high coincidence rates [4].

14.2.1 AMOC relief and lineshape function

Since both 511 keV photons resulting from a 2γ-annihilation event transmit equivalent information, one photon may be used to determine the age of the annihilating positron and the other for the correlated measurement of the momentum of the annihilating positron-electron pair by measurement of the

Doppler shift of the photon energy. Data acquisition is performed by a two-parameter multichannel analyzer [1-4].

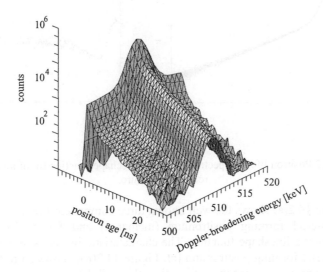

Figure 14.1 AMOC relief of methanol at room temperature.

The AMOC results can be visualized by an AMOC relief (Figure 14.1), where the number of counts is plotted logarithmically versus the positron age and the energy of the annihilation radiation. Summing all counts along the age axis results in the time-integrated Doppler-broadening photon lineshape, summing along the energy axis results in the usual lifetime spectrum.

For the visualization of the correlation between Doppler broadening and positron age it is instructive to compute the age-dependent *lineshape function* $S^t(t)$ [Figure 14.2(b)] by determining separately for each time channel of the AMOC relief (see Figure 14.1) the lineshape parameter S (defined as the ratio of the number of counts in the center of the annihilation line to the number of counts in the entire line [10] and plotting it versus the positron age. A large S parameter corresponds to a narrow Doppler broadening of the annihilation line, a small one to a broad line. The absolute value of S has no immediate physical significance since the width of the center region of the annihilation lines is deliberately chosen to get S parameter values of about 0.5 in order to achieve maximum sensitivity against small variations of the linewidth.

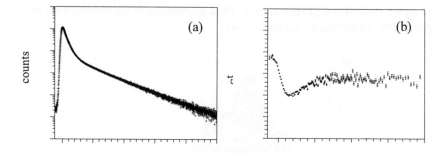

Figure 14.2 Positron lifetimespectrum (a) and lineschape function (b) of methanol at room temperature.

Figure 14.2(a) shows the well known three component lifetime spectrum of a material forming positronium (methanol) and Figure 14.2(b) the corresponding lineshape function. The characteristic lineshape function $S^t(t)$ of a material forming positronium [cf. Figure 14.2(b)] shows large S^t values at short positron ages, resulting from the self-annihilation of slowed down p-Ps. This is followed by a steep decrease of $S^t(t)$. At higher positron age the level corresponding to the rather broad momentum distribution of the electrons which annihilate with the positrons of o-Ps via the pick-off process is reached. In cases where free positrons annihilate with electrons of a different momentum distribution than the o-Ps via pick-off process the annihilation of free positrons shows up as a change in $S^t(t)$ in a time range around 1 ns. This effect is clearly visible in the lineshape function of methanol [see Figure 14.2(b)].

14.2.2 Analysis of AMOC data

In order to take advantage of the full information contained in the AMOC data we use a two-dimensional fitting procedure: A two-dimensional model function representing the number of counts as a function of positron age and energy of the annihilation quanta is fitted to the "raw" AMOC relief without prior data reduction. On the age axis, each positron state is represented by an exponential decay function convoluted with the time resolution function of the spectrometer the linewidth σ_t of which is also a fitting parameter of the procedure. Transitions between different positron states, as they take place in chemical reactions, are calculated from adequate rate equation models and lead to changes in the exponential decay functions describing the age-dependence. This changes are known as "chemical quenching". Therefore, we are able to extract the transition rates K_{ij}, i.e., the chemical reaction rates, directly from the fitting procedure.

On the energy axis, one Doppler-broadened Gaussian lineshape corresponds to each positron state. The annihilation with core electrons is taken into account by a broad Gaussian line of low amplitude. The energy resolution function which is determined simultaneously by measuring the linewidth of the 497 keV ^{103}Ru photon line is used in order to get the deconvoluted linewidth σ_i of the annihilation lines. This deconvoluted linewidth may be used to calculate the mean kinetic energy of the annihilating positron-electron pairs [11] if we assume that the velocity distribution of these pairs is described by or at least not too far away from a Maxwell velocity distribution. This seems to be a valid assumption at least for positronium in liquids which are known to build bubbles. Using this assumption we may calculate the mean kinetic energy of p-Ps as [11].

$$E_{p-Ps}^{kin} = 3 \frac{\sigma_{p-Ps}^2}{m_e c^2}, \tag{1}$$

where m_e is the electron mass and c the velocity of light. If we generalize this relation for the annihilation from other positron states such as free positrons or o-Ps, we have to mention that in both cases the momentum of the annihilating positron-electron pair comes mainly from the electron which is, in general, bound to a molecule. In these cases, the positron can be regarded as completely thermalized. Since only one particle, the electron, contributes to the momentum distribution we can calculate its mean kinetic energy according to [11]

$$E_{e^-}^{kin}(e^+) = 6 \frac{\sigma_{e^+}^2}{m_e c^2} \text{ and } E_{e^-}^{kin}(o-Ps) = 6 \frac{\sigma_{o-Ps}^2}{m_e c^2} \tag{2}$$

from the deconvoluted Gaussian linewidths obtained from the two-dimensional data analysis.

354 *Principles and Applications of Positron and Positronium Chemistry*

The mean kinetic energies of annihilating electrons may provide information on the electronic surroundings where the positrons undergo annihilation.

Table 14.1 shows the relevant parameters obtained from a two-dimensional data analysis in summary.

Table 14.1 Parameters obtained from the two-dimensional fitting procedure.

Parameter	Description
$\tau_i = 1/\lambda_I$	Lifetime (inverse of annihilation rate) of the positron state i
I_i	Intensity of the positron state i
K_{ij}	Transition rate between state i and state j
σ_t	Linewidth of the time resolution function
$E^{kin}(i)$	Mean kinetic energy of the annihilating positron-electron pair or
σ_i	deconvoluted linewidth of the positron state i

14.2.3 Analysis of the young age broadening of positronium

In all positronium-forming materials examined so far by the AMOC technique, juvenile broadening arising from the delayed slowing-down of positronium [11-14] was observed (see also section 3): Positronium which is formed at an early stage of the slowing-down process cannot loose its kinetic energy via ionization of molecules or creation of excitons (since it is an electrically neutral particle), but via the less effective phonon excitation process. Therefore, a considerable amount of positronium is not thermalized when undergoing annihilation. Its kinetic energy results in a large Doppler broadening visible at young positron ages. This effect is clearly visible in the lineshape functions of positronium forming materials in a decrease towards negative positron ages near time-zero. This juvenile broadening was taken into account by a simple two-state rate-equation model [11, 14]. This model approximates the Ps ernergy loss by allowing for transitions between epithermal and slowed-down Ps. Two additional parameters are added to the fitting procedure by this model, viz. a thermalization energy E_0 from which the slowing-down process starts, and a slowing-down time t_{th}. It was shown that this simplified model serves well for the analysis of positron-chemistry measurements as the positron and positronium reactions take place in a different time regime than the young age broadening (see Figure 14.3).

14.3 Time domain observations in positron and Ps chemistry

Positrons are widely used as non-destructive mobile probes in condensed matter, e.g., for the study of vacancy-like defects in crystals. In many isolating materials such as organic liquids or polymers, positronium (Ps), the bound state of a positron and an electron is at our disposal as an additional, electrically neutral probe. With some justification, Ps can be regarded as the lightest isotope of hydrogen. It may undergo chemical reactions analogous to those of hydrogen (e.g., oxidation or complex formation). In contrast to most other branches of reaction chemistry, in which information on the reactions is primarily derived from studying educts and products, positronium chemistry offers the possibility to observe the behavior of one of the reaction partners, viz. the positron or positronium, rather directly on time scales between 10^{-11} s and 10^{-8} s. Here, AMOC makes it possible to watch a chemical reaction in the time domain by observing the vanishing of one of the reaction partners during the reaction via annihilation. The chemical reactions observable by AMOC are: oxidation and complex formation of positronium, reactions of positrons, and spin conversion of positronium. Figure 14.3 gives a schematic overview of the influence these reactions have on the positron lifetime spectra and the lineshape functions extracted from an AMOC measurement. As can be seen in Figure 14.3, most reactions lead to a shortening of the longest lifetime component which is known as "chemical quenching" whereas the lineshape function allows the reaction to be identified easily, since it is characteristic for each reaction.

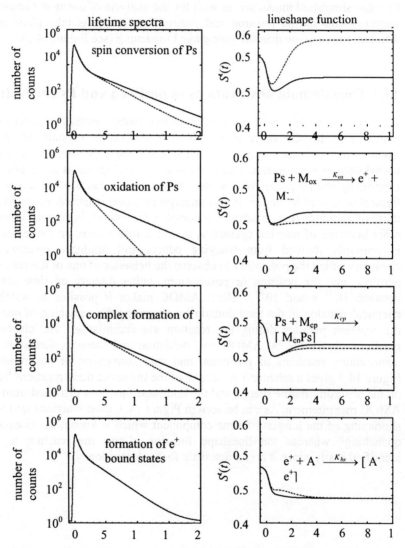

Figure 14.3 Schematic lifetime spectra (right hand side) and lineshape functions $S^t(t)$ (left hand side) of different chemical reactions observable by AMOC (solid lines denote the pure solvents, dashed lines show the reactions).

The fact that positrons annihilate and, therefore, only few positrons are present in the sample at the same time gives us reaction equations which are easy to handle. Back reaction are not to be considered in most cases.

We have used the AMOC method so far to study chemical reactions of positrons, oxidation reactions of positronium, as well as the spin conversion of positronium in the presence of organic radicals. In all cases, the method was well suited for the time-domain observation of the reactions. In this contribution, we will confine ourselves to the reactions of positrons, as they take place in liquid solutions of halides.

14.3.1 Chemical reactions of positrons

In 1974, O.E. Mogensen and V.P. Shantarovich concluded from ACAR measurements on aqueous solutions of sodium chloride [15] that in this system a fourth positron state (in addition to p-Ps, free positrons and o-Ps) was formed, which they identified as an e^+Cl^- bound state. The existence of such a bound state found support in several Doppler broadening and ACAR investigations on aqueous and non-aqueous solutions of halides and pseudo-halides [16-20]. Since the commonly used expression "bound state of positrons" may lead to confusion with positrons bound in Ps, we refer to this "fourth" state as positron molecules and characterize it by the subscript M.

The more recent theoretical work on the binding between Ps and the halides agrees that there is a substantial positive binding energy E_{PsA}, at least for A=F, Cl, and Br. According to Schrader, Yoshide, and Iguchi [21, 22] the binding energies for the isolated molecules are given by $E_{PsF} = 1,98$ eV, $E_{PsCl} = 1,91$ eV, $E_{PsBr} = 1,14$ eV. Mogensen [17] claimed good agreement between the theoretical prediction of Farazdel and Cadé [23] for the one-dimensional ACAR curves and the measurements. This has been referred to as "a part of what is perhaps the most striking example of agreement between calculation and experiment in positron chemistry". However, the agreement did not include PsF. It was supposed that PsF may not be stable in aqueous solutions due to hydration effects—certainly a surprising conclusion in view of the fact that the theory had predicted the highest binding energy for PsF. The AMOC method provides a possibility to investigate this subject without the need of using Ps inhibitors [see, e.g., 24] which may affect the the microstructure of water strongly when used in high concentrations.

A survey of our AMOC measurements on aqueous solutions of sodium halides may be obtained by computing the lineshape function $S^t(t)$ from the AMOC histogram. In Figure 14.4 the lineshape functions of pure water and a

0.01 M aqueous solution of NaBr are shown for comparison. The lineshape function of pure water is characteristic for positronium-forming materials: The large S parameter at short positron ages corresponds to the intrinsic annihilation of p-Ps. At old ages, at which only o-Ps survives because of its long lifetime, the S parameter is low due to the broad momentum distribution of the electrons annihilating with the positrons via pick-off. In the time range around 1 ns the lineshape function is dominated by the annihilation of free positrons which, in the case of water, have an S parameter close to that of o-Ps. When sodium bromide is added to the solvent, the lineshape function is significantly enhanced at ages about 1 ns compared to the pure solvent, as can be seen in Figure 14.4. This behavior results from the narrow annihilation lineshape of the positron-molecule state PsBr. As the effect appears just in the time region of the fre-positron annihilation, it is clear that the positron-molecule state originates from the free- positron state.

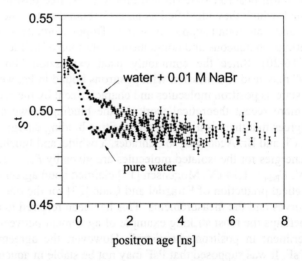

Figure 14.4 Lineshape functions of water and an 0.01 M aqueous solution of NaBr.

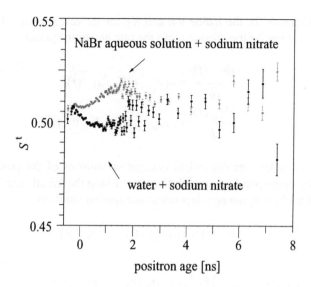

Figure 14.5 Lineshape functions of a 3 M NaNO$_3$ aqueous solution with and without 0.01 M NaBr.

When Ps formation is suppressed by adding a high quantity of a Ps inhibitor such NaNO$_3$ to the solvent, the lineshape function becomes essentially a flat line (Figure 14.5). (The slight curvature visible between time zero and 2 ns in Figure 14.5 is due to a very small residual amount of Ps.) This behavior indicates the presence of only one positron state, which can be ascribed to the free positrons, i.e., to positrons not forming positronium. When 0.01 M NaBr is added to the inhibitor solution, an increase of $S^t(t)$ is observed in the time region around 1ns. We can see that the reaction of positrons and halide ions is still taking place when Ps is absent.

14.3.2 The model

The simplest way of describing the reaction of free positron with halide ions is the reaction equation

$$e^+ + A^- \xrightarrow{K_M} PsA \qquad (3)$$

where A⁻ stands for the halide ion and K_M for the reaction rate. The reaction equation (3) can be transformed into the rate equation system

$$\frac{dn_{e^+}(t)}{dt} = -(\lambda_{e^+} + K_M)n_{e^+}(t), \qquad (4)$$

$$\frac{dn_M(t)}{dt} = -\lambda_M n_M(t) + K_M n_{e^+}(t), \qquad (5)$$

where $n_i(t)$ is the age-dependent occupation number of the positron state i and λ_i the corresponding annihilation rate. Using the initial conditions $n_e^+(0) = N$ and $n_M(0) = 0$, the age-dependent occupation numbers

$$n_{e^+}(t) = N \exp(-(\lambda_{e^+} + K_M)t) \qquad (6)$$

$$n_M(t) = \frac{K_M N}{\lambda_{e^+} + K_M - \lambda_M}(\exp(-\lambda_M t) - \exp(-(\lambda_{e^+} + K_M)t)) \quad (7)$$

are obtained. N is the number of positrons which do not annihilate from a positronium state. Annihilation from the para- and the orthopositronium state is described by two exponential decay functions.

14.3.3 Influence of the halide ion size

Table 14.2 Annihilation parameters of PsF, PsCl, PsBr, PsI, and Ps yield.

Solution	τ_M [ps]	σ_M [keV]	K_M [10^9/s]	I_{Ps}
H₂O	--	--	--	0.369
0.01 M NaF aq. solution	1236	1.08	0.18 ± 0.04	0.316
0.01 M NaCl aq. solution	898	0.92	0.36	0.250
0.01 M NaBr aq. solution	622	0.84	0.33	0.304
0.01 M NaI aq. solution	625	0.86	0.39	0.279
typ. statistical error	15	0.01	0.02	0.002

Table 14.2 shows the annihilation and reaction parameters of the positron-molecule states PsA of 0.01 M aqueous solutions of NaF, NaCl, NaBr, and NaI, as well as for pure water. The two-dimensional model functions based on equations (6) and (7) were shown to describe the AMOC

measurements well for the halide solutions. The annihilation parameters of free positrons, p-Ps and o-Ps for the solutions do not differ within accuracy of the measurement from their values in pure water for this solute concentration.

14.3.4 Positron-molecule lifetimes

The lifetimes of the positron-molecule states τ_M (see Table 14.2) are considerably longer than that of the free positrons ($\tau_e^+ = 400$ ps). τ_M is longest for PsF, which has the smallest number of electrons in its shell, and, hence, the lowest electron density, and decreases with increasing number of electrons. Seeger and Banhart [25] give an upper limit for the lifetime τ_∞ of positron states where no o-Ps is involved:

$$\tau_\infty^{-1} = \frac{1}{4}\tau_p^{-1} + \frac{3}{4}\tau_o^{-1}. \tag{8}$$

For water, with $\tau_p = (140 \pm 2)$ ps and $\tau_o = (1886 \pm 2)$ ps, we get $\tau_\infty = 580$ ps. The positron-molecule lifetimes exceed this limit for all halides examined. As data analyses with positron-molecule lifetimes fixed below this limit were definitely not successful, we are confessed that the long lifetimes are a hard experimental fact. However, we have not yet come to a satisfactory explanation what makes the positrons survive this long in the neighborhood of the halide ions. Possible explanations could involve the spin similar to the effect in o-Ps, or electron depletion due to charge effects in the hydration sphere around the halide ions.

14.3.5 Reaction rates

The reaction rates are the same for the formation of PsCl, PsBr, and PsI within the accuracy of the measurement. The reaction rate of PsF is somewhat smaller and has a larger statistical error (see Table 14.2). The determination of K_M is less accurate for PsF than for the other halides since the annihilation lineshape from this state is very similar to that of the free positrons. If one tentatively uses $K_M = 0.3 \cdot 10^9$/s as a fixed parameter in the two-dimensional analysis, the quality of the fit does not deteriorate compared to $K_M = 0.18 \cdot 10^9$/s. It may, therefore, be concluded that the reaction rate is independent of the halide size. This result shows that the size and the mass of the halide ions does not play a major role in the reaction process. We assume a possible reason for this behavior is that the positrons are much more mobile than the halide ions.

14.3.6 Existence of the PsF molecule

The linewidth of annihilation from the free-positron state is $\sigma_e^+ = 1.08 \pm 0.01$ keV. It is very similar to σ_M in the NaF solution. The similarity of the lineshapes explains the fact why the PsF state could never be detected by ACAR and Doppler-broadening measurements. In lifetime measurements the PsF component hides beneath the o-Ps component which has a similar lifetime. This is a case where the two-dimensional data analysis shows its great advantage: As the Doppler broadening of each positron state is determined in its "own" time regime even positron states with similar features may be seperated from each other. Moreover, a tentative fitting procedure with only the three positron states as in pure water did not come to a satisfactory result with the AMOC histogram of the NaF solution.

14.4 Observation of positronium slowing down

Beam-based AMOC allowed time-dependent observations of positronium slowing-down in condensed matter for the first time. For all Ps-forming solids and liquids investigated so far [11-14] a decrease in the $S^t(t)$ values has been found at young positron ages as shown for poly(isobutylene) (PIB) in Figure 14.6. This juvenile broadening has never been observed in materials for which there is no evidence for Ps formation (e.g., see Al results, Figure 14.7). When positrons form Ps, the ionization of molecules and the creation of excitons and electron-hole pairs as effective mechanisms of losing kinetic energy ceases to operate. In materials with optical phonon branches the dominate slowing-down process for Ps is thought to be the transfer of kinetic energy to the lattice via collision with optical phonons [13, 26].

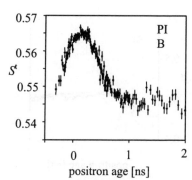
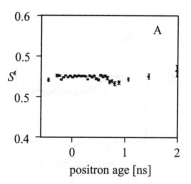

Figure 14.6 Lineshape functions $S^t(t)$ of poly(isobuthylene) (PIB) at room temperature.

Figure 14.7 Lineshape function of aluminum at room temperature.

The AMOC data of the juvenile broadening have been analysed in terms of a two state model [11] which approximates the Ps energy loss via interaction with phonons by allowing for transitions between epi-thermal Ps having a kinetic energy E_0 and slowed down Ps (cf. Sec. 1.2). A better analysis of AMOC data on Ps slowing-down is suggested by Seeger [27]. Room temperature slowing-down times t_{th} between 10 ps and 40 ps and initial kinetic energies E_0 of Ps between 3 eV and 6 eV are obtained by this procedure for materials with optical phonon [3]. The values obtained for t_{th} have a higher reliability than those for E_0, which appear to be more model-dependent.

In positronium forming materials without optical phonons much longer Ps slowing-down times should be found. This may be tested by AMOC measurements in solid rare gases, crystallizing in the face centered cubic (fcc) structure which, being a Bravais lattice, does not have optical phonon branches.

In Ar and even more pronounced, in Kr and Xe, the lineshape functions $S^t(t)$ show indeed a clear shift of the juvenile Doppler broadening to higher positron ages (Figure 14.8).

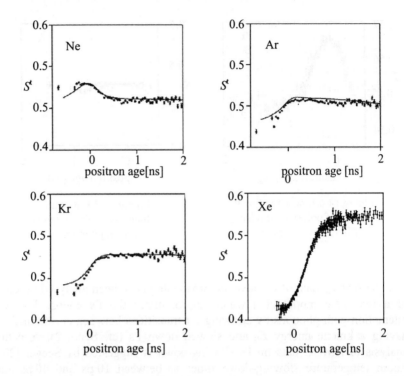

Figure 14.8 Lineshape functions $S^t(t)$ of solid rare gases. Data points with error bars are calculated from the AMOC data, solid lines from the two-dimensional model function [11].

The narrowing annihilation photo line of slowed-down p-Ps resulting in a high S parameter at low positron ages as shown for PIB in Figure 14.6 has vanished completely (cf. Figure 14.8) showing that most of the p-Ps annihilates from an epi-thermal state. This indicates Ps slowing-down times t_{th} in the order of the p-Ps lifetime. The analysis based on the two-state model mentioned above (solid lines in Figure 14.8) yields slowing down times t_{th} of 125 ps to 250 ps for solid Ar, and of 400 ps to 600 ps for solid Kr [3, 4, 11]. They do not differ very much from those for the liquid states. The Ps slowing-down times in solid (cf. Figure 14.8) and liquid Xe are estimated to be in the range of 300 ps to 600 ps [4]. The rather long Ps slowing-down times t_{th} observed in Ar, Kr, and Xe are in agreement with the idea that Ps cannot lose energy by generating electron-hole pairs and, since in the rare gases studied there are no optical phonon branches, the Ps energy can only be transferred to acoustic phonons which is significantly less effective than the transfer to optical phonons.

The lineshape function $S^c(t)$ of solid Neon (Figure 14.8) shows a maximum similar to PIB (Figure 14.6). The slowing-down times t_{th} according to the two state model analysis are in the range of 20 ps to 40 ps [3, 4, 11] and thus are similar to t_{th} in materials with optical phonons. The process responsible for the shorter slowing-down times in Ne is not yet fully understood and requires further study.

The observation of delayed Ps slowing-down in *all* positronium forming liquids and solids investigated so far shows unambiguously that for positronium formation no special sites are required [28, 29].

14.5. Evidence for the existence of two kinds of o-Ps in condensed rare gases

Positron-lifetime and AMOC spectra were measured on Ne, Ar, and Kr in the liquid and in the solid states [4, 11]. The measured o-Ps lifetimes of 2.4 ns in solid Ar at 16 K and of 2.1 ns in solid Kr at 50 K are accounted for pick-off annihilation of o-Ps. The much longer o-Ps lifetimes in the liquids (15.7 ns in Ne at 26 K, 7.0 ns in Ar at 87 K and 5.7 ns in Kr at 120 K) and are explained by o-Ps annihilation in long-lived self-localized states, the so-called "Positronium Bubbles" [30].

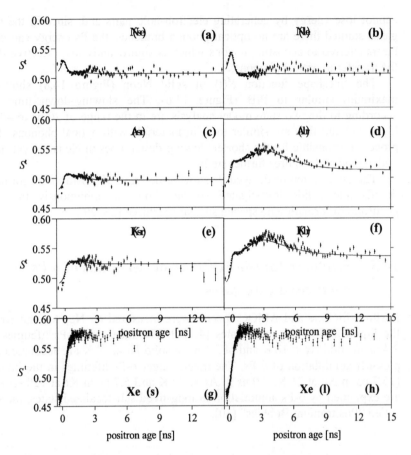

Figure 14.9 Lineshape functions $S^t(t)$ calculated from AMOC data and model functions (solid lines) for condensed rare gases: Solid Ne at 15.2 K (a), liquid Ne at 26.0 K (b), solid Ar at 83.3 K (c), liquid Ar at 86.3 K (d), solid Kr at 50.0 K (e), liquid Kr at 120.0 K (f), solid Xe at 150 K (g), and liquid Xe at 163 K (h).

The lineshape functions of Ar and Kr in the liquid states only showed a surprising maximum at positron ages of about 3 ns [Figs.14.5(d) and 14.5(f)]. Analysis of the AMOC data obtained on liquid Ar and Kr revealed for the first time that the positronium bubbles are formed from an additional delocalized, metastable o-Ps state [4, 11]. The lifetime of the metastable o-Ps states in the liquids were found to be about the same as the o-Ps lifetimes in the solids in the vicinity of the melting point. The transition rate to the longer-lived and apparently more stable o-Ps bubble states in liquid Ar and Kr is about $3 \cdot 10^8$ s^{-1} [11]. A lower limit of the height of the energy barrier between the two different o-Ps states of about 10^{-1} eV is estimated by assuming that the barrier is overcome by overbarrier jumps with an attempt frequency of 10^{14} s^{-1}. A bump in the lineshape function of liquid Ne may also be possibly visible at positron ages of about 4 ns to 6 ns [Figure 14.9(b)] indicating that a metastable Ps state may be formed in Ne, too. In liquid Xe no significant effect was observed in the lineshape function [cf. Figure 14.9(h)]. The S parameters of the two o-Ps states in Xe may be very similar. However, an additional positron lifetime component found in Xe in the liquid state only [31] indicates that two distinct o-Ps states are also present in liquid Xe.

We can conclude that in all liquid rare gases investigated positronium is formed first before it is trapped into self-localized "positronium bubble" states. Seeger emphasized that AMOC has the potential to solve the important question "where an how positronium is formed" also for more complex positronium formers, e.g., polymers [28].

Acknowledgements

The authors would like to thank all of those who were involved in the Stuttgart MeV positron beam. This paper would not have been possible without the work of all of them. In particular we would like to thank Prof. A. Seeger for his continuous support, Dr. P. Bandžuch for providing us with the Xe data, Prof. H. D. Carstanjen and Dr. J. Major for many fruitful discussions, and Aleksandar Puzic for his help in preparing the manuscript.

References

[1] H. Stoll, P. Wesolowski, M. Koch, K. Maier, J. Major, and A. Seeger, *Materials Science Forum* **105-110**, 1989 (1992).
[2] H. Stoll, M. Koch, K. Maier, and J. Major, *Nucl. Instr. and Meth.* B, **56-57**, 582 (1991).

[3] H. Stoll, P. Castellaz, S. Koch, J. Major, H. Schneider, A. Seeger, and A. Siegle, *Materials Science Forum* **255-257**, 92 (1997).
[4] H. Stoll, P. Bandzuch, and A. Siegle, *Materials Science Forum* **363-365**, 547 (2001).
[5] H. Stoll: *MeV Positron Beams*, in: Positron Beams and Their Applications, ed. P. Coleman (World Scientific Singapore, 2000) p. 237.
[6] W. Bauer, K. Maier, J. Major, H.-E. Schaefer, A. Seeger, H. D. Carstanjen, W. Decker, J. Diehl, and H. Stoll, *Appl. Phys.* **B 43**, 261 (1987).
[7] W. Bauer, J. Briggmann, H. D. Carstanjen, W. Decker, J. Diehl, K. Maier, J. Major, H.-E. Schaefer, A. Seeger, H. Stoll, and R. Würschum: *The Stuttgart Positron Beam, its Performance and Recent Experiments*, in: Positron Annihilation, eds. L. Dorikens-Vanpraet, M. Dorikens, and D. Segers (World Scientific, Singapore, 1989) p. 579.
[8] W. Bauer, J. Briggmann, H. D. Carstanjen, S. H. Connel, W. Decker, J. Diehl, K. Maier, J. Major, H.-E. Schaefer, A. Seeger, H. Stoll, and E. Widmann, *Nucl. Instr. and Meth.* **B 50**, 300 (1990).
[9] K. Maier and R. Myllylä: *Positron Lifetime Spectrometer with $\beta^+\gamma$–Coincidences*, in: Positron Annihilation, eds. R. H. Hasiguti and K. Fujiwara (Japan Institute of Metals, Sendai, 1979) p. 829.
[10] I.K.MacKenzie, J.A.Eady, R.R.Gingerich, *Phys. Lett.* **33A**, 279 (1979).
[11] Siegle: Positronenzerstrahlung in kondensierter Materie—eine Untersuchung mit der Methode Lebensalter-Impuls-Korrelation (AMOC), Dr. rer. Nat. Thesis, Universität Stuttgart, (Cuvillier, Göttingen, ISBN 3-89712-129-8, 1998).
[12] H. Stoll, M. Koch, U. Lauff, K. Maier, J. Major, H. Schneider, A. Seeger, and A. Siegle, *Appl. Surf. Sci.* **85**, 17 (1995).
[13] Seeger, *Appl. Surf. Sci*, **85**, 8 (1995).
[14] H. Schneider, A. Seeger, A. Siegle, H. Stoll, P. Castellaz, J. Major, *Appl. Surf. Sci.* **116**, 145 (1997).
[15] O.E. Mogensen, V. P. Shantarovich, *Chem. Phys.* **6**, 100 (1974)
[16] G. Duplâtre, J.Ch. Abbé, A. G. Maddock, A. Haessler, *Radiat. Phys. Chem.* **11**, 199 (1978).
[17] O. E. Mogensen, *Chem. Phys.* **37**, 139 (1979).
[18] J. Talamoni, J. Ch. Abbé, G. Duplâtre, A. Haessler, *Chem. Phys.* **58**, 13 (1981).
[19] G. Duplâtre, J. Ch. Abbé, A. G. Maddock, A. Haessler, *J. Chem. Phys.* **72**, 89 (1980).
[20] J. Ch. Abbe, G. Duplatre, A. G. Maddock, and A. Haessler, *Rad. Phys. Chem.* **15**, 617 (1980).
[21] D. M. Schrader, T. Yoshida, K. Iguchi, *Phys. Rev. Lett.* **68**, 3281 (1992).
[22] D. M. Schrader, T. Yoshida, K. Iguchi, *J. Chem. Phys.* **98**, 7185 (1993).
[23] D.A. Farazdel, P. E. Cadé, *J. Chem. Phys.* **66**, 2612 (1977)
[24] J. R. Andersen, N. J. Pedersen, O. E. Mogensen, and P. Christensen, *Chem. Phys. Lett.*. **63**, 171 (1979).

[25] Seeger and F. Banhart, *Helv. Phys. Acta* **63**, 403 (1990).
[26] Seeger, *J. Phys.: Condensed Matter* **10**, 10465 (1998).
[27] Seeger, Radiation Physics and Chemistry **58**, 411 (2000).
[28] Seeger, Radiation Physics and Chemistry **58**, 503 (2000).
[29] Seeger, *Materials Science Forum* **363-365**, 257 (2001).
[30] R.A. Ferrell, *Phys. Rev.* **108**, 167 (1957).
[31] P. Bandžuch, private communication.

Problem

The following reaction equation describes the spin conversion of positronium:

$$pPs(oPs) + M \uparrow \xrightarrow{K_{sc}} \frac{1}{4}pPs + \frac{3}{4}oPs + M \downarrow$$

where M denotes a paramagnetic molecule.
This equation can be transformed into the rate equation system:

$$\frac{dn_p(t)}{dt} = -\left(\lambda_p + \frac{3}{4}K_{sc}\right)n_p(t) + \frac{1}{4}K_{sc}n_o(t)$$

$$\frac{dn_o(t)}{dt} = -\left(\lambda_o + \frac{1}{4}K_{sc}\right)n_o(t) + \frac{3}{4}K_{sc}n_p(t)$$

where $n_p(t)$, $n_o(t)$ are the time dependent occupation numbers of the para- and orthopositronium state, respectively, λ_p and λ_o their annihilation rates and K_{sc} the spin conversion rate. The initial conditions are:

$$n_p(0) = n_o(0)/3$$

$$n_p(0) + n_o(0) = I_{Ps}$$

Use the Laplace transformation to calculate the occupation numbers $n_p(t)$ and $n_o(t)$. K_{sc} can be regarded as time independent.

Answer to problem

$$n_p(t) = \frac{I_{Ps}}{16\gamma}\{[\gamma - \lambda_p - 3(\lambda_o + K_{sc})]\exp(-s_1 t) + [\gamma + \lambda_p + 3(\lambda_o + K_{sc})]\exp(-s_2 t)\}$$

$$n_o(t) = \frac{3I_{Ps}}{16\gamma}\{[\gamma - \lambda_o - 3(\lambda_p + K_{sc})]\exp(-s_1 t) + [\gamma + \lambda_p + 3(\lambda_o + K_{sc})]\exp(-s_2 t)\}$$

where

$$s_{1,2} = (\lambda_p + \lambda_o + K_{sc} \mp \gamma)/2$$

$$\gamma := \left[(\lambda_p - \lambda_o + K_{sc}/2)^2 + 3K_{sc}^2/4\right]^{1/2}$$

Appendix: Free-volume Data in Polymeric Materials (R.T.)

Ying Li, Renwu Zhang and Y.C. Jean

Department of Chemistry, University of Missouri – Kansas City, Kansas City, MO 64110, USA

In PAL, we use the relationship between the o-Ps lifetime (τ_3 in ns) and free-volume radius R (in Å) [103] to calculate R:

$$\tau_3^{-1} = 2\left[1 - \frac{R}{R+1.66} + \frac{1}{2\pi}\sin\left(\frac{2\pi R}{R+1.66}\right)\right] \ (ns^{-1}),$$ and estimate free-volume fraction, f_v from o-Ps intensity (I_3 in %) and free-volume hole radius (in Å) [109] as: $f_v(\%) = 0.0018 I_3(\%) \left\langle \frac{4}{3}\pi R^3 \right\rangle$.

Polymer name	o-Ps Lifetime τ_3(ns)	Free-Volume Radius (Å)	o-Ps Intensity I_3(%)	Free-Volume Fraction (%)	Longer o-Ps Lifetime τ_4 (ns)	Longer o-Ps Intensity I_4 (%)	Ref	Comments
1,2-Polybutadiene (1,2-PB)	2.45 ± 0.01	3.23 ± 0.01	36.5 ± 0.6	9.27 ± 0.2			31	
3-nitropropionic acid	1.04 ± 0.06	1.72 ± 0.06	3.40 ± 0.6	0.13 ± 0.03			11	
Bisphenol-A (BA)	1.81 ± 0.02	2.67 ± 0.02	19.1 ± 0.6	2.73 ± 0.11			35	
Bisphenol-A (BA)	1.72 ± 0.02	2.58 ± 0.03	23.8 ± 0.3	3.07 ± 0.11			55	
Cellulose acetate	1.85 ± 0.02	2.71 ± 0.03	24.2 ± 0.9	3.62 ± 0.12			8	

cis-1,4-polybutadiene (Mw 810000)	2.61 ± 0.05	3.35 ± 0.01	39.4 ± 0.07	11.2 ± 0.23		58	45K, τ_3 = 1.1, I_3 = 24% 75K, τ_3 = 1.2, I_3 = 24% 200K, τ_3 = 1.7, I_3 = 25%
cis-1,4-polyisoprene	2.24 ± 0.01	3.06 ± 0.01	28.0 ± 0.23	6.05 ± 0.17		16	150K τ_3 = 0.72, I_3 = 34% 205K τ_3 = 0.690, I_3 = 28%
Copolymer (4-vinyl pyridine methyl methacrylate)	2.48 ± 0.01	3.25 ± 0.01	40.8 ± 0.2	10.6 ± 0.21		24	
Cresol novolac (CR)	1.90 ± 0.01	2.76 ± 0.01	18.0 ± 0.60	2.84 ± 0.12		36	
Cresol novolac (CR)	2.03 ± 0.02	2.88 ± 0.03	21.3 ± 0.20	3.82 ± 0.10		55	
Crysene	1.50 ± 0.02	2.34 ± 0.03	4.30 ± 0.20	0.410 ± 0.03		8	
Crytex (polymer sulfure)	1.14 ± 0.05	1.87 ± 0.07	7.50 ± 1.00	0.370 ± 0.09		10	77K, τ_3 = 0.7, I_3 = 7.5%
Diglycidyl ether of bisphenol A-4,4'-diamino-diphenyl-methane (DGEBA-DDM)	2.20 ± 0.01	3.02 ± 0.01	13.4 ± 0.30	2.80 ± 0.08		44	
Diglycidyl ether of bisphenol A-hexamethylene diamine (DGEBA-HMDA)	2.06 ± 0.01	2.90 ± 0.01	12.5 ± 0.3	2.31 ± 0.07		44	
Diglycidyl ether of resorcinol-hexamethylene diamine (DGERO-HMDA)	1.58 ± 0.02	2.43 ± 0.02	10.2 ± 0.3	1.10 ± 0.06		44	
Diglycidyl ether of resorcinol-4,4'-diaminodiphenyl-methane (DGERO-DDM)	1.74 ± 0.01	2.60 ± 0.01	12.4 ± 0.3	1.64 ± 0.06		44	
Epoxy resin	1.82 ± 0.01	2.68 ± 0.01	25.3 ± 0.3	3.66 ± 0.08		32	
Epoxy resin (DGEBA/DBH/ DAB)	1.4 -2.7 ± 0.02	2.22 ± 0.12	18-25 ± 0.1	1.48 ± 0.05		103	78°C - 230°C composition variation, distribution.

Epoxy (TGDDM/DGEBA /DDS)	1.4-1.8 ± 0.1	2.22 ± 0.12	14-25 ± 0.2	1.15 ± 0.19			107	Composition variation, chain extension.
Hexafluoro bisphenol-A polycarbonate (6FPC)	2.65 ± 0.01	3.38 ± 0.01	27.9 ± 0.1	8.13 ± 0.10			42	
Hexamethyldisiloxane (HMDSiO)	3.23 ± 0.05	3.78 ± 0.01	27.5 ± 0.5	11.19 ± 0.16			66	
Hydroxybenzoic acid/hydroxynapht hoic acid (75/25) (HBA/HNA)	1.15 ± 0.03	1.89 ± 0.03	8.7 ± 0.3	0.44 ± 0.04			47	100°C, $\tau_3 = 1.40, I_3 = 12\%$ 200°C, $\tau_3 = 1.75, I_3 = 21\%$ 300°C, $\tau_3 = 1.82, I_3 = 23\%$
Lucite	1.55 ±0.01	2.39 ± 0.01	27.9 ± 1.8	2.89 ± 0.22			8	
Lucite	2.5 ± 0.2	3.27 ± 0.12	21.0 ± 2.0	5.53 ± 0.23			9	
Naphthalene	0.93 ± 0.03	1.54 ± 0.05	19.1 ± 1.1	0.52 ± 0.08			8	
Nylon	1.39 ± 0.04	2.21 ± 0.05	22.2 ± 1.0	1.80 ± 0.20			8	
Nylon	1.63	2.48	18.2	2.10			14	
Nylon-6	1.71 ± 0.08	2.57 ± 0.07	15.6 ± 1.0	1.99 ± 0.30			6	
Nylon-6	1.72 ± 0.047	2.58 ± 0.04	20.7 ± 1.3	2.68 ± 0.30			22	
Para nitrophenol	1.42 ± 0.14	2.24 ± 0.16	3.0 ± 0.8	0.26 ± 0.12			11	
Paraffin	2.5 ± 1	3.27 ± 0.59					2	
Paraffin wax	0.90 ± 0.10	1.48 ± 0.18	5.5 ± 1.5	0.13 ± 0.06	3.20 ± 0.10	19.0 ± 1.0	6	Long o-Ps free volume radius is 3.76Å, fractional free volume is 7.62%.
Phenantrene	0.94 ± 0.06	1.55 ± 0.11	8.5 ± 0.2	0.24 ± 0.05			8	
p-Hydroxyenzoicacid/ poly(ethylene terephthalate) (HBA/PET)	1.45 ± 0.01	2.28 ± 0.01	4.00 ± 0.3	0.36 ± 0.03			38	100°C, $\tau_3 = 1.70, I_3 = 7.0\%$ 150°C, $\tau_3 = 1.90, I_3 = 14.5\%$ 200°C, $\tau_3 = 2.04, I_3 = 22.5\%$
Plastic Scintilator	1.98	2.83	23.0	3.93			5	For amorphous materials

Material							Ref	Notes
Naton	± 0.04	± 0.03	± 1	± 0.30				
Plexiglas (PMMA)	1.40	2.22					1	
Plexiglass (PMMA)	2.50 ± 1	3.27 ± 0.01					2	
Poly styrene-butadiene-styrene (Casting solvent: Methyl Ethyl ketone)	2.020	2.87					19	Mw = 124000
Poly styrene-butadiene-styrene (Casting solvent: Carbon tetrachloride)	2.381	3.17					19	Mw = 124000
Poly styrene-butadiene-styrene (Casting solvent: Ethyl acetate)	1.835	2.69					19	Mw = 124000
Poly styrene-butadiene-styrene (Casting solvent: toluene)	2.248	3.07					19	Mw = 124000
Poly((ethylene glycol)n dimethacrylate)	1.83 ± 0.03	2.69 ± 0.02	13.7 ± 0.3	2.01 ± 0.01			80	
Poly(1,4-cyclohexylenedimethylene terephthalate)	1.77 ± 0.02	2.63 ± 0.03	27.8 ± 0.2	3.80 ± 0.10			84	
Poly(2,6-dimethyl-1,4-phenylene oxide) (PPO)	2.68 ± 0.03	3.40 ± 0.02	33.0 ± 0.3	9.81 ± 0.23			81	343K, τ_3 = 2.72, I_3 = 34% 403K, τ_3 = 2.82, I_3 = 35%. 423K, τ_3 = 2.86, I_3 = 36%
Poly(2,6-dimethylphenylene oxide) (PMPO)	3.12 ± 0.06	3.71 ± 0.03	18.0 ± 0.6	6.92 ± 0.38			59	
Poly(2,6-diphenylphenylene oxide) (PPPO)	2.36 ± 0.03	3.16 ± 0.02	26.0 ± 1	6.17 ± 0.35			59	
Poly(2-methyl-6-allyl-1,4-phenylene oxide - co-2,6-dimethyl-1,4-phenylene-oxide)	2.68 ± 0.04	3.40 ± 0.02	18.0 ± 0.5	5.35 ± 0.25			59	

Appendix 377

Poly(3-methylthiophene)(doped)(P3MTd)	2.25 ± 0.7	3.07 ± 0.47	5.00 ± 0.3	1.09 ± 0.56		25	
Poly(3-methylthiophene)(neutral)(P3MTd)	1.55 ± 0.06	2.39 ± 0.06	0.50 ± 0.2	0.05 ± 0.02		25	
Poly(4-methy-1-pentene) (PMP)	2.66 ± 0.02	3.39 ± 0.01	27.0 ± 0.3	7.92 ± 0.16		63	
Poly(4-methylpentene-1-co-☐-olefine) (P4MP-C)	2.56 ± 0.01	3.31 ± 0.01	19.2 ± 0.6	5.27 ± 0.19		31	
Poly(aryl-ether-ether-ketone) (PEEK)	1.73 ± 0.02	2.59 ± 0.02	21.5 ± 0.4	2.81 ± 0.11		45	
Poly(aryl-ether-ether-ketone) (PEEK)	1.79 ± 0.02	2.65 ± 0.02	23.5 ± 0.2	3.29 ± 0.09		103	Crystalinity 0-30%
Poly(aryl-ether-ether-ketone) (PEEK)	1.78 ± 0.02	2.64 ± 0.02	18.0 ± 0.2	2.49 ± 0.08		106	Stretched, ellipsoidal model
Poly(benzoxazole) (PBO)	3.24 ± 0.02	3.79 ± 0.01	22.0 ± 0.3	9.00 ± 0.18		81	
Poly(butadiene) (Mw 20000)	2.72 ± 0.02	3.43 ± 0.01	36.2 ± 0.5	11.04 ± 0.25		69	100K, $\tau_3 = 1.28$, $I_3 = 23\%$ 200K, $\tau_3 = 1.72$, $I_3 = 27\%$ 350K, $\tau_3 = 2.72$, $I_3 = 36\%$
Poly(butyl methacrylate) (PMBA)	2.35 ± 0.01	3.15 ± 0.01	27.0 ± 0.6	6.36 ± 0.18		31	
Poly(Chlorotrifluro ethylene) (Mw 250000)	1.74 ± 0.02	2.60 ± 0.02	4.62 ± 0.2	0.61 ± 0.04		54	
Poly(di-n-hexylsilane) (PDHS)	2.55 ± 0.02	3.31 ± 0.01	19.0 ± 0.5	5.18 ± 0.19		76	
Poly(ether ure-thane)-LiClO4	1.89 ± 0.02	2.75 ± 0.02	11.4 ± 0.5	1.78 ± 0.11		34	100K, $\tau_3 = 1.12$, Intensity = 11%
Poly(ethyl methacrylate) (PEMA)	2.20 ± 0.02	3.02 ± 0.01	32.9 ± 0.3	6.87 ± 0.16		21	
Poly(ethylene terephthalate)	1.75 ± 0.03	2.61 ± 0.03	12.0 ± 0.3	1.60 ± 0.09		27	Crystalinity : 50% $\tau_3 = 1.80$, $I_3 = 9\%$. 20% $\tau_3 = 1.76$, $I_3 = 10\%$
Poly(ethylene	1.64 ±	2.49	21.2	2.48		84	

terephthalate)	0.02	± 0.02	± 0.5	± 0.12				
Poly(ethylene terephthalate) (PET)	1.60 ± 0.01	2.45 ± 0.01	12.6 ± 0.3	1.40 ± 0.05		31		
Poly(ethylene terephthalate) (PET) (crystallinity 0%, Mw 50000)	1.77 ± 0.02	2.63 ± 0.02	12.2 ± 0.2	1.67 ± 0.06		85		Crystallinity 20%, τ_3 =1.77, I_3=10%. Crystallinity 30%, τ_3 = 1.73, I_3 = 10%. Crystallinity 55%, τ_3 = 1.77, I_3 = 9%.
Poly(isobutyl methacrylate) (PIBMA)	2.16 ± 0.02	2.99 ± 0.01	24.9 ± 0.3	5.02 ± 0.13		21		
Poly(methyl acrylate) (PMA)	1.80 ± 0.01	2.66 ± 0.01	30.5 ± 0.6	4.32 ± 0.13		31		
Poly(methyl methacrylate)	1.93 ± 0.01	2.78 ± 0.01	24.6 ± 0.2	4.00 ± 0.07		94		
Poly(methyl methacrylate) (PMMA)	1.96 ± 0.02	2.81 ± 0.02	32.1 ± 0.5	5.38 ± 0.17		21		
Poly(methyl methacrylate) (PMMA) (Mw 280000)	2.06 ± 0.03	2.90 ± 0.02	11.8 ± 0.5	2.18 ± 0.14		50		
Poly(methyl methacrylate) (PMMA) (Mw 120000)	1.86 ± 0.02	2.72 ± 0.02	33.4 ± 0.5	5.05 ± 0.17		75		
Poly(methyl methacrylate) (PMMA) (Mw 38000)	1.85 ± 0.02	2.71 ± 0.02	12.3 ± 0.5	1.84 ± 0.11		77		Mw 68000 τ_3 = 1.96, I_3 = 16% Mw 110000 τ_3 = 1.95, I_3 = 16%
Poly(methyl methacrylate) (PMMA) (Mw 75000)	1.89 ± 0.02	2.75 ± 0.02	28.3 ± 0.5	4.42 ± 0.16		72		
Poly(methyl methacrylate) (PMMA) (Mw 90000)	1.91 ± 0.02	2.76 ± 0.02	28.5 ± 0.3	4.54 ± 0.13		64		
Poly(methyl-n-propylsilane) (PMPrS)	2.45 ± 0.02	3.23 ± 0.01	19.3 ± 0.5	4.90 ± 0.18		76		
Poly(n-butyl methacrylate) (PNBMA)	2.23 ± 0.02	3.05 ± 0.01	33.9 ± 0.4	7.25 ± 0.18		21		

Poly(phenylene oxide)(PPO)	1.78 ± 0.14	2.64 ± 0.12	15.38 ± 2.65	2.13 ± 0.66	3.06 ± 0.13	15.8 ± 2.97	60	
								Long o-Ps free volume radius is 3.67Å, fractional free volume is 6.0%.
Poly(1-trimethylsilyl-1-propyne) (PTMSP)	1.52 ± 0.01	2.36 ± 0.01	5.58 ± 0.05	0.55 ± 0.01	5.58 ± 0.02	5.63 ± 0.5	60	
								Long o-Ps free volume radius is 5.00Å, fractional free volume is 5.1%.
Poly(1-trimethylsilyl-1-propyne) (PTMSP)	1.87 ± 0.29	2.73 ± 0.24	5.4 ± 0.5	0.83 ± 0.29	5.88 ± 0.06	33.1 ± 0.7	79	
								Long o-Ps free volume radius is 5.13Å, fractional free volume is 34%. In N_2, τ_3 = 2.68, I_3 = 4.4% τ_4= 10.9, I_3 = 34%
Poly(1-trimethylsilyl-1-propyne) (PTMSP)	2.53 ± 0.02	3.29 ± 0.01	11.3 ± 0.1	3.04 ± 0.06	12.54 ± 0.04	20.2 ± 0.1	74	
								Long o-Ps free volume radius is 7.21Å, fractional free volume is 57%.
Poly(vinyl acetate)	2.78 ± 0.02	3.48 ± 0.01	27.6 ± 0.4	8.74 ± 0.20			82	
poly(vinyl acetate) (PVAC)	1.89 ± 0.01	2.75 ± 0.01	23 ± 0.6	3.59 ± 0.13			31	
Poly(vinyl acetate) (PVAC) (Mn 260000)	1.92 ± 0.02	2.77 ± 0.02	27.8 ± 0.3	4.48 ± 0.13			64	
Poly(vinyl alcohol) (PVA)	1.40 ± 0.01	2.22 ± 0.01	9.1 ± 0.3	0.75 ± 0.04			39	50°C, τ_3 = 1.50, I_3= 9% 90°C, τ_3 = 1.48, I_3 = 9% 150°C, τ_3 = 1.73, I I_3= 8%
Poly(vinyl alcohol) (PVA)	1.59 ± 0.03	2.44 ± 0.03	17.0 ± 0.3	1.86 ± 0.10			57	

380 *Principles and Applications of Positron and Positronium Chemistry*

Polymer								Notes
Poly(vinyl chlorate)	1.73 ± 0.01	2.59 ± 0.01	5.4 ± 0.3	0.71 ± 0.07			40	
Poly(vinyl chlorate)	2.13 ± 0.02	2.96 ± 0.01	10.2 ± 0.5	2.00 ± 0.07			86	50°C, $\tau_3 = 2.22$, $I_3 = 9\%$ 100°C, $\tau_3 = 2.24$, $I_3 = 10\%$, 140°C, $\tau_3 = 2.63$, $I_3 = 9\%$.
Poly[(2-dimethylamino)-ethyl ethacrylate]-l-polyisobutylene	2.54 ± 0.02	3.30 ± 0.01	19.9 ± 0.2	5.39 ± 0.11			61	
poly[(ethylene glycol) 23-dimethacrylate]	2.43 ± 0.02	3.21 ± 0.01	25.5 ± 0.2	6.38 ± 0.12			98	
Poly[1-phenyl-2-[p-(triisopropylsilyl)phenyl] acetylene] (PPrSiDPA)	2.65 ± 0.05	3.38 ± 0.03	16.7 ± 0.3	4.86 ± 0.21	7.21 ± 0.06	25.8 ± 0.4	79	Long o-Ps free volume radius is 5.65Å, fractional free volume is 35%. In N_2, $\tau_3 = 3.30$, $I_3 = 16\%$ $\tau_4 = 9.38$, $I_3 = 28\%$
Poly[1-phenyl-2-[p-(triphenylsilyl)phenyl] acetylene] (PPhSiDPA)	2.01 ± 0.08	2.86 ± 0.06	21.1 ± 2.2	3.71 ± 0.62	3.12 ± 0.01	15.6 ± 2.5	79	Long o-Ps free volume radius is 3.71Å, fractional free volume is 6.0%.
Polybithiophene (doped) (PBTd)	1.10 ± 0.3	1.81 ± 0.45	1.00 ± 0.3	0.05 ± 0.01			25	
Polybithio-phene (neutral) (PBTn)	1.45 ± 0.15	2.28 ± 0.17	1.30 ± 0.6	0.12 ± 0.08			25	
Bisphenol A – Polycarbonate	2.04 ± 0.03	2.88 ± 0.02	31.2 ± 0.6	5.65 ± 0.24			78	350K, $\tau_3 = 2.17$, $I_3 = 38\%$ 400K, $\tau_3 = 2.23$, $I_3 = 40\%$. 450K, $\tau_3 = 2.58$, $I_3 = 39\%$
Bisphenol A - Polycarbonate (BPA-PC)	2.13 ± 0.05	2.96 ± 0.04	34.5 ± 0.5	6.78 ± 0.34			65	
Bisphenol-A polycarbonate	2.10 ± 0.02	2.94 ± 0.01	28.0 ± 0.6	5.36 ± 0.19			28	
Bisphenol-A poly-	2.02	2.87	33.0	5.86			29	

carbonate	± 0.03	± 0.01	± 0.6	± 0.25			
Bisphenol-A poly-carbonate	2.10 ± 0.01	2.94 ± 0.02	32.7 ± 0.2	6.25 ± 0.08		42	
Polycarbonate	2.00 ± 0.02	2.85 ± 0.02	40.0 ± 0.5	6.97 ± 0.20		20	
Polycarbonate	2.14 ± 0.01	2.97 ± 0.01	29.0 ± 1.0	5.75 ± 0.24		30	75K $\tau_3 = 1.81$, $I_3 = 22\%$.
Polycarbonate	2.04 ± 0.05	2.88 ± 0.04	21.0 ± 0.4	3.80 ± 0.22		33	
Polycarbonate	2.17 ± 0.01	3.00 ± 0.01	31.5 ± 0.3	6.41 ± 0.11		43	
Polycarbonate	2.11 ± 0.01	2.95 ± 0.01	29.2 ± 0.1	5.64 ± 0.06		92	
Polycarbonate	2.12 ± 0.02	2.96 ± 0.01	27.3 ± 0.2	5.32 ± 0.12		87	
Polycarbonate (Mw 58000)	2.08 ± 0.03	2.92 ± 0.01	14.3 ± 0.5	2.69 ± 0.15		51	
Polycarbonate (Mw 105000)	1.75 ± 0.01	2.61 ± 0.01	15.0 ± 0.2	2.01 ± 0.05		101	50-450K
Polycarbonate (PC 1151)	2.11 ± 0.01	2.95 ± 0.01	32.4 ± 0.3	6.25 ± 0.10		63	
Polyethylene (LDPE)	2.44 ± 0.01	3.22 ± 0.01	27.0 ± 0.2	6.80 ± 0.09		63	
Polyethersulphone	2.19 ± 0.02	3.01 ± 0.01	32.0 ± 0.5	6.59 ± 0.19		20	
Polyethylene	2.40 ± 0.3	3.19 ± 0.01	29.0 ± 5.0	7.09 ± 2.45		3	
Polyethylene	2.04 ± 0.06	2.88 ± 0.01	29.5 ± 0.4	5.34 ± 0.32		8	
Polyethylene	2.20 ± 0.2	3.02 ± 0.01	22.0 ± 2.0	4.59 ± 1.04		9	
Polyethylene	1.22	1.99	10.8	0.64	2.67	16.9	13	Long o-Ps free volume radius is 3.4Å, fractional free volume is 5.0%.
Polyethylene	1.03	1.70	9.40	0.35	2.61	21	14	Long o-Ps free volume radius is 3.35Å, fractional free volume is 6.0%.
Polyethylene	2.62 ± 0.02	3.36 ± 0.01	27.1 ± 0.4	7.74 ± 0.19		82	

Polyethylene	2.59 ± 0.02	3.34 ± 0.01	21.9 ± 0.5	6.13 ± 0.20			91	
Polyethylene (branched)	2.48 ± 0.02	3.25 ± 0.01	28.1 ± 0.5	7.29 ± 0.21			68	100K, $\tau_3 = 1.4$, $I_3 = 33\%$. 200K, $\tau_3 = 1.8$, $I_3 = 23\%$ 350K, $\tau_3 = 2.85$, $I_3 = 34\%$
Polyethylene (HDPE)	2.51 ± 0.02	3.28 ± 0.01	20.1 ± 0.5	5.33 ± 0.19			70	100K, $\tau_3 = 2.6$, $I_3 = 11\%$ 200K, $\tau_3 = 2.4$, $I_3 = 14\%$ 250K, $\tau_3 = 2.0$, $I_3 = 14\%$
Polyethylene (HDPE) (Mw 4000000)	2.62 ± 0.03	3.36 ± 0.02	19.4 ± 0.6	5.54 ± 0.25			49	
Polyethylene (high density)	2.52 ± 0.03	3.28 ± 0.02	21.2 ± 0.3	5.66 ± 0.17			26	
Polyethylene (LDPE)	2.78 ± 0.02	3.48 ± 0.01	26.2 ± 0.1	8.30 ± 0.11			90	
Polyethylene (LDPE) (Mw 40000)	2.55 ± 0.03	3.31 ± 0.02	21.1 ± 0.3	5.75 ± 0.17			49	
Polyethylene (linear)	2.28 ± 0.02	3.09 ± 0.01	21.2 ± 0.5	4.73 ± 0.17			68	100K, $\tau_3 = 1.3$, $I_3 = 26\%$ 200K, $\tau_3 = 1.8$, $I_3 = 23\%$ 350K, $\tau_3 = 2.5$, $I_3 = 24\%$
Polyethylene (low density)	2.66 ± 0.03	3.39 ± 0.02	26.0 ± 0.3	7.63 ± 0.20			26	
Polyethylene (Mn 20000)	1.30 ± 0.02	2.09 ± 0.03	37.8 ± 0.7	2.62 ± 0.14			48	
Polyethylene (Mw 100000)	1.20 ± 0.01	1.96 ± 0.01	6.20 ± 0.1	0.35 ± 0.01	2.92 ± 0.02	28.7 ± 0.3	93	Long o-Ps free volume radius is 3.57Å, fractional free volume is 9.9%.
Polyethylene (PE)	2.25 ± 0.01	3.07 ± 0.01	25.6 ± 0.6	5.57 ± 0.17			31	
Polyethylene oxide/NH4ClO4 (80:20) (Mw 600000)	1.66 ± 0.02	2.51 ± 0.02	11.7 ± 0.5	1.40 ± 0.09			53	

Polyethylene terephthalate (Mw 18000)	1.91 ± 0.02	2.76 ± 0.02	15.8 ± 0.2	2.52 ± 0.08			97	340K, $\tau_3 = 2.0, I_3 = 20\%$ 400K, $\tau_3 = 2.3, I_3 = 22\%$.
Polyethylene (HDPE)	2.34 ± 0.02	3.14 ± 0.01	20.4 ± 0.2	4.77 ± 0.10			63	
Polyethylene (LDPE)	2.72 ± 0.01	3.43 ± 0.01	26 ± 0.5	7.93 ± 0.19			35	
Polyethylene (low-density)	2.57 ± 0.02	3.32 ± 0.01	28.2 ± 0.2	7.79 ± 0.13			62	
Polyhydroxamic acid	2.20 ± 0.01	3.02 ± 0.01	15.8 ± 0.3	3.30 ± 0.08			95	
Polyisobutylene	2.18	3.01	43.0	8.82			15	Quenched from RT. to -196°C with no annealing (at -60°c) above Tg -100°C, $\tau_3 = 1.3, I_3 = 34\%$ -90°C, $\tau_3 = 1.3, I_3 = 32\%$. -80°C, $\tau_3 = 1.4, I_3 = 31\%$ -70°C, $\tau_3 = 1.4, I_3 = 32\%$. -65°C, $\tau_3 = 1.5, I_3 = 32\%$ -60°C, $\tau_3 = 1.5, I_3 = 31\%$. -50°C, $\tau_3 = 1.6, I_3 = 32\%$. -40°C, $\tau_3 = 1.7, I_3 = 35\%$ -20°C, $\tau_3 = 1.8, I_3 = 39\%$. 0°C, $\tau_3 = 2.0, I_3 = 42\%$
Polyoxymethylene	2.55 ± 0.01	3.31 ± 0.01	23.0 ± 1.0	6.27 ± 0.30			30	75K $\tau_3 = 1.3, I_3 = 20\%$
Polypropylene	0.84 ± 0.10	1.37 ± 0.20	0.08 ± 0.07	0.00 ± 0.00	2.50 ± 0.10	21.4 ± 1.0	6	Long o-Ps free volume radius is 3.27Å, fractional free volume is 5.6%.
Polypropylene	2.36 ± 0.03	3.16 ± 0.02	5.13 ± 0.16	1.22 ± 0.06			26	
Polypropylene No. 6056	2.20 ± 0.10	3.02 ± 0.07	6.26 ± 0.84	1.31 ± 0.26			11	

Polypropylene No. 6056 (treated with HNO3)	1.05 ± 0.08	1.74 ± 0.13	0.16 ± 0.07	0.01 ± 0.00		11	
Polypropylene (K1011)	2.21 ± 0.02	3.03 ± 0.01	5.01 ± 0.13	1.05 ± 0.04		63	
Polystyrene	2.50 ± 1	3.27 ± 0.59	100	26.31		2	
Polystyrene	2.30	3.11	36.0	8.15		3	-196°C, $\tau_3 = 1.7$.
Polystyrene	2.20 ± 0.2	3.02 ± 0.14	100	20.87		7	
Polystyrene	1.60 ± 0.01	2.45 ± 0.01	45.2 ± 0.1	5.01 ± 0.07		8	
Polystyrene	2.07	2.91	33.2	6.18		14	
Polystyrene (Mw 280000)	2.10 ± 0.02	2.94 ± 0.01	34.0 ± 0.2	6.50 ± 0.13		99	
Polystyrene	1.86 ± 0.01	2.72 ± 0.01				18	Mw= 104, $\tau_3 = 2.4$. Mw= 600, $\tau_3 = 2.2$. Mw= 970, $\tau_3 = 2.0$. Mw= 5000, $\tau_3 = 1.9$ Mw = $5.0 * 10^4$, $\tau_3 = 1.9$. Mw = $2.145 * 10^5$, $\tau_3 = 1.9$.
Polystyrene	2.06 ± 0.02	2.90 ± 0.01	42.9 ± 0.5	7.91 ± 0.21		68	
Polystyrene (Mn 5700)	1.65 ± 0.03	2.50 ± 0.03	38.5 ± 0.8	4.56 ± 0.25		48	
Polystyrene (Mn 5730)	2.21 ± 0.02	3.03 ± 0.01	38.2 ± 0.5	8.04 ± 0.21		52	
Polystyrene (Mn 85000)	2.03 ± 0.01	2.88 ± 0.01	41.3 ± 0.3	7.41 ± 0.11		63	
Polystyrene (Mw 230000)	2.05 ± 0.01	2.89 ± 0.01	18.2 ± 0.1	3.33 ± 0.04		96	Mw 105000, $\tau_3 = 2.0$, $I_3 = 19\%$. Mw 80000, $\tau_3 = 2.0$, $I_3 = 15\%$ Mw 70000, $\tau_3 = 2.0$, $I_3 = 13\%$
Polystyrene (Mw 600000)	2.03 ± 0.03	2.88 ± 0.02	16.2 ± 0.5	2.91 ± 0.16		50	
Polystyrene (PS)	1.90 ± 0.01	2.76 ± 0.01	34.5 ± 0.6	5.44 ± 0.14		31	
Polystyrene (PS)	2.03 ± 0.03	2.88 ± 0.02	35.0 ± 0.6	6.28 ± 0.26		78	

Material								Notes
Polystyrene (PS) Mw=400K	2.05 ±0.01	2.89 ± 0.01	35.8 ±0.16	6.54 ± 0.09			37	10°C, $\tau_3 = 2.0$, $I_3 = 35\%$, 100°C, $\tau_3 = 2.2$, $I_3 = 36\%$, 130°C, $\tau_3 = 2.5$, $I_3 = 37\%$.
Polysulphone	2.18 ± 0.02	3.01 ± 0.01	32.0 ± 0.5	6.57 ± 0.19			20	
Polysulphone	2.00 ± 0.1	2.85 ± 0.08	17.5 ± 0.1	3.05 ± 0.26			108	Temperature dependence 20-240°C. CO_2 exposed, live relaxation
Polyurethane	2.42 ± 0.02	3.21 ± 0.01	19.4 ± 0.4	4.82 ± 0.15			83	100K, $\tau_3 = 1.3$, $I_3 = 17\%$. 200K, $\tau_3 = 1.4$, $I_3 = 19\%$ 250K, $\tau_3 = 1.7$, $I_3 = 20\%$
Polyurethane	1.70 ± 0.1	2.56 ± 0.09	27.8 ± 0.1	3.50 ± 0.39			89	
Polyurethane	2.00 ± 0.02	2.85 ± 0.02	26.5 ± 0.2	4.62 ± 0.11			100	Depth profile.
Polyvinyl alcohol	1.78 ± 0.01	2.64 ± 0.01	21.4 ± 0.1	2.96 ± 0.04			88	
Polyvinyl chloride	1.25 ± 0.03	2.03 ± 0.04	9.50 ± 0.7	0.60 ± 0.08			8	
Polyvinyl chloride	1.9 ± 0.2	2.76 ± 0.16	26.0 ± 2.0	4.10 ± 1.04			9	
Polyvinyl Chloride	0.64 ± 0.01	0.85 ± 0.02	26.7 ± 0.2	0.13 ± 0.01			17	Bulk-polymerization
Polyvinyl Chloride	0.69 ± 0.01	1.00 ± 0.02	26.0 ± 0.2	0.20 ± 0.02			17	Suspension-polymerization
Polyvinyl chloride	1.74 ± 0.01	2.60 ± 0.01	6.30 ± 0.1	0.83 ± 0.02			56	
Polyvinyl chloride	1.88	2.74	4.70	0.73 ± 0.32			14	
Polyvinylidene-fluoride (PVDF)	2.12 ± 0.02	2.96 ± 0.01	14.2 ± 0.5	2.77 ± 0.14			71	200K, $\tau_3 = 1.6$, $I_3 = 14\%$
Polyvinyl-toluene	1.78 ± 0.05	2.64 ± 0.04	35.8 ± 0.5	4.96 ± 0.32			8	
Pyrene	1.25 ± 0.05	2.03 ± 0.07	1.90 ± 0.1	0.12 ± 0.02			8	
Sulphur (ortho-hombic)	1.03 ± 0.05	1.70 ± 0.08	9.00 ± 1.0	0.34 ± 0.08			10	77K, $\tau_3 = 0.7$, $I_3 = 9.0\%$
Polytetrafluoro-ethylene (PTFE)	1.75 ± 0.02	2.61 ± 0.02	10.1 ± 0.5	1.35 ± 0.09	4.12 ± 0.02	16.5 ± 0.5	73	Long o-Ps free volume radius is

Teflon	1.20 ± 0.5	1.96 ± 0.68	100	5.67 ± 6.06			2	4.30Å, fractional free volume is 9.9%.
Teflon	2.25 ± 0.03	3.07 ± 0.02	19.0 ± 2	4.13 ± 0.52			4	30°C, τ_3 = 3.1. -78°C, τ_3 = 1.4. -196°C, τ_3 =1.3.
Teflon	3.22 ± 0.07	3.77 ± 0.03	21.6 ± 0.7	8.75 ± 0.49			8	
Teflon	3.50 ± 0.2	3.95 ± 0.08	17.0 ± 2.0	7.88 ± 1.39			9	
Teflon	2.90	3.56	17.0	5.78			9	Normal (D = 2.190 g/ml)
Teflon	2.90 ± 0.1	3.56 ± 0.05	17.0	5.78 ± 0.92			10	77K, τ_3 = 1.5, I_3 = 13%
Teflon	3.54 ± 0.03	3.97 ± 0.01	19.7 ± 0.3	9.30 ± 0.22			12	
Teflon	1.70 ± 0.1	2.56 ± 0.09	8.4 ± 0.9	1.06 ± 0.23	4.11 ± 0.07	14.2 ± 0.9	12	Long o-Ps free volume radius is 4.3Å, fractional free volume is 8.5%.
Teflon	1.23	2.00	17.50	1.06	4.33	14.4	13	Long o-Ps free volume radius is 4.41Å, fractional free volume is 9.3%.
Teflon	1.22	1.99	12.10	0.72	4.14	17.3	14	Long o-Ps free volume radius is 4.31Å, fractional free volume is 10.4%.
Teflon (Annealed)	2.20	3.02	14.0	2.92			9	Annealed (D = 2.205 g/ml)
Teflon (Annealed)	2.20 ± 0.1	3.02 ± 0.07	13.0 ± 2.0	2.71 ± 0.60			10	77K, τ_3 = 1.2, I_3 = 12%
Teflon (Quenched)	3.3	3.82	17.0	7.17			9	Quenched (D = 2.150 g/ml)
Teflon (Quenched)	3.3 ± 0.1	3.82 ± 0.04	17.0 ± 2.0	7.17 ± 1.08			10	77K, τ_3 = 1.8, I_3 = 12%.

Tetramethyl Bisphenol-A polycarbonate (TMPC)	2.32 ± 0.6	3.12 ± 0.38	31.0 ± 0.6	7.13 ± 2.77		29	
Tetramethyl Bisphenol-A polycarbonate (TMPC)	2.41 ± 0.01	3.20 ± 0.01	30.7 ± 0.2	7.57 ± 0.09		42	
Tetramethyl Bisphenol-A polycarbonate (TMPC) Mw 67000	2.43 ± 0.01	3.21 ± 0.01	23.5 ± 0.2	5.88 ± 0.08		105	Blends with PC
Tetramethylbiphenol (TMB)	1.88 ± 0.02	2.74 ± 0.02	21.5 ± 0.6	3.32 ± 0.15		36	
Tetramethylbiphenol (TMB)	1.78 ± 0.02	2.64 ± 0.02	22.9 ± 0.3	3.17 ± 0.10		55	
Tetramethylbisphenol-A polycarbonate (Mw 43000)	2.37 ± 0.02	3.17 ± 0.01	24.5 ± 1	5.86 ± 0.31		51	50K, $\tau_3 = 2.2, I_3 = 20\%$
Tetramethylhexafluorobisphenol-A polycarbonate (TM6FPC)	3.10 ± 0.01	3.70 ± 0.00	29.0 ± 0.1	11.03 ± 0.08		42	Blends PC (TMPC) FPC
trans-1,4-polyisoprene	2.33	3.13	32.0	7.42		16	150K $\tau_3 = 1.4, I_3 = 28\%$, 205K $\tau_3 = 1.5, I_3 = 25\%$
Trihydroxyphenylmethane (THPM)	1.73 ± 0.02	2.59 ± 0.02	19.1 ± 0.2	2.49 ± 0.08		55	
Trihydroxypoly(ethylene oxide copropylene oxide)	2.64 ± 0.03	3.37 ± 0.02	23.3 ± 0.5	6.75 ± 0.24		41	
Trihydroxypoly(ethylene oxide copropylene oxide) (3:1) (Mw 5000)	2.63 ± 0.03	3.37 ± 0.02	23.1 ± 0.3	6.65 ± 0.18		46	
Trimethyl cyclohexane polycarbonate (TMC-PC)	2.69 ± 0.05	3.41 ± 0.03	34.3 ± 0.5	10.26 ± 0.39		65	
Tris-hydroxyphenylmethane (THPM)	2.12 ± 0.01	2.96 ± 0.01	17.0 ± 0.6	3.31 ± 0.14		36	
XAD-11 (amide)	3.18 ± 0.04	3.75 ± 0.02	12.6 ± 0.6	5.00 ± 0.31		23	
XAD-2 (styrene-divinylbenzene)	3.78 ± 0.04	4.11 ± 0.01	11.4 ± 0.6	5.97 ± 0.38		23	
XAD-4 (styrene-divinylbenzene)	4.62 ± 0.04	4.56 ± 0.01	6.60 ± 0.5	4.71 ± 0.39		23	
XAD-7 (acrylic es-	2.71	3.43	11.1	3.36		23	

ter)	± 0.04	± 0.02	± 0.6	± 0.24			
XAD-8 (acrylic ester)	3.94 ± 0.04	4.20 ± 0.01	11.3 ± 0.6	6.32 ± 0.40		23	

References

1. Millett, W.E. (1951) "A preliminary measurement of the decay of positron in Plexiglas". *Phys. Rev.* **82**, 336.
2. Depenedetti, S., Richings, H.J. (1952) "The half-life of positrons in condensed materials". *Phys. Rev.* **85**, 377.
3. Bell, R. E., Graham, R. L. (1953) " Time Distribution of Positron Annihilation in Liquids". *Phys. Rev.* **90**, 644.
4. Fabri, G., Germaqnoli, E., Raudone, G. (1963) "Positronium decay in Teflon. Influence of lattice transition". *Phys. Rev.* **130**, 204.
5. Hsu, F.H., Wu, C.S. (1967) "Correlation between decay lifetime and angular distribution of positron annihilation in the plastic scintillator naton". *Phys. Rev. Lett.* **18**, 889.
6. Tao, S. J., Green, J. J. (1965) " A third positron lifetime in solid polymers". *Phys. Soc.* **85**, 463.
7. Thosar, B. V., Kulkarni, V. G., Lagu, R. G., Chandra, G. (1965) "The study of lifetimes of positronium in polyethylene irradiated by gamma rays". *Phys. Lett.* **19**, 201.
8. Bertolaccini, M., Bussolati, C., Zappa, L. (1965) "Three-quantum annihilation of positrons in metals and insulators". *Phys. Rev.* **193**, 697.
9. Chandra, G., Kulkarni, V. G., Lagu, R.G., Thosar, B. V. (1968) "Quenching of triplet positronium in condensed media". *Phys. Lett.* **25**, 368.
10. Kulkarni, V. G.,Lagu, T. G., Chandra, G., Thosar, B. V. (1969) "The lifeimtes of positrons in oxides of arsenic and antimony" Proc. *Indian Acad.Sci. A.* **70**, 107.
11. Ogata, A., Tao, S. J. (1970) "*ortho*-Positronium annihilation in nitric acid treated polypropylene". *J. Appl. Phys.* **41**, 4261.
12. Jain, P. C., Bhatnagar, S., Gupta, A. (1970) "Positron annihilation in some high polymers". *J. Phys. C.* **5**, 2156.
13. Kerr, D. P. (1974) "Positron Annihilation in Teflon and Polyethylene". *Can. J. Phys.* **52**, 935.
14. Bertolaccini, M., Bisi, A., Gambarini, G., Zappa, L. (1974) "Relaxed positronium in polymers". *J. Phys. C. Solid State Phys.* **7**, 3827.
15. Stevens, J.R., Rows, R.M. (1973) "Evidence for structure in polyisobutylene from positron lifetimes". *J. Appl. Phys.* **44**, 4328.

16. Hsu, F. H., Tseng, P. K., Chuang, S. Y., Chong, Y. L. (1979) "Temperature dependence of positron lifetime in trans- and cis-1,4-Polyisoprene". *Proc. 5th Int. Conf. Positron Annihilation.* 581.
17. Hsu, F. H., Tseng, P. K., Chuang, S. Y., Chong, Y. L. (1978) "Temperature and phase dependence of positron lifetimes in polyvinyl chloride and polyisoprene". *Chin. J. Phys.* **16**(4), 196.
18. West, D. H.D., MxBrierty, V.J.M., Delaney, C.F.G. (1975) "Positron decay in polymers: molecular weight dependence in polysterene." *Appl. Phys.* **7**, 171.
19. Djermouni, B., Ache, H.J. (1980) "Effect of casting solvent on the properties of styrene-butadiene-styrene block copolymers studied by positron annihilation techniques". *Macromolecules,* **13**, 168.
20. Malhotra, B.D., Pethrick, R.A. (1983) "Positronium annihilation studies of polycarbonate, polyethersulphone and polydulphone". *Eur. Polym. J.* **19**(6) 457.
21. Malhotra, B.D., Pethrick, R.A. (1983) "Positron annihilation studies of the glass-rubber transition in Poly(alkyl methacrylate). *Macromolecules* **16**, 1175.
22. Singh, J.J., Clair, T.L. (1984) "Moisture dependence of positron annihilation spectra in Nylon-6".*Nucl. Instr. and Meth. in Phys. Res.* **221**, 427.
23. Venkateswaran, K., Cheng, K.L., Jean, Y.C. (1984) "Application of positron annihilation to study the surface properties of porous resins". *J. Phys. Chem.* **88**, 2465.
24. Malhotra, B.D., Pethrick, R.A (1985) "Positron annihilation in 4-vinyl pyridine methyl methacrylate copolymer". *Poly. Comm.* **26**, 14.
25. Nicolau, Y.F., Moser, P. (1993) "Study of free volume and crystallinity in polybithiophene and poly(3-methylthiophene)". *J. Poly. Sci. B: Polymer Physics.* **31**, 1529.
26. Suzuki, T., Miura, T., Oki, Y., Numajiri, M., Kondo, K., Ito, Y. (1995) "Positron irradiation effects on polypropylene and polyethylene studied by positron annihilation". *Radiat. Phys. Chem.* **45**(4), 657.
27. Xie, L., Gidley, D.W., Hristov, H.A., Yee, A.F. (1994) "Positronium formation in semicrystalline poly(ethylene terephthalate)". *Polymer.* **35**(1), 14.
28. Kluin, J.E., Vleeshouwers, S., McGervey, J.D., Jamieson, A.M., Simha, R. (1992) "Temperature and time dependence of free volume in bisphenol-A polycarbonate studied by positron lifetime spectroscopy". *Macromolecules.* **25**, 5089.
29. Kluin, J.E., Yu, Z., Vleeshouwers, S., McGervey, J.D., Jamieson, A.M., Simha, R., Sommer, K. (1993) "Ortho-positronium lifetime studies of free volume in polycarbonates of different structures: Influence of hole size distributions". *Macromolecules.* **26**, 1853.

30. Kristiak, J., Kristiakova, K., Sausa, O., Bandzuch, P., Bartos, J. (1993) "Temperature dependence of free volume distributions in polymers studied by position lifetime spectroscopy". *Journal De Physique IV.* 265.
31. Okamoto, K., Tanaka, K., Katsube, M., Kita, H., Ito, Y. (1993) "Free volume holes of rubbery polymers probed by positron annihilation". *Bull. Chem. Soc.* **66**, 61.
32. Suzuki, T., Oki, Y., Mumajiri, M., Miura, T., Kondo, K. (1993) "Positron annihilation and polymerization of epoxy resins". *Polymer.* **34**(7), 1361.
33. Kristiak, J., Bartos, J., Fristiakova, K., Sausa, O., Bandzuch, P. (1994) "Free-volume microstructure of amorphous polycarbonate at low temperatures determined by positron-annihilation-lifetime spectrospcopy". *Phys. Rev. B.* **49**(10), 6601.
34. Peng, Z.L., Wang, B., Li, S.Q., Wang, S.J. (1994) "Free volume and ionic conductivity of poly(ether urethane)-LiClO4 polymeric electrolyte studied by positron annihilation". *J. Appl. Phys.* **76**(12), 1.
35. Zhang, Z., Ito, Y. (1991) "Microstructure of gamma-ray irradiated polyethylenes studied by positron annihilation". *Radiat. Phys. Chem.* **38**(2), 221.
36. Suzuki, T., Oki, Y., Numajiri, M., Miura, T., Kondo, K., Shiomi, Y., Ito, Y. (1993) "Novolac epoxy resin and positron annihilation". *J. Appl. Poly. Sci.* **19**, 1921.
37. Yu, Z., Yahsi, U., McGervey, J.D., Jamieson, A.M., Simha, R. (1994) "Molecular weight-dependence of free volume in polystyrene studied by positron annihilation measurements". *J. Poly. Sci. B.* **32**, 2637.
38. Uedono, A., Sadamonto, R., Kaqano, T., Tanigawa, S., Uryu, T. (1995) "Free volumes in liquid-crystalline main-chain polymer probed by positron annihilation". *J. Poly. Sci. B: Poly. Phys.* **33**, 891.
39. Abdel-Hady, E.E., El-Sayed, A.M.A. (1995) "Free volume home distributions of polymers via the positron lifetime techniques". Polymer degradation and stability. **47**, 369.
40. Borek, J., Osoba, W. (1995), "Free volume in plastified poly(vinyl chlorate". *Acta Physica Polociva,* A. **88**, 91.
41. Forsyth, M., Meakin, P., MacFarlane, D.R., Hill, A.J. (1993) "Free volume and conductivity of plasticized polyether-urethane solid polymer electrolytes". *J. Phys. Condens. Matter* **7**, 7601.
42. Jean, Y.C., Yuan, J.-P., Liu, J., Deng, Q., Yang, H. (1995) "Correlations between gas permeation and free volume hole properties probed by positron annihilation spectroscopy". *J. Poly. Sci.:Part B: Poly. Phys.***33**, 1.
43. Xie, L., Gidley, D.W., Hristov, H.A., Yee, A.F. (1995) "Evolution of nanometer voids in polycarbonate under mechanical stress and thermal expansion using positron spectroscopy". *J. Poly. Sci: Part B: Poly. Phys.* **33**, 77.

44. Yang, Liu, Hristov, H.A., Yee, A.F., Gidley, D.W., Bauchiere, D., Halary, J.L., Monnerie, L. (1995) "Changes of the hole volume in model epoxy networks". *Polymer*, **36**(21), 3997.
45. Kobayashi, Y., Haraya, K., Hattori, S. (1994) "Evaluation of polymer free volume by positron annihilation and gas diffusivity measurements". *Polymer* **35**(5), 925.
46. Forsyth, M., Meakin, P., MacFarlane, D.R., Hill, A.J. (1993) "Positron annihilation lifetime spectroscopy as a probe of free volume in plasticized solid polymer electrolytes". *Electrochimica Acta*, **40**(13), 2349.
47. MxCullagh, D.M., Yu, Z., Jamieson, A.M., Blackwell, J., McGervey, J.D. (1995) "Positron annihilation lifetime measurements of free volume in wholly aromatic copolyesters and blends". *Marcromolecules*. **28**, 6100.
48. Uedono, A., Kawano, T., Tanigawa, S., Ban, M., Kyoto, M., Uozumi, T. (1996) "Study of relaxation processes in polyethylene and polystyrene by positron annihilation". *J. Poly. Sci. Part B: Poly. Phys.* **24**, 2145.
49. Suzuki, T., Oki, Y., Numajiri, M., Miura, T., Kondo, K., Ito, Y. (1992) "Positron annihilation in irradiated and unirradiated polyethylenes". *J. Poly. Sci. Part B: Poly. Phys.* **30**, 517.
50. Hristov, H.A., Bolan, B., Yee, A.F., Xie, L., Gidley, D.W. (1996) "Measurement of hole volume in amorphous polymers using positron spectroscopy". *Macromolecules*. **29**, 8507.
51. Bartos, J., Kristiakova, K., Sausa, O., Kristiak, J. (1996) "Free volume microstructure of tetramethylpolycarbonate at low temperatures studied by positron annihilation lifetime spectroscopy: a comparison with polycarbonate". *Polymer*, **37**(15), 3397.
52. Uedono, A., Kawano, T., Wei, L., Tanigawa, S., Ban, M., Kyoto, M. (1995) "Free volume in polystyrene probed by positron annihilation". *Journal de Physique III*, **5**, C1-199.
53. Haldar, B., Singru, R.M., Maurya, K.K., Chandra, S. (1996) "Temperature dependence of positron-annihilation lifetime, free volume, conductivity, ionic mobility, and number of charge carries in a polymer electrolyte polyethylene oxide complexed with NH_4ClO_4". *Phys. Rev. B*. **54**, 7143.
54. Ramachandra, P., Ramani, R., Ravichandran, T.S.G., Ramgopal, G., Gopal, S., Ranganathaiah, C., Murthy, N.S. (1996) "Free volume study of poly(chlorotrifluoroethylene) using positron annihilation spectroscopy as a mircoanalytical tool". *Polymer*, **37**(15), 3223.
55. Suzuki, T., Oki, Y., Numajiri, M., Miura, T., Kondo, K., Shiomi, Y., Ito, Y. (1996) "Free – volume characteristics and water absorption of novolac expoxy resins investigated by positron annihilation". *Polymer* **37**(14), 3025.
56. Borek, J., Osoba, W. (1996) "Free volume in plasticized polyvinyl chloride". *J. Poly. Sci. Part B: Poly. Phys.* **34**, 1903.

57. El-Samahy, A.E., Abdel-Rehim, N., El-Sayed, A.M.A. (1996) "Temperature dependence of free-volume holes in poly(vinyl alcohol) studied by positron annihilation technique". *Polymer* **37**(19), 4413.
58. Bartos, J., Bandzuch, P., Suasa, O., Kristiakova, K., Kristiak, J., Kanaya, T., Jenninger, W. (1997) "Free volume mircostructure and its relationship to the chain dynamics in cis-1,4-poly(butadiene) as seen by positron annihilation lifetime spectroscopy". *Macromolecules*, **30**, 6906.
59. Alentiev, A., Drioli, E., Gokzhaev, M., Golemme, G., Ilinich, O., Volkov, V., Yampholskii, Y. (1998) "Gas permeation properties of phenylene oxide polymers". *J. Membr. Sci.* **138**(1), 99.
60. Shanarovich, V.P., Azamatova, Z.K., Novikov, Y.A., Yampolskii, Y.P. (1998) "Free-volume distribution of high permeability membrane materials". *Macromolecules*, **31**, 3963.
61. Suvegh, K., Domjan, A., Vanko, G., Ivan, B., Vertes, A. (1998) "Free volume and swelling dynamics of the poly[(2-dimethylamino)ethyl methacrylate]-l-polyisobutylene amphiphilic network by positron annihilation investigations". Macromolecules, 31, 7770.
62. Kobayashi, Y., Wang, C.L., Hirata, K., Zheng, W., Zhang, C. (1998) "Effects of composition and external electric fields on positron formation in a polymer blend system". *Phys. Rev. B.* **38**(9), 5384.
63. Wang, C.L., Hirata, K., Kawahara, J., Kobayashi, Y. (1998) "Electric-field dependence of positronium formation in liquids and polymers". *Phys. Rev. B*. **58**(22), 864.
64. Wang, C.L., Hirata, T., Maurer, F.H.J., Eldrup, M., Pedersen, N. J. (1998) "Free- volume distribution and positronium formation in amorphous polymers: temperature and positron-irradiation-time dependence". *J. Chem. Phys.* **108**(11), 4654.
65. Bohlen, J., Wolff, J., Kirchhelm, R. (1999) "Determination of free-volume and hole number density in polycarbonates by positron lifetime spectroscopy". Marcromolecules, 32, 3766.
66. Wang, C.L., Kobayashi, Y., Togashi, H., Kato, K., Hirotsu, T., Hirata, K., Suzuki, R., Ohdaira, T., Mikado, T. (1999) "Plasma-polymerized hexamthyldisiloxane films characterized by variable-energy positron lifetime spectroscopy". *J. Appl. Poly. Sci.* **74**(10), 2522.
67. Suzuki, N., Takamori, A., Baba, J., Matsuda, J., Hyodo, T., Okamoto, Y., Miyagi, H. (2000) "Positron annihilation study of high impact polystyrene". *Radiation Physics and Chemistry.* **58**, 593.
68. Hsu, F.H., Choi, Y.J., Hadley Jr, J.H. (2000) "Temperature dependence of positron annihilation lifetime spectra for polyethylene: positron irradiation effects". *Radiation Physics and Chemistry.* **58**, 473.
69. Sausa, O., Zrubcova, J., Band, P., Kristiak, J., Bartos, J. (2000) "A study of time dependence of ortho-positronium annihilation in a poly(butadiene) at

different temperatures: a meaning of I_3 parameters". *Radiation Physics and Chemistry.* **58**, 479.
70. Suzuki, T., Ito, Y., Kondo, K., Hamada, E., Ito, Y. (2000) "Radiation effect on positronium formation in low-temperature polyethylene". *Radiation Physics and Chemistry.* **58**, 485.
71. Murakami, H. (2000) "Changes in free volume of polyvinylidene fluoride". *Radiation Physics and Chemistry.* **58**, 531.
72. Schmidt, M., Maurer, F.H.J. (2000) "Ortho-positronium lifetime and intensity in pressure-densified amorphous polymers". *Radiation Physics and Chemistry.* **58**, 535.
73. Ito, Y. (2000) "Positronium as a probe of sorption mechanism in polytetrafluoroethylene". *Radiation Physics and Chemistry.* **58**, 551.
74. Bi, J., Simon, G.P., Yamasaki, A., Wang, C.L., Kobayashi, Y., Griesser, H.J. (2000) "Effects of solvent in casting of poly(1-trimethylsilyl-1-propyne) membranes". *Radiation Physics and Chemistry.* **58**, 563.
75. Huang, C.M., Yuan, J.-P., Cao, H., Zhang, R., Jean, Y.C., Suzuki, R., Ohdaira, T., Nielsen, B. (2000) "Positron annihilation studies of chromophore-doped polymers". *Radiation Physics and Chemistry.* **58**, 571.
76. Tashiro, M., Pujari, P.K., Seki, S., Honda, Y., Nishijima, S., Tagawa, S. (2000) "Studies on thermally induced structural changes in silicon based polymers by positron annihilation". *Radiation Physics and Chemistry.* **58**, 587.
77. Mohamed, H.F.M., Abd-Elsadek, G.G. (2000) "Positron annihilation investigation on poly(methyl methacrylate)". *Radiation Physics and Chemistry.* **58**, 597.
78. Bohlen, J., Kirchheim, R. (2001) "Macroscopic volume changes versus changes of free volume as determined by positron annihilation spectroscopy for polycarbonate and polystyrene". *Macromolecules* **34**, 4210.
79. Yampolshii, Y.P., Korikov, A.P., Shantarovich, V.P., Nagai, K., Freeman, B.D., Masuda, T., Teraguchi, M., Kwak, G. (2000) "Gas permeability and free volume of highly branched substituted polyacetylene". *Macromolecules* **34**, 1788.
80. Reiche, A., Dlubek, G., Weinkauf, A., sandner, B., Fretwell, H.M., Alan, A.A., Fleischer, G., Rittig, F., Karger, J., Meyer, W. (2000) "Local free volume and structure of polymer gel electrolytes on the basis of alternating copolymers". *J. Phys. Chem. B.* **104**, 6397.
81. Tanaka, K., Kawai, T., Kita, H., Okamoto, K., Ito, Y. (2000) "Correlation between gas diffusion coefficient and positron annihilation lifetime in polymers with rigid polymer chains". *Macromolecules,* **33**, 5513.
82. Dlubek, G., Lupke, T., Stejny, J., Alam, M.A., Arnold, M. (2000) "Local free volume in ethylene-vinyl acetate copolymers: a positron lifetime study". *Macromolecules,* **33**, 990.

83. He, C., Dai, Y., Wang, B., Wang, S., Wang, G., Hu, C. (2001) "Temperature dependence of free volume in cross-linked polyurethane studied by positrons". *Chin. Phys. Lett.* **18**(1), 123.
84. Hill, A.J., Winhold, S., Stack, G.M., Tant, M.R. (1996) "Effect of copolymer composition on free volume and gas permeability in poly(ethylene terephthalate) –poly(1,4 cyclohexylenedimethylene terephthalate) copolymers". *Eur. Polym. J.* **32**(7), 843.
85. Xie, Li, Gidley, D.W., Hristov, H.A., Yee, A.F. (1994) "Positronium formation in semicrystalline poly(ethylene terephthalate)". *Polymer*, **32**, 3861.
86. Al-Waradawi, I.Y., Abdel-Hady, E.E. (1997) "Positron annihilation lifetime study of pure and treated polyvinyl chloride". *Materials Science Forum.* **255-257**, 366.
87. Chen, Z., Suzuki, T., Kondo, K., Uedono, A., Ito, Y. (2001) "Free volume in polycarbonate studied by positron annihilation effects of free radicals and trapped electrons on positronium formation". *Jpn. J. Appl. Phys.* **40**, 5036.
88. Muramatsu, H., Matsumoto, K., Minekawa, S., Yagi, Y., Sasai, S. "ortho-Positronium annihilation parameters in polyvinyl alcohol with various degree of polymerization, saponification and crystallinity". *Radiochim acta*, **89**, 119.
89. Zhang, R., Chen, H., Cao, H., Huang, C.-M., Mallon, P.E., Li, Y., He, Y., Sandreczki, T.C., Jean, Y.C., Suzuki, T., Ohdaira, T. (2001) "Degradation of polymer coating systems studied by positron annihilation spectroscopy. IV. Oxygen effect of UV irradiation". *J. Poly. Sci. Part B: Poly. Sci.* **39**, 2035.
90. Hamada, E., Oshima, N., Katoh, K., Suzuki, T., Kobayashi, H., Kondo, K., Kanazawa, I., Ito, Y. (2001) "Application of a pulsed slow-positron beam to low-density polyethylene film". *Acta physica polonica A.* **99**, 373.
91. Osoba, W. (2001) "Influence of the isothermal annealing on the free volume changes in thermo-shrunken polyethylene by positron annihilation". *Acta physica polonica A.* **99**, 447.
92. Suzuki, T., Kondo, K., Hamada, E., Ito, Y. (2001) "Positron irradiation effects on positronium formation in polycarbonates during a positron annihilation experiments". *Acta physica polonica A.* **99**, 515.
93. Dlubek, G., Stejny, J., Lupke, T., Bamford, D., Petters, K., Hubner, C., Alam, M.A., Hill, M.J. (2001) "Free-volume variation in polyethylenes of different cystallinities: positron lifetime, density, and X-rays studies". *J. Poly. Sci. Part B: Poly. Sci.* **40**, 65.
94. Blomquist, B., Helgee, B., Maurer, F.H.(2001) "Synthesis and characterization of copolymers of methyl methacrylate and 11-(4-ethoxyazobenzene – 4'-oxy) undecyl methacrylate". *Marcromolecules Chem. Phys.* **202**, 2742.

95. Mazzroua, A., Mostafa, N., Gomaa, E., Mohsen, M. (2001) "The use of positron annihilation lifetime technique to study the effect of doping metal salts on polyhydroxamic acid polymers". *J. Appl. Poly. Sci.* **81**, 2095.
96. Mohamed, H.F.M. (2001) "Study on polystyrene via positron annihilation lifetime and Doppler broadened techniques". *Polymer*, **42**, 8013.
97. Consolati, G.,Quasso, F. (2001) "An experimental study of the occupied volume in polyethylene terephthalate". *J. Chem. Phys.* **114**, 2825.
98. Bamford, D., Dlubek, G., Reiche, A., Alam, M.A., Meyer, W., Galvosas, P., Rittig, F. (2001) "The local free volume, glass transition, and ionic conductivity in a polymer electrolyte: a positron lifetime study". *J. Chem. Phys.* **115**, 7260.
99. Jean, Y.C., Zhang, R., Cao, H., Yuan, J.P., Huang, C.M., Nielsen, B., Asoka-Kuwai, P. (1997) "Glass transition of polystyrene near the surface studied by positron annihilation spectroscopy". *Phys. Rev. B* **56**, 8459.
100. Cao, H., Zhang, R., Yuan, J.P., Huang, C.M., Jean, Y.C., Suzuki, K., Ohdaira, T. (1998) "Free-volume hole model for position formation in polymers surface studies". *J. Phys. Cond. Mart.* **10**. 10429.
101. Sandreczki, T.C., Hong, X., Jean, Y.C. (1996) "Sub-glass transition temperature annealing of polycarbonate studied by positron annihilation spectroscopy". *Macromolecules.* **29**, 4015.
102. Jean, Y.C., Sandreczki, T.C., Ames, D.P. (1986) "Positron annihilation in anneal-cured epoxy polymers". *J. Polymer. Sci. Phys.* **24**, 1247.
103. Nakanishi, H., Jean, Y.C., Smith, E.G., Sandreczki, T.C. (1989) "Positron formation at free volume – the amorphous region of semicrystalline PEEK". *J. Poly. Sci. B.* **27**, 1419.
104. Deng, Q., Zandie H.G., Jean, Y.C. (1992) "Free volume distribution of an epoxy polymer probed by positron annihilation by positron annihilation spectroscopy". *Macromolecules.* **25**, 1090.
105. Liu, J., Jean, Y.C., Yang, H. (1995) "Free volume properties of polymer blends by positron annihilation spectroscopy: Miscibility". *Macromolecules.* **28**, 5774.
106. Jean, Y.C., Shi, H. (1994) "Positron lifetime in an ellipsoidal free-volume hole of polymers". *J. Non-cryst. Solids.* **806**, 172.
107. Jean, Y.C., Deng, Q., Nguyen, T.T. (1995) "Free-volume hole properties in thermosetting plastics studied by positron annihilation spectroscopy: chain extension". *Macromolecules.* **28**, 8840.
108. Yuan, J-P., Cao, H., Hellmeth, E.W., Jean, Y.C. (1998) "Subnano hole properties of CO_2 exposed polysulphone studied by positron annihilation spectroscopy". *J. Poly. Sci., B. Polym, Phys.* **36**, 3049.
109. Wang, Y.Y., Nakanishi, H., Jean, Y.C. and Sandreczki, T.C. (1990) "Pressure Dependence of Positron Annihilation in Epoxy Polymers," *J. Poly. Sci. B* **28**, 1431

Index

A

Ab initio, 84, 156
2D-ACAR, 223, 224
ACAR, 38, 44, 56, 57, 58, 59, 225, 226, 230, 241, 271, 282, 316, 357, 362
activation energy
    for bound state formation, 97
    for defect formation, 109
    of intrinsic molecular defect, 110
    of partial inhibition, 79
AMOC, 56, 204, 241, 253, 349, 350, 351, 352, 354, 355, 356, 357, 360, 362, 363, 364, 365, 366, 367, 368
amorphous, 266
    crystal interface, 266
    crystalline interface, 267
    material, 265
    phase, 266, 267, 268
    polymers, 260, 265, 286, 291
    Si, 235
    silicon, 236, 249
    SiO2, 174, 235, 240, 241, 242, 243, 244, 245, 250
    state in polymers, 254
amorphous polymers
    polymers, 174, 260
amorphous region of the polymer
    of a polymer, 266
amorphous regions, 265, 266
    coefficient of thermal expansion of the, 260
angular correlation, 73
    curve, 57
    experiments, 74, 203
    measurements, 203
    of annihilation radiation, 37, 38, 241, 271, 316
    of radiation, 225
    resolution of experiments, 213
annealing
    for recovery of Ps intensity, 273
    H_2 atmoshere, 333
    in a-Si
        H, 237
    in O_2, 336
    O_2 atmoshere, 333, 335
    of defects, 239
    of metals, 59
    of MgO embedded with Au, 329, 334
    of polymers, 273
    of polypropylene, 273
    of porogen, 239, 339
    study of, 111
    temperature dependence, 236
annihilation rate
    and diffusion constant, 178
    calculation, 316
    calculation of, 317
    contribution of atoms to, 219
    due to wall bounces, 185
    from oxygen, 219
    in open porosity, 188
    of a positron states, 258
    of surface trapped positrons, 317
    of the triplet state, 186
    partial positron, 318
    self-, 185
annihilation rates
    determination, 331
    extraction of, 188

in metals and semiconductors, 316
spin-averaged, 195
anti-inhibition, 81, 82, 83
atomic force microscopy (AFM), 292
Auger electron spectroscopy, 3, 64, 326
Auger Electron Spectroscopy (AES), 311

B

beam
    area, 63
    bunching, 63
    electrostatically guided, 186
    experiments, 64
    intensity, 63
    LINAC, 63, 64
    MeV beam, 45, 63
    monoenergetic, 167, 174, 186
    monoenergetic pulsed positron, 239
    plusing, 186
    pulsed, 205
    pulsed positron, 241
    pulses, 63
    size, 205
    transport, 60
    variable energy monoenergetic, 167
beam based lifetime measurements, 186
beam based positron lifetime, 198
beam line, 59, 65
beam measurements, 40
beam-based AMOC, 64, 350
beam-based microscopes, 64
beams
    brightness enhancement of, 62
    facility-based, 41
    laboratory bases, 205
    laboratory-based, 61

magnetic transport of, 62
plasma-generated positron, 65
polarized positron, 63
pulsed, 186
Variable positron implantation energy, 281
binding energy (energies), 17, 18, 21, 23, 24, 25, 27, 28, 29, 32, 33, 123, 124, 125, 126, 132, 133, 144, 153, 158, 162, 174, 176, 310, 316, 357, 358
Blob
    model, 117, 129, 131, 134, 138, 140, 141, 144
    formation energy, 119
    size (radius), 121, 131
    expansion, 121
Born amplitude, 119
Born cycle, 32
Born-Haber cycle, 125, 126, 127, 128
Born-Oppenheimer approximation, 5, 18, 119
bubble model, 89

C

cavity, 89
    mirrors, 243
    sizes, 257, 283
    structure, 5
    volume, 246
cavity size
    in polystyrene, 267
charge
    build up in polymers, 272
    transfer, 219, 220, 221
charge density, 106, 314, 317
charge effects, 361
charged defects, 109
chemical analysis, 3
    of atomic defects, 3

chemical changes during degradation, 292
Chemical composition and the S parameter, 290
chemical effects
   the spur model, 86
chemical environment, 56, 271
   of voids, 271
chemical equilibria, 77, 106
chemical properties, 104
chemical quenching, 353, 355
chemical rate constant, 275
chemical reaction
   observed by AMOC, 355
chemical reaction rates, 353
chemical reactions
   of positrons, 56, 357
   of Ps, 275, 355
   Ps, 2
Chemical reactions
   of positronium, 350
   of positrons, 357
Chemical sensitivity, 271
chemical signature of metal, 202
chemical structure
   of polymers, 264
CI calculation, 159
Coulomb attraction, 118, 121, 124, 126, 132, 133, 144
Coulomb interaction, 121, 131
Coulomb repulsion, 20, 123, 152
configuration interaction (CI), 23, 24
continuous lifetime analysis, 258
cross section
   for positronium formation, 175
   of three-photon annihilation, 2
crystal
   amorphous interface, 266
   faces, 265
   NaI(Tl), 43
   NaI(Tl), 65
   polyethylene, 266
   Ps trapped at the interface, 266
   single crystal Cu, 313
   structure in polymers, 265
crystals
   defect free single, 259
   MgO, 330
   MgO (100) single, 330
   polypropylene, 266

D

Debye-Huckel theory, 122
Debye screening, 122
Debye radius, 122
defect
   concentration, 109, 255, 283, 293
   dynamics, 52
   environment, 333
   formation, 109, 110, 111
   -free bulk material, 55
   near-surface, 65
   parameter, 302
   size, 293
   spectroscopy, 60
defect size
   size, 255, 283
defects
   atomic, 3, 235
   bond-breaking-type, 243
   created by undulator photons, 245
   extrinsic, 86, 87, 109, 110
   flux-pinning, 224
   formation of, 91
   fraction of, 286
   intrinsic, 87
   near pore surfaces, 249
   near-surface, 281
   non-equilibrium, 59
   structural, 222
   study of, 38, 52, 224, 249
   sub-nanometer, 286
   trapping at, 226
   vacancy type, 212
   vacancy-like, 355

vacancy-type, 239
deformation
   of polymers probed by PAL, 270
   of PTFE, 270
delocalized
   metastable o-Ps state, 367
   positron, 222, 316
   states of the electron, 82
Density Function Theory (DFT), 24
depth profiles
   by Rutherford backscattering, 330
   in polymeric materials, 284
   lifetimes, 196
   of coatings, 291
   of free volume, 300
   of porous materials, 173
   of vacancy clusters, 335
detrapping, 197, 216, 229
diffusion
   barriers, 183, 202
   coefficient, 101, 175, 268
   coefficients, 84, 268, 285, 286
   constant, 178
   controlled process, 97
   -controlled rate, 100, 102, 103
   -controlled reaction, 88, 99, 100, 101
   controlled reaction of Ps in polymers, 276
   equation, 177
   metal ion or atom in-, 183
   of gases through a polymer matrix, 268
   of o-Ps from crystalline to amorphous phase, 266
   of the positron and Ps back to the surface, 285
   positronium, 100
   probability, 340
   Ps, 100, 102, 276, 286, 340, 345, 346
diffusion barrier
   barrier, 183

diffusion length, 174, 175, 176, 178, 196, 286, 339, 341, 342
   and pore interconnectivity, 339
   of positron, 174, 178
   of positron in amorphous polymers, 286
   of positronium, 174
   of positronium in pores, 175
   of positrons in porogens, 346
   of positrons in solid gas, 60
   of psoitron undergoing Brownian motion, 286
   of o-Ps, 342
diffusion model, 84, 287, 340
diffusion path
   of Ps, 339
Dirac delta function, 153
Dissociation thresholds, 22
DMC, 22, 29, 31
doping
   chromophore, 276
   experiments, 212
   ionic KIO4, 86
   of Cu-O planes, 218
   of PMMA, 276
   rare earth and transition metal, 226
Doppler broadening, 351
   correlation with positron age, 351
   experiments, 54
   for pure Au, 333
   lineshape, 216, 222
   lineshape parameter, 222
   lineshape parameter S, 335
   measurement, 52, 202
   measurements, 204, 319
   of positron states, 362
   on aqueous and non aqueous solutions, 357
   parameters, 38
   shift, 363
   two-detector coincidence (2D-DBAR), 332

Doppler broadening of annihilation radiation (DBAR), 330
Doppler broadening of energy spectra (DBES), 282
Doppler broadening spectroscopy, 37, 38, 40, 52, 53, 64
Doppler broadening technique, 202
Doppler broadening techniques, 202, 204

E

EI mass spectrum, 161
electric dipole polarizability, 152
electron
  Auger, 3, 64, 309, 310, 311, 312, 318, 326
  bombardment, 60
  capture, 77, 83
  densities, 256
  densitiy, 256
  density, 3, 83, 86, 95, 180, 185, 259, 314, 315, 316, 317, 331, 332, 333, 336, 361
  depletion, 361
  gas, 314, 315, 317
  mass, 353
  momentum, 52, 54, 212, 213, 283, 290, 335
  radius, 316
  rest mass, 52
  scavenger, 77, 86
  scavengers, 78, 86, 87
  scavenging, 76, 77, 86, 108
  solvated, 75, 77, 84, 87
  spectroscopy, 3
  surface, 285
  transfer, 219, 220, 222, 336
  wave function, 5, 173, 185
  work functions, 311
  yield, 83, 107
Electron Affinity, 14

Electron induced Auger Electron Spectroscopy (EAES), 311
electron spin resonance (ESR), 292, 299
electron-doped superconductor $Nd_{2-x}Ce_xCuO_4$, 220
electron-hole pairs, 44, 362, 365
electronic structure, 54, 211, 214, 226, 335
Electronic transfer, 336
electron-positron correlation effect, 317
electron-positron momentum density, 59
electron-positron overlap function, 218, 220
electrons
  core, 54, 56, 202, 213, 309, 311, 317, 319, 353
  momentum distribution of, 352, 358
  near the Fermi surface, 213
  secondary, 45, 64, 186, 311, 312
  solvated, 75, 107
ellipsometric porosimetry, 184

F

Fermi surface, xiv, 57, 212, 225, 226, 230
Feshbach resonances, 157
Feynman diagram, 153
Fourier transform infrared and Raman spectroscopy (FTIR), 292
fractional free volume, 259, 262, 267, 270, 375, 379, 380, 381, 382, 383, 385, 386
free volume, xi, 5, 6, 86, 87, 90, 109, 250, 254, 255, 256, 257, 258, 259, 260, 261, 262, 263, 267, 268, 270, 271, 274, 288, 293, 300, 301,

302, 303, 304, 375, 379,
380, 381, 382, 383, 385,
386, 389, 390, 391, 393,
394, 395
free-electron laser experiments, 243
Fullerenes, xiv, 227

G

gas permeability, 268, 394
glass transition temperature, 259, 260, 395
glassy polymers, 269
gloss, 299, 302

H

Hartree-Fock (HF) method, 23, 27, 30

I

impurity centres in the crystal lattice, 42
inhibition, 78, 82, 83, 85, 88, 275
    constant, 75, 77, 80, 82, 83, 86, 107, 276
    constants, 75, 76, 79
    effect, 276
    in polymers, 275, 277
    of positronium formation, 74
    partial, 75, 77, 78, 79, 80, 81, 82, 87, 88
    strong, 86
    total, 75, 76, 79, 80, 82, 84, 107
Ionization energy, 153
Ionization potential, 27, 28, 153, 160

J

Jastrow factor, 22

L

lifetime
    analysis, 169, 187, 188, 192, 198, 199, 258, 259, 349
    depth profile, 204
    depth profiled, 202
    Depth profiled, 201
    depth profiles, 196
    distribution, 199, 259
    distributions, 188, 191, 200, 259
    experiments, 40, 223
    free positron, 96
    in terms of positron trapping behaviour, 216
    mean values, 188, 189, 190, 192, 199, 205
    measurement, 37, 38
    measurements, 63, 71, 186, 202, 204, 205, 214, 220, 224, 228, 282, 335, 362
    of cation vacancy, 110
    of positronium states, 3, 186
    of solvated positron, 76
    o-Ps, 3, 86, 87, 91, 92, 99, 195, 246, 255, 256, 257, 262, 263, 264, 265, 267, 274, 275, 287, 288, 289, 342
    pick-off, 3, 236, 282, 283
    positron, 37, 40, 44, 49, 50, 56, 66, 192, 193, 198, 201, 214, 220, 221, 223, 224, 227, 228, 229, 236, 239, 241, 246, 249, 250, 288, 289, 330, 331, 334, 335, 342, 343, 349, 355, 367
    p-Ps, 91, 99, 364
    quenching, 77
    shortening, 239

spectra, 51, 188, 236, 239, 241, 246, 258, 265, 266, 331, 333, 334, 342, 355, 356
spectrometers, 50
spectrometry, 49, 54
spectroscopy, 40, 65, 73, 236, 249, 250, 251, 253, 255, 257, 330
start signal, 186
studies, 45
Lifetime
  depth profiles, 204
lifetime of the bound-states
  of bound-states, 81
lifetimes
  continuous distribution of, 201
  discrete, 201
  distributions of, 201
  in polymers, 185
  measurement, 185
  measurement of, 51
  measurement of the mean, 49
  of bound-states, 96
  of positronium states, 74
  of positron-molecule states, 361
  o-Ps, 109, 110, 257, 266, 267, 272, 338, 342, 365, 367
  positron-molecule, 361
lifetimes of all e^+ bound-states
  of all e+ bound-states, 87
Linear Energy Transfer (LET), 119, 146

M

mean free path, 66, 186, 195
MELT, 52, 169, 188, 191, 201, 259
metal indiffusion experiments, 202
microbeam, 64
microelectronic devices, 246, 337
microelectronics industry, 338
moderation efficiency, 60, 66
moderators, x, 59, 60

mono-energetic positrons beam, 38
Monte-Carlo calculations, 84
multichannel analyser, 49

N

nodal surface, 22, 23
nuclear magnetic resonance (NMR), 292

O

Onsager radius, 82, 86, 109
o-Ps out- diffusion probability, 340
Ore gap, 123, 124, 125, 127, 128, 129, 144
Ore mechanism, 134
Ore model, 123
out- diffusion, 340, 341
out diffusion probability, 340
out-diffusion model, 340

P

PAES, 3, 59, 64, 309, 310, 311, 312, 313, 318, 319, 321, 322, 323, 324, 325
percolation, 171, 175, 190, 194, 338, 346
photodegradation, 293, 294
photoelectron spectroscopy, 161
pick-off annihilation, 3, 90, 180, 181, 236, 241, 242, 256, 271, 282, 283, 365
PI mass spectrum, 161
plasmon resonance frequency shift, 337
polymer blends, 271, 395
Polarization potential, 152
polymer stereoregularity, 268
porosity, xiii, 167, 168, 169, 170, 171, 172, 173, 177, 179,

        180, 181, 182, 184, 188,
        190, 191, 192, 193, 194,
        196, 197, 198, 199, 203,
        204, 205, 338
positron
    lifetimes, 38, 51, 258, 282, 316,
        331, 343
    wave function, 5, 54, 313, 316,
        319, 325
    positron beam, 7, 41, 54, 59, 60, 61,
        63, 64, 66, 169, 205, 236,
        239, 241, 243, 249, 250,
        281, 282, 292, 293, 309,
        367
    positron beam technique, 7
    positron beams, 3, 37, 56, 59, 62, 63,
        65, 173, 186, 253, 277,
        281
        MeV beam, 56
    Positron beams, 59, 68
    positron diffusion length, 66, 174,
        178, 346
    positron spur, 84, 85, 276
    positron state
        Doppler broadening of, 362
    positron states
        evolution of, 349
        first-principles calculations of,
            313
        transitions between, 56, 353
        transitions of, 350
        upper limit for the lifetime of,
            361
    positron-electron correlation
        potential, 314, 315
    positronium, 173
        annihilates, 173
        annihilation, 168, 170, 171, 172,
            173, 174, 176, 185, 192, 203,
            271
        annihilation measurements, 169
        annihilation rates, 178, 370
        atom, 185, 186
        binding energy of, 174, 176

bubble, 367
bubbles, 365, 366
chemical reaction, 355
chemical reaction of, 275
chemical reactions of, 350, 355
chemistry, 1, 3, 6, 94, 277, 349,
    355
de Broglie wavelength of, 185
diffusion, 100, 174, 175, 177,
    187
diffusion length, 196, 345
diffusion length of, 174
epithermal, 350
escape, 171, 175, 178, 180, 187,
    188, 193
formation, 73, 86, 167, 173, 175,
    176, 180, 191, 235, 239, 245,
    249, 253, 256, 260, 272, 275,
    338, 341, 350, 352, 365
fraction, 189, 190
inhibition of formation, 275
intensity, 243
lifetime, 171, 185, 191, 192, 258,
    342, 343
lifetime spectroscopy, 342, 343
lifetimes, 195
measurements, 168
oxidation reactions of, 357
pick off, 173
quenching of, 275
reactions, 95
size, 255
slowing down, 362, 365
slowing-down of, 354
spin conversion of, 355, 357, 370
spin states, 186
states, 94, 185, 282, 360
thermalization, 185
thermalized, 354
trapping, 170, 174, 181, 185,
    195, 202, 204, 338, 339
velocity distribution of, 203
wave function, 185